**GEOTECHNICS
FOR
BUILDING PROFESSIONALS**

GEOTECHNICS
FOR
BUILDING PROFESSIONALS

J A CHARLES
PhD, DSc(Eng), FREng, FICE

Foreword by
PROFESSOR J B BURLAND
FRS, FREng, FICE, FIStructE

BRE

BRE is committed to providing impartial and authoritative information on all aspects of the built environment for clients, designers, contractors, engineers, manufacturers, occupants, etc. We make every effort to ensure the accuracy and quality of information and guidance when it is first published. However, we can take no responsibility for the subsequent use of this information, nor for any errors or omissions it may contain.

BRE is the UK's leading centre of expertise on building and construction, and the prevention and control of fire. Contact BRE for information about its services, or for technical advice, at:

BRE, Garston, Watford WD25 9XX
Tel: 01923 664000
Fax: 01923 664098
email: enquiries@bre.co.uk
www.bre.co.uk

Details of BRE publications are available from:

www.brebookshop.com
or
IHS Rapidoc (BRE Bookshop)
Willoughby Road
Bracknell RG12 8DW
Tel: 01344 404407
Fax: 01344 714440
email: brebookshop@ihsrapidoc.com

Published by BRE Bookshop

Index compiled by Linda M Sutherland

Requests to copy any part of this publication should be made to:

BRE Bookshop,
Building Research Establishment,
Watford WD25 9XX
Tel: 01923 664761
Fax: 01923 662477
email: brebookshop@emap.com

BR 473
© Copyright BRE 2005
First published 2005
ISBN 1 86081 727 0

CONTENTS

Foreword		viii
About the author		x
Preface		xi
Acknowledgements		xi
Abbreviations		xii
Notation		xiii
1 Introduction		1
1.1	Foundations and geotechnics	1
1.2	Building professionals and geotechnical specialists	2
1.3	Scope of the book	3
2 Regulations for foundations and geotechnics		6
2.1	Introduction	6
2.2	Planning constraints	8
2.3	Building regulations	8
2.4	Contaminated land	11
2.5	Codes and standards	13
2.6	Insurance and litigation	14
3 Types of ground		19
3.1	Introduction	19
3.2	Soil formation	20
3.3	Soil composition – solid particles	22
3.4	Soil composition – phase relationships	23
3.5	Coarse soil	26
3.6	Fine soil	28
3.7	Organic soil	32
3.8	Rock	33
3.9	Fill	35
3.10	Groundwater	37
4 Soil behaviour		43
4.1	Introduction	43
4.2	Stress and strain	44
4.3	Effective stress and pore water pressure	45
4.4	Groundwater flow	47
4.5	Deformation and compressibility	51
4.6	Shear strength	57
4.7	Partially saturated soil	62
4.8	Shrinkage and swelling of clay	63
4.9	Collapse compression on wetting	66
4.10	Compaction of partially saturated fine soil	68
5 Ground assessment		75
5.1	Introduction	75
5.2	Ground investigation	79
5.3	Desk study	81
5.4	Walk-over survey	84

5.5	Direct investigation	86
5.6	Laboratory measurement of ground properties	90
5.7	In-situ measurement of ground properties	91
5.8	Geophysical measurement of ground properties	98
6 Foundation design		**104**
6.1	Introduction	104
6.2	Types of foundation	105
6.3	Acceptable ground movements	111
6.4	Basic features of design	113
6.5	Bearing resistance of shallow foundations	115
6.6	Bearing resistance of deep foundations	118
6.7	Use of elastic theory in settlement calculations	120
6.8	Settlement of shallow foundations in response to loading	124
6.9	Settlement of deep foundations in response to loading	127
6.10	Foundations for low-rise building extensions	128
6.11	Ground chemistry	129
7 Foundations on difficult ground		**134**
7.1	Introduction	134
7.2	Soft natural ground	135
7.3	Loose natural ground	137
7.4	Fill	137
7.5	Shrinkable clay and the effect of trees	139
7.6	Unstable slopes	148
7.7	Underground mines and cavities	154
8 Brownfield land		**163**
8.1	Introduction	163
8.2	Types of brownfield site	165
8.3	Systems at risk	170
8.4	Ground-related hazards	171
8.5	Identification of contamination	176
8.6	Strategies for building on brownfield sites	178
8.7	Foundations and services	182
9 Ground treatment		**189**
9.1	Introduction	189
9.2	Vibrated stone columns	191
9.3	Impact compaction	194
9.4	Preloading	197
9.5	Excavation and refilling	199
9.6	Drainage	202
9.7	Soil mixing	203
9.8	Grouting	204
9.9	Remediation of contaminated land	206
9.10	Specification and quality management	209
10 Foundation movement and damage		**213**
10.1	Introduction	213
10.2	Causes of building damage	214

10.3	Types of foundation movement	215
10.4	Causes of foundation movement	217
10.5	Investigating damage	223
10.6	Diagnosis	226
10.7	Uses of monitoring	229
10.8	Structural monitoring	230
10.9	Foundation and ground movement monitoring	232
11 Remedial treatment and underpinning		**236**
11.1	Introduction	236
11.2	Mitigating the cause of ground movement	236
11.3	Soil stabilisation	238
11.4	Repairing or strengthening the superstructure	238
11.5	Remedial underpinning	239
11.6	Mass concrete underpinning	240
11.7	Pier-and-beam underpinning	241
11.8	Pile-and-beam and piled-raft underpinning	243
11.9	Mini-piling underpinning	243
11.10	Design of underpinning	247
11.11	Partial underpinning	250
12 Ancillary works		**253**
12.1	Hardcore	253
12.2	Soakaways	257
12.3	Drains and sewers	262
12.4	Small embankments	267
12.5	Small retaining walls	268
12.6	Freestanding walls	271
Glossary		275
References		298
Index		311

FOREWORD

Having been Head of the Geotechnics Division at the Building Research Station from 1972 to 1980, it is a particular pleasure for me to be invited to write this foreword. Written for the non-specialist, this book is a distillation of the large reservoir of expertise, experience and guidance on foundations for low-rise buildings which has accumulated at BRE over many years. This experience is based on research into the actual behaviour of buildings founded on a wide range of difficult ground conditions, studies of the effectiveness of various ground treatment methods and detailed examinations of the causes and severity of damage in hundreds of buildings.

For many, indeed most, building professionals and structural engineers, geotechnics is something of a black art. The ground is mysterious: every trench, hole and excavation opened up is a journey into an uncharted underworld. When problems with foundations occur, rumours abound – underground streams, old wells, hidden cavities, previous uses of the site and so on. On one occasion, I visited the site of a foundation excavation in Salisbury, the bottom of which had turned into a soft slurry owing to upward seepage of water, and was told by an engineer that the whole of Salisbury was floating on liquid chalk!

Why is it that geotechnics comes across as a mysterious and inaccessible subject? For most building materials, such as steel, cement, concrete and brickwork, the stiffness and strength can be specified and remain reasonably fixed and dependable. For soils, however, the stiffness and strength depends on the stress acting on them and the water pressures acting within them, and these can change significantly during and after construction. In most cases, the professional cannot specify the founding material but has to work with what nature has provided; and nature is seldom straightforward.

To understand the methodology of the geotechnical engineer, it is helpful to compare it with the approach adopted by a structural engineer or building professional working on an ancient building. A detailed inspection is essential to discover what materials were used and how the building was constructed. Special attention is given to detecting signs of weakness and defects since these can dominate behaviour. Historical records can reveal the history of construction and give important clues as to the building's overall response.

This book demonstrates that the geotechnical engineer works in exactly the same way. First, discover what is there: the ground profile and groundwater conditions at the site and how they vary across it. For low-rise buildings, highly sophisticated testing is rarely required; it is usually necessary to know only what the soil types are (fine or coarse material) and how dense they are (density is an indicator of stiffness and strength). Particular emphasis must be placed on identifying previous uses of the site, potential hazards and difficult ground. Much can be learned from local experience and records such as aerial photographs and old maps. Skimping on the investigation can lead to very costly and time-consuming corrective measures, just as they can for an ancient building. So Chapter 5 Ground assessment is fundamental to the whole book and is a 'must-read'.

For the professional who wishes to become more informed about the mysteries of soil mechanics, Chapter 3 Types of ground and Chapter 4 Soil behaviour provide a useful introduction to the subject. But a deep understanding of soil mechanics is not needed to follow the logic behind the choice of the various types of foundations given in Chapter 6, the approaches required for dealing with difficult ground conditions outlined in Chapter 7, the assessments of risks associated with building on brownfield land described in Chapter 8, and the most common methods of ground treatment in Chapter 9.

One of the most difficult and emotive problems that the building professional has to deal with is that of foundation movement and damage together with remedial treatment. BRE has carried out research in these areas for many years and is respected internationally for its wise and balanced guidance. Chapters 10 and 11 draw together this guidance in a concise but comprehensive manner.

One of the great attractions of the book is that it informs the practitioner about when expert advice is necessary and where to obtain it. It contains comprehensive lists of relevant regulations, standards, BRE publications and authoritative papers. This book deserves to find its way on to the bookshelves of most building professionals and structural engineers.

Professor John Burland, FRS, FREng, FICE, FIStructE

ABOUT THE AUTHOR

Dr Andrew Charles is a distinguished geotechnical engineer. He worked in BRE's Geotechnics Division from 1967 to 2002 and is now a BRE Associate. He has had extensive involvement in many aspects of geotechnical engineering and developed particular expertise in the behaviour of fill materials and the effectiveness of ground treatment techniques.

He has published more than 150 technical papers and was awarded the Geotechnical Research Medal of the Institution of Civil Engineers for 2000 and the British Geotechnical Society Prize for 1993. In 1985 Dr Charles was awarded the George Stephenson Medal by the Institution of Civil Engineers and he was elected a Fellow of the Royal Academy of Engineering in 1999. His work on fill materials culminated in the publication Building on fill: geotechnical aspects.

He was chairman of the BSI committee for Methods of test for soils for civil engineering purposes (BS 1377) from 1980 to 2003 and is technical editor for the European Standard on Ground treatment by deep vibration.

PREFACE

Of all the components that make up a modern building or civil engineering structure, the ground is often the least well known and appreciated and a lack of attention to foundation design and construction can have undesirable and expensive consequences. For low-rise buildings the situation is particularly troublesome for two reasons:
- many important decisions concerning ground and foundations are made by people who normally have had little or no training in geotechnical engineering;
- the near-surface soils that support most low-rise buildings frequently do not receive the attention that their complex behaviour deserves.

In an endeavour to improve matters, the Building Research Establishment (BRE) has, over many years, published guidance on various aspects of foundations for low-rise buildings in Digests, Information Papers, Good Building Guides and Reports. This book draws together this guidance. Although the material has been written by specialists, it has been written primarily for building professionals who are not geotechnical specialists.

The idea of bringing the published BRE guidance together in the form of a book was first discussed by Richard Driscoll, then Head of Geotechnics Division at BRE, and Andrew Charles. The time that has elapsed between that first discussion and the appearance of the book can be attributed to several factors, by no means the least of these has been that the task has proved to be a good deal more arduous than originally anticipated!

While this book principally focuses on the application of geotechnics to foundations, the geotechnical work at BRE has never been narrowly confined to building as such, but has had a breadth commensurate with that of the construction industry.

ACKNOWLEDGEMENTS

The work of those colleagues who prepared the Digests and Reports which have formed the source documents for this book is gratefully acknowledged. Many of the publications, such as those in the Digest series, were not attributed to authors, so it is impossible to identify and acknowledge individual contributions. In the initial phase of preparing the book, drafts of a number of chapters were written by Tim Freeman, a former member of BRE Geotechnics Division. Responsibility for the remaining chapters was divided amongst several members of staff but much of the drafting was done by Andrew Charles with assistance from Richard Driscoll. This approach led to significant differences in style and some missing material! Subsequently, Mike Crilly prepared a draft chapter on damage to foundations and underpinning.

On retirement from BRE at the end of 2002, Andrew Charles was commissioned to complete the work on the book. This involved revising the layout, rewriting the earlier material into a common style with appropriate updating, and filling in some missing sections. In the final phase of the work, a number of colleagues assisted by reviewing sections and chapters as follows: Richard Driscoll (Chapters 1 to 7 and 10 to 12), John Powell (Chapter 5 and Section 12.2), Paul Tedd (Chapter 8) and Ken Watts (Chapter 9).

ABBREVIATIONS

The abbreviation BRE is used in the book to describe the Building Research Station/Building Research Establishment.

AGS	Association of Geotechnical and Geoenvironmental Specialists
ASCE	American Society of Civil Engineers
BS	British Standard
BSI	British Standards Institution
CEN	European Committee for Standardisation
CIRIA	Construction Industry Research and Information Association
CLEA	Contaminated land exposure assessment model
CPT	Cone penetration test
CPTU	Cone penetration test using piezocone
CSW	Continuous surface wave method
DD	Draft for development (BSI)
DEFRA	Department of the Environment, Food and Rural Affairs
DP	Dynamic probing
DPC	Damp-proof course
DPM	Damp-proof membrane
EA	Environment Agency
EN	European standard
ENV	European pre-standard
GBG	BRE Good Building Guide
GRG	BRE Good Repair Guide
HDPE	High density polyethylene
HSE	Health and Safety Executive
ICE	Institution of Civil Engineers
ISO	International Standards Organisation
ISSMGE	International Society for Soil Mechanics and Geotechnical Engineering
LCR	Standard Land Condition Record
MDD	Maximum dry density
MDPE	Medium density polyethylene
NHBC	National House-Building Council
OCR	Overconsolidation ratio
pfa	Pulverised-fuel ash
PCB	Polychlorinated biphenyl
PGA	Peak ground acceleration
PPG	Planning policy guidance note
prEN	European pre-standard
SASW	Spectral analysis of surface waves
SPT	Standard penetration test
VOC	Volatile organic compound
WRc	Water Research Centre

NOTATION

a_{50} internal surface area of soakaway below 50% of effective depth, excluding base area
c_u undrained shear strength
c_v coefficient of consolidation
e void ratio
e_o initial void ratio
f infiltration rate
f_s sleeve friction in cone penetration test
f_s shape and rigidity factor in settlement calculation
f_d depth factor in settlement calculation
g gravity $g = 9.81$ m/s^2
h height of sample
h_w depth of water table below ground level
i hydraulic gradient
k coefficient of permeability or hydraulic conductivity
m_v coefficient of volume compressibility
n porosity (%)
p line load
p mean principal total stress
p' mean principal effective stress
p_k suction
q deviator stress $(\sigma_1 - \sigma_3)$
q applied pressure or gross bearing pressure
q bearing resistance or ultimate bearing capacity
q_a allowable bearing pressure
q rate of flow of water
q_c cone resistance in cone penetration test
q_d dynamic point resistance in dynamic probing
q_u unconfined compressive strength
s settlement
s' effective stress path parameter in triaxial test $0.5\,(\sigma'_1 + \sigma'_3)$
s total stress path parameter in triaxial test $0.5\,(\sigma_1 + \sigma_3)$
t time
t stress path parameter in triaxial test $0.5\,(\sigma_1 - \sigma_3)$
u_a pore air pressure
u_w pore water pressure
v discharge velocity $v = (q/A)$
v_s actual seepage velocity
w water content (%)
w_{opt} optimum water content (%)
w_P plastic limit (%)
w_L liquid limit (%)
z depth below ground level or foundation level
z_e depth to which ground is significantly stressed by applied load

A cross-sectional area
A impermeable area drained to a soakaway
B width of foundation
C_c compression index
C_R relative compaction
C_U coefficient of uniformity $C_U = D_{60}/D_{10}$
C_α coefficient of secondary compression $C_\alpha = \Delta\varepsilon_v/\log(t_2/t_1)$
D constrained modulus
D depth of foundation
D significant depth of desiccation
D_{10} particle size such that 10% of particle size distribution by mass is finer than D_{10}
D_{60} particle size such that 60% of particle size distribution by mass is finer than D_{60}
D duration of design storm
E Young's modulus
E_u Young's modulus under undrained conditions
E total applied energy in dynamic compaction
G shear modulus
H height of wall, tree or fill in embankment
H height of fall of weight in dynamic compaction
I_C consistency index $I_C = 1 - I_L$
I_D density index (relative density)
I_L liquidity index $I_L = (w - w_P)/I_P$
I_P plasticity index $I_P = w_L - w_P$
I_S index for strength of rock
K bulk modulus
K coefficient of earth pressure
K_a active earth pressure coefficient
K_o coefficient of earth pressure at rest
K_p passive earth pressure coefficient
L length of foundation, wall or other object
M mass of weight used in dynamic compaction
M mass of hammer in dynamic probing
N SPT blow count
N_c bearing capacity factor
N_γ bearing capacity factor
N_q bearing capacity factor
P point load
R total rainfall in design storm
R_f friction ratio in cone penetration test
S_f water shrinkage factor
S_r degree of saturation of a soil; ratio of volume of water to volume of pores (%)
T time factor for one-dimensional consolidation
V volume
V_a volume of air voids or air voids expressed as a percentage of total volume
V_s volume of solids
V_w volume of water voids
W weight in dynamic compaction

α	total stress skin friction factor; ratio of shear stress on pile shaft to undrained shear strength
α	tilt of building
α	logarithmic creep compression rate parameter
β	effective stress skin friction factor, ratio of shear stress on pile shaft to effective vertical stress
γ	unit weight
γ_d	dry unit weight
γ'	submerged unit weight
γ_s	unit weight of solid particles
γ_w	unit weight of water $\gamma_w = 9.81$ kN/m^3
δ	angle of friction at fill/ pile interface
ε	strain
ε_v	vertical strain
ν	Poisson's ratio
ρ	bulk density
ρ_d	dry density
ρ_{dmax}	maximum dry density
ρ_{dmin}	minimum dry density
ρ_s	particle density
ρ_w	density of water $\rho_w = 1.0$ Mg/m^3
σ	total stress
σ_h	horizontal total stress
σ_v	vertical total stress
σ_1	major principal total stress
σ_3	minor principal total stress
σ'	effective stress
σ'_h	horizontal effective stress
σ'_p	preconsolidation pressure
σ'_v	vertical effective stress
σ'_1	major principal effective stress
σ'_3	minor principal effective stress
τ	shear stress
τ_f	shear strength or maximum shear stress
ϕ'	effective angle of shear resistance
ϕ'_{cv}	constant volume angle of shear resistance
ϕ'_r	residual angle of shear resistance
Δ	relative deflection – maximum vertical displacement relative to straight line connecting two points
Δ	increment

CHAPTER 1: INTRODUCTION

The foundations will affect how the other elements of a building perform and so can be a major factor in determining whether or not the building achieves its intended purpose with satisfactory long-term performance. However, decisions concerning the foundations for low-rise buildings are often made by people involved in planning and other aspects of the building development who have little formal training in geotechnical engineering. This book has been written for the benefit of such people and their non-specialist professional advisors; these include architects, surveyors, loss adjusters, planners, insurance underwriters and property developers, most of whom are likely to have only a limited understanding of the engineering of the ground. The book brings together a large quantity of material relevant to building foundations published by BRE over many years. This introductory chapter defines the scope of the book.

1.1 Foundations and geotechnics

The building industry was largely untouched by developments in science and technology in the nineteenth century. Indeed, up to 1914 building remained unaffected by such developments to an extent that was greater than for most other forms of industrial activity. Doubtless such a state of affairs, with its reliance on practical experience and its hostility to innovation, would have inspired Kipling to write the opening lines to *A truthful song*[1]:

> *I tell this tale, which is strictly true,*
> *Just by way of convincing you*
> *How very little, since things were made,*
> *Things have altered in the building trade.*

Following the First World War, the Building Research Station[2] was formed to remedy this unhelpful situation by advancing the scientific knowledge of building materials and applying this science to building. It was the first organisation of its kind in the world.

The investigation of the causes and cures of defects in building materials and structures can be difficult but understanding the behaviour of the ground on which the building is placed can, in some situations, present even more problems. Yet the stability and life of any building depend on the strength and deformation of the ground on which it is founded. If the ground fails or deforms substantially, the building on it will fail or be significantly impaired regardless of how well the building has been designed and built.

It has been said that soil is both the oldest and the most complex construction material. Of all the components that make up a building or civil engineering construction, the ground will usually be the least well known and appreciated. It is often a case of 'out of sight, out of mind' and the originator of the scientific basis of much of modern ground engineering, Karl Terzaghi[3], pointed out that *'there is no glory attached to the foundations'*[4].

> **Box 1.1 Some basic terminology**
> The study of the engineering behaviour of the ground is variously described as *ground engineering, geotechnical engineering, geotechnics* or *geomechanics*. Although these terms have different nuances, they are often used interchangeably and may be taken to encompass the utilisation or alteration of the behaviour of the ground for man's construction activities. *Ground* includes both soils and rocks; *soil mechanics* and *rock mechanics* are commonly used to describe subsets of the more general terms.

However, lack of attention to foundation design and construction can have undesirable and expensive consequences. Some early work on ground behaviour (see Box 1.1 for terminology) began at the Building Research Station[5] in the late 1920s, but it was another ten years or more before really substantial progress was made.

Modern geotechnical engineering can be dated from the formation of the International Society for Soil Mechanics and Foundation Engineering – now known as the International Society for Soil Mechanics and Geotechnical Engineering (ISSMGE) – which held its first conference at Harvard University in the USA in 1936. One British applied scientist, Leonard Cooling[6], attended this historic event. Dr Cooling was a major catalyst for much of the study of the behaviour of the ground that began at the Building Research Station, which from 1926 was located at Garston in Hertfordshire. By 1937, a well-equipped laboratory had been established at Garston and in the following years many geotechnical problems were investigated. Until the end of the Second World War, the Building Research Station remained the location of much of the geotechnical activity in the United Kingdom. In 1972, the Building Research Station formed the major constituent element of the newly formed BRE.

The book draws heavily upon the work done by Cooling and those who followed at BRE. It contains a compilation of material relevant to building foundations produced over the 75-year history of geotechnics at BRE and a great debt is owed to the many outstanding geotechnical engineers who have work at BRE. Throughout the book there are references to their achievements and to other important publications. The references are not required reading since some of them are fairly specialised but they should be of interest and value to the reader who wants more detail. An account of the full range of geotechnical work carried out at BRE *Seventy-five years of building research: geotechnical aspects* (Charles et al, 1996) has been published in the Proceedings of the Institution of Civil Engineers.

1.2 Building professionals and geotechnical specialists
Using tried and tested empirical rules, a vast amount of building was successfully conducted before there was any great awareness of geotechnical engineering. However, the well-tried and tested methods of the past are no longer adequate, for several reasons:
❏ In an industrialised country like Great Britain, much of the good land has already been built upon. Growing use is made of marginal and brownfield land hitherto considered too poor to use for building. Such sites can contain a wide variety of geotechnical and geoenvironmental hazards.

INTRODUCTION

❐ Modern construction is far more sensitive to foundations movements than are buildings constructed in Victorian Britain. Where foundation problems occur, they can be expensive to rectify and litigation is likely.

❐ The majority of houses are now owner-occupied and most owners are very conscious of the market value of their home. Despite the occasional collapse of property values, owners continue to perceive their homes as major financial investments, with good reason. Consequently, any perceived threat to the value of the home, and subsidence is one of the principal hazards, is highly emotive.

With this background, it is not sensible to neglect the important processes involved in the investigation of the ground before construction begins. Extensive and costly problems can be encountered both during and after construction because of inadequate ground information. The Institution of Civil Engineers has highlighted the need for geotechnical specialists to be involved at the earliest possible stage of a project[7].

Many important decisions concerning foundations for low-rise buildings are made by people who have had little or no training in geotechnical engineering. Such people may be involved in planning decisions that have far-reaching consequences for foundation designs and subsequent structural performance; they may be responsible for the maintenance of buildings that suffer from subsidence, and they may need some knowledge of foundation engineering to assess insurance risks. Property developers will require an understanding of the financial implications of developing a particular site. This book has been written for the benefit of such people and the non-specialist professionals who advise them; they include architects, surveyors, loss adjusters, planners and insurance underwriters, who may have only a limited understanding of the engineering of the ground.

Although geotechnical engineering is a specialised subject requiring considerable experience as well as deep knowledge, building professionals will benefit from a better understanding of the most important elements of geotechnics. This should enable them to better appreciate the limitations of their knowledge, to ask the geotechnical specialist informed questions, and to have greater confidence in the answers they receive.

1.3 Scope of the book
The book concentrates primarily on the way in which the ground is used to support building foundations. It deals principally with low-rise buildings (not more than three storeys in height) and much of the emphasis is placed on near-surface ground conditions. Although the material contained in this book has been written by specialists, an attempt has been made to see geotechnical engineering through the eyes of a non-specialist. The book has been structured to take the reader through a logical sequence of information, from the viewpoint of someone who is confronted, perhaps for the first time, with having to consider, say, a new housing development on a site about which little is known.

Chapter 2 discusses some important aspects of planning, the law relating to foundation issues and insurance against subsidence and landslip. It encourages realistic expectations for the performance of foundations that represent good value for money. Where perfect performance is required, the price is likely to be excessive.

The wide variety of ground conditions encountered in the United Kingdom is discussed in Chapter 3, together with an account of some behavioural peculiarities and an indication of how these might influence the design process. An understanding of the behaviour of soil is fundamental to appreciating the load-carrying characteristics and compressibility of the ground, and hence the safe and economical design of foundations. Chapter 4 gives a simple outline of the relevant principles of soil mechanics.

Chapter 5 introduces the crucial subject of ground investigation. In practice, events may start with planning considerations and move on to legal and insurance matters; only then does attention begin to focus on finding out what is in the ground. By the time the ground is adequately appraised, it may be too late; a site may have been bought cheaply without realising that expensive ground treatment will be required. Unforeseen ground conditions can lead to delays, expensive re-design and costly modifications on site. The importance of an adequate, appropriate site investigation is emphasised. Inexpensive information can be overlooked while quite costly site investigations can include unnecessary sampling and testing, deriving ground properties that are not used subsequently by the designer.

Chapters 6 and 7 describe the design and performance expectations for different types of foundations for low-rise buildings on commonly encountered ground conditions. Their relative merits are assessed and problems which are known to have occurred are outlined. Chapter 6 deals with basic foundation design, including bearing capacity and settlement due to loading. The problems encountered with some common types of difficult ground are considered in Chapter 7.

With a target that 60% of all new houses should be built on brownfield land[8], the particular hazards encountered on such sites, which are a legacy from previous use, require careful examination. Chapter 8 discusses the physical and chemical hazards in simple terms and gives guidance on the various solutions offered by geotechnical and geoenvironmental engineering.

In some situations ground treatment will be required before building development can safely proceed. Chapter 9 identifies the most commonly used ground treatment techniques and discusses their advantages and limitations.

Over the past 30 years, there has been a substantial growth in subsidence claims submitted to insurers. These claims have arisen from a variety of ground problems.

INTRODUCTION

Chapter 10 describes the types of foundation movement that cause cracks to appear in buildings. Investigating foundation damage and determining its cause are key processes. The role of ground and structural monitoring is described. BRE is a leading authority on ground movement problems and the chapter constitutes a distillation of many earlier publications. Continuing foundation movement may require geotechnical engineering to effect a remedy and appropriate remedial action is described in Chapter 11, with particular emphasis on the uses and limitations of underpinning.

Not all geotechnical engineering associated with low-rise building developments concerns the ground beneath the foundations of a building; Chapter 12 describes some other common building operations where geotechnics has a significant role.

In each chapter, certain detailed information, which has not been put in the body of the text, can be found in notes which are placed at the end of the chapter. References to other publications are listed at the end of the book. Generally, the Harvard system of references has been adopted which is based on the name of the first author. BRE and BSI publications are listed separately at the end of the reference section.

Notes

1. In *A truthful song* Rudyard Kipling (1865-1936) asserted that, apart from glazing and plumbing, the building trade had remained pretty much unchanged since the days when the Pharaohs were building the pyramids! Kipling was a leading novelist, poet, and short-story writer, who in 1907 became the first British writer to receive the Nobel Prize for literature.

2. The history of the Building Research Station has been described by a former Director of Building Research, Dr F M Lea, subsequently Sir Frederick Lea, in Science and Building *(Lea, 1971)*.

3. Karl Terzaghi (1883-1963) is the undisputed 'father' of soil mechanics. The publication of his book *Erdbaumechanik* in 1925 laid the foundation of the subject and established its importance within building and civil engineering.

4. In his paper on *The influence of modern soil studies on the design and construction of foundations* which he presented at the Building Research Congress in London in 1951, Terzaghi commented that *'On account of the fact that there is no glory attached to the foundations, and that the sources of success or failure are hidden deep in the ground, building foundations have always been treated as step-children; and their acts of revenge for the lack of attention can be very embarrassing'*.

5. The Building Research Station was founded in 1921 and was initially located in East Acton.

6. Dr L F Cooling joined what was then the Building Research Station in 1927; from 1933 he was the leader of a newly-formed group working on soil problems. For many years this group was known as the Geotechnics Division of BRE.

7. In 1993 the Site Investigation Steering Group of the Institution of Civil Engineers published *Site investigation in construction* in four parts.

8. A national target has been set that by 2008, 60% of additional housing should be on previously developed land and through conversions of existing buildings (Department of the Environment, Transport and the Regions, 2000a). Maximising the re-use of previously developed land should promote urban regeneration and minimise the amount of greenfield land used for development.

CHAPTER 2: REGULATIONS FOR FOUNDATIONS AND GEOTECHNICS

Total foundation failures are rare but inadequate foundation performance is not uncommon and may render a building unfit for its intended purpose even where there is no danger of structural collapse. It is appropriate therefore to review the ways in which controls and safeguards are imposed so that buildings are located on stable ground and do minimal damage to the local environment. Many of these safeguards are beneficial to the developer, as well as the owner of the building, in helping to avoid expensive mistakes in the ground. This chapter deals with planning constraints, building regulations, special features of regulations for contaminated land, codes and standards, and insurance and litigation. Most of the content is specific to the UK and, as regulatory regimes change more rapidly than does soil behaviour, could soon become out-of-date! Nevertheless, this subject will be of considerable importance to those engaged in building developments.

2.1 Introduction

It has been pointed out[1] that *'Virtually every structure is supported by soil or rock. Those that aren't either fly, float or fall over'*! Fortunately most structures that are built on soil and rock do not fall over. Nevertheless, the consequences of a foundation failure, which results in the collapse of a building, are so great that certain rules have been imposed on engineers and builders to ensure that the health and safety of the occupants, and public safety generally, are not jeopardised by poor design or construction practices. Indeed, the present paucity of collapsing buildings in the UK may be at least partly attributable to the development over several centuries of a legal framework controlling building works[2].

The need for a legal system which can protect the purchaser of a house from an incompetent builder has long been appreciated. The famous code of laws of Hammurabi[3], the king of Babylon, included a law that if a man built a house for another so carelessly that it fell down and killed the owner; the housebuilder was to be put to death[4]! Since it is doubtful that Hammurabi would have bothered drafting laws for hypothetical or very rare situations, it can be concluded that it was not uncommon for houses to fall down owing to the carelessness of builders in Babylon nearly four thousand years ago.

In a chilly northern climate, the provision of adequate shelter for man and his goods comes second only to the provision of food as a necessity for the maintenance of life. It is not surprising, therefore, that for many centuries there have been regulations to control building work in England. The most basic requirements are that the building should be structurally stable and provide protection from the weather. However, the Rebuilding Act of 1667, which was enacted after the Great Fire of London had destroyed 80% of the city in September 1666[5], was largely concerned with minimising fire risk by such features as substantial party walls, which could prevent a fire spreading to adjacent houses.

REGULATIONS FOR FOUNDATIONS AND GEOTECHNICS

Although total foundation failure is a very rare event, inadequate foundation performance may render a building unfit for its intended purpose even where there is no danger of collapse. With growing public awareness of possible negative impacts of construction on the environment, planning constraints are becoming more onerous and are encouraging builders to redevelop brownfield[6] sites where a wide range of ground problems may be encountered. It is therefore appropriate to review of the ways in which controls and safeguards are imposed to ensure that buildings and other constructions are located on stable ground and do minimal damage to the local environment.

It would be optimistic to assume that all building professionals are enthusiastic about regulations. Benjamin Franklin[7] famously asserted that '*in this world nothing can be said to be certain, except death and taxes*' and some hard-pressed building practitioner might well feel that regulations would make an appropriate third item on Franklin's list. However, many of these regulatory requirements are beneficial to the developer in helping to avoid expensive mistakes in the ground and they certainly provide a much needed safeguard for the individual house purchaser.

There is one respect in which Chapter 2 is unlike the other chapters in this book. While textbooks on geotechnics occasionally need to be updated because of advances in the understanding of soil behaviour, the materials that they describe are not normally subject to change. This permanence is in marked contrast to the rapidly changing world of regulations and legislation. It was recognised that in a chapter on regulations for foundations, the information would only be applicable to the UK and even there it could become out-of date quite soon. Nevertheless, the regulatory context within which foundations are designed and built is of importance and it was considered that this facet of the subject should not be excluded.

In the course of obtaining approval to proceed with a building scheme, the developer requires planning permission and the structural and geotechnical designs need to satisfy the Building Regulations to ensure the safety and health of people in and around the buildings. Whatever the size and nature of the buildings and their intended use, consideration of the ground conditions should begin at the earliest stages in the planning of a development. Problems can occur when a developer has obtained outline planning permission but discovers at the detailed design stage that unforeseen foundation costs will be incurred in complying with Building Regulations.

It is important that all parties involved in a building development are aware of their particular responsibilities to ensure that the works are carried out in a safe manner. The Construction (Design and Management) Regulations 1994 are relevant[8].

2.2 Planning constraints

The British planning system operates in the public interest to control development. There is a long history of such controls; in Elizabethan times an attempt was made to limit the growth of London by prohibiting building except on existing foundations. However, such controls often have unexpected and unwelcome effects. By the eighteenth century the well-meant policy of previous centuries had produced illegally built warrens out of sight of the authorities[9]. Such housing had no amenities and presented major problems for law enforcement. This shows that those drafting such laws need to consider carefully not only what they wish to achieve but also the actual effect that the legislation is likely to have.

Planning permission is required for any development of land and the planning system is designed to facilitate the provision of homes and buildings in a manner that is compatible with sustainable development. The primary legislation governing the planning process is contained in a number of Acts of Parliament[10].

The fundamental requirement is that development must have prior permission[11]. Development is defined as *'the carrying out of building, engineering, mining or other operations in, on, over or under land, or the making of any material change in the use of any buildings or other land'*. Certain works and uses do not come within the definition, such as *'the use of buildings or land within the curtilage of a dwelling-house for any purpose incidental to the enjoyment of the dwelling-house as such'*. When making an application for planning permission, the developer is only required to prove the case for his development if it would cause demonstrable harm to the interests of conservation. The onus is on the planning authorities to give reasons for refusing permission or for imposing conditions on the proposal.

A series of Planning Policy Guidance Notes[12] describe Government policies on different aspects of planning. PPG14 examines the effects of ground instability on development and land use. It provides guidance on geological and geotechnical constraints on planning and development, including natural and man-made cavities, unstable slopes and landslides, ground compression and subsidence. The guidelines aim to ensure that development is suitable and that physical constraints are taken into account at all stages of the planning process.

2.3 Building regulations

The Public Health Act of 1875 gave local authorities the power to enact by-laws with respect to the structure of walls, foundations, roofs and chimneys of new buildings for the purposes of stability, fire prevention and health. While these by-laws served a very useful function, it was appreciated that they could be a barrier to innovation and in 1936 the Building Research Board commented that *'There is too frequently in byelaws a tendency to specify particular forms of construction, forms usually based on particular materials ....*

REGULATIONS FOR FOUNDATIONS AND GEOTECHNICS

The ideal is undoubtedly that the byelaws should state the performance required leaving open the methods by which the requisite standard is to be attained...'[13].

In the 1960s, building regulations were introduced applicable to the whole country; the Scottish Building Regulations were issued in 1963 and those for England and Wales in 1964. The Building Research Board was pleased to note that '*The form of the new regulations ... stems from the principles first put forward by this Board in 1936 and amplified in later years. Progressively the building byelaws, and now the building regulations which are replacing them, have reflected more and more the philosophy enunciated by our predecessors*'[14]. The building regulations are operated independently of the planning system and plans may be approved prior to, in parallel with, or after determination of a planning application.

With certain specified exceptions, all building work, as defined in the regulations, in England and Wales is governed by building regulations. A separate system of building control applies in Scotland and in Northern Ireland. The Building Act 1984 consolidated most of the earlier primary legislation relating to buildings. There was a major revision of the building control system in England and Wales in 1985. The Schedule 1 Requirements were written in general terms requiring reasonable standards of health and safety. Reference was made to supporting Approved Documents, which could be complied with in various ways. Builders and developers are required by law to obtain building control approval in the form of an independent check that the building regulations have been complied with.

The Building Regulations 2000[15] impose requirements on people carrying out building works in England and Wales. There are 24 regulations arranged in five parts and there are three schedules. Schedule 1 of the Regulations presents the technical requirements, expressed in functional terms, to which building work must comply. The regulations are concerned principally with safe construction and the provision of services to and from buildings such that the '*health, safety, welfare and convenience of persons in and about the building*' are secured. The regulations impose minimum standards relating to design, workmanship and materials.

The technical requirements in Schedule 1 are mostly set out in a functional form for building design and construction and they are presented in 13 parts. Each part of Schedule 1 is supported by an Approved Document, which gives guidance on how to comply with the requirements in some of the more common building situations. The series of Approved Documents (Box 2.1) give some technical detail and refer to British Standards and guidance material such as BRE publications. Compliance must be shown with the Building Regulations, but not necessarily the contents of the Approved Documents, which give practical guidance on meeting the requirements of the regulations.

There is no obligation to adopt a particular solution contained in an Approved Document if it is preferred to meet the relevant requirement in some other way. However, compliance with the Building Regulations is deemed to have been satisfied when the guidance contained in Approved Documents is followed.

> **Box 2.1 Approved Documents in which geotechnics has some relevance** [16]
> A: Structure
> C: Site preparation and resistance to contaminants and moisture
> H: Drainage and waste disposal

Part A of Schedule 1 to the Building Regulations, *Structure*, is concerned with the strength, stability and resistance to deformation of the building. Requirements A1 and A2 have some geotechnical significance. Requirement A1 (Box 2.2) stipulates that buildings are constructed so that all loads are transmitted to the ground safely and without causing ground movement that will impair the stability of another building.

Requirement A2 (Box 2.3) stipulates that the stability of the building is not impaired by ground movement caused by certain processes which are not directly associated with the loading applied to the ground by the building. In connection with requirement A2b, attention is drawn to PPG14.

> **Box 2.2 Requirement A1: loading**
> (1) The building shall be constructed so that the combined dead, imposed and wind loads are sustained and transmitted by it to the ground:
> (a) safely; and
> (b) without causing such deflection or deformation of any part of the building, or such movement of the ground, as will impair the stability of any part of another building.
>
> (2) In assessing whether a building complies with the above regard shall be had to the imposed and wind loads to which it is likely to be subjected in the ordinary course of its use for the purpose for which it is intended.

Practical guidance on meeting the requirements of Part A of Schedule 1 is in Approved Document A[17], which gives guidance on the design of traditional strip foundations, including minimum strip foundation widths. It is concerned with the provision of adequate width and strength to limit to acceptable levels the settlement of foundations on ground that is compressible and laterally variable in its compressibility.

Part C of Schedule 1 to the Building Regulations, *Site preparation and resistance to contaminants and moisture*, addresses the hazards posed by contaminants in the ground or groundwater. Requirement C1 (Box 2.4) deals with preparation of the site and resistance to contaminants and requirement C2 (Box 2.5) resistance to moisture.

> **Box 2.3 Requirement A2: ground movement**
> The building shall be constructed so that ground movement caused by:
> (a) swelling, shrinkage or freezing of the subsoil; or
> (b) land-slip or subsidence (other than subsidence arising from shrinkage), in so far as the risk can be reasonably foreseen,
> will not impair the stability of any part of the building.

REGULATIONS FOR FOUNDATIONS AND GEOTECHNICS

The Building Regulations are concerned principally with the health and safety of people in and about buildings. The performance requirements of the regulations could be inferred from the words '... *ground movement ... will not impair the stability of any part of the building.*' However, all materials deform under stress and it is a matter of degree. Therefore, in assessing compliance with the regulations, judgement has to be exercised about the impact of any deformation and instability arising from ground conditions and the provision of foundations. A decision on what constitutes an impairment of stability that poses a threat to health and safety, rests with the building control service[18], who are responsible for implementing the regulations. Developers should ascertain the approach of the building control service in advance of an application for inspection.

> **Box 2.5 Requirement C2: resistance to moisture**
> The floors, walls and roof of the building shall adequately protect the building and people who use the building from harmful effects caused by:
> (a) ground moisture;
> (b) precipitation and wind-driven spray;
> (c) interstitial and surface condensation.

2.4 Contaminated land

Ground contamination can impact on a building development. The contamination may present a hazard for workers during building and could have adverse effects on the built environment. Occupiers or owners of contaminated land could, in some circumstances, be required to remediate the ground. Development on land affected by contamination[19] will normally be regulated through the planning and development control regimes[20]. Planning policy guidance note PPG 23 deals with planning and pollution control.

> **Box 2.4 Requirement C1: preparation of site and resistance to contaminants**
> **(1)** The ground to be covered by the building shall be reasonably free from any material that might damage the building or affect its stability, including vegetable matter, top-soil and pre-existing foundations.
> **(2)** Reasonable precautions shall be taken to avoid danger to health and safety caused by contaminants on or in the ground covered or to be covered by the building, and any land associated with the building.
> **(3)** Adequate sub-soil drainage shall be provided if it is needed to avoid:
> (a) the passage of ground moisture to the interior of the building;
> (b) damage to the building, including damage through the transport of water-borne contaminants to the foundations of the building.
> **(4)** For the purposes of this requirement, 'contaminant' means any substance which is or may become harmful to persons or buildings including substances which are corrosive, explosive, flammable, radioactive or toxic.

Land contamination is a material consideration within the planning regime and a planning authority has to consider the potential implications of contamination when it is developing structure or local plans, and when considering individual applications for planning permission. The objectives of the planning system in respect of land affected by contamination are:
❏ to encourage the redevelopment and beneficial reuse of previously developed land, both to bring about the social, economic and environmental regeneration of that land and surrounding area, and to

reduce unnecessary development pressures on greenfield sites (although some greenfield sites may also be contaminated);
❐ to ensure, that any unacceptable risks to human health, buildings and the environment from contamination are identified and properly dealt with, as new development and land-uses proceed.

The only Approved Document directly concerned with contaminated land is associated with Part C of Schedule 1 to the Building Regulations, *Site preparation and resistance to contaminants and moisture*, although Regulation 7 of the Building Regulations on materials and workmanship and Part A of Schedule 1 on structure may also be relevant.

Part IIA of the Environmental Protection Act 1990, which was introduced by Section 57 of the Environment Act 1995, inserted new powers to order remediation of contaminated land[21]. A remediation notice may be served on an appropriate person, requiring that person to clean up contaminated land. Land is considered to be contaminated only where it appears to the authority that, by reason of substances in, on or under the land, significant harm is being caused or there is a significant possibility of such harm being caused; or pollution of controlled waters is being, or is likely to be, caused. Harm is defined by reference to harm to the health of living organisms or other interference with the ecological systems of which they form a part and, in the case of man, is stated to include harm to property. Where building development is proposed, risk assessment should consider a broad spectrum of potential receptors, which may be vulnerable to different hazards. The regulatory regime requires local authorities to identify contaminated sites and determine the appropriate remediation requirements. An objective is '*to reduce or remove perceived blight associated with land contamination by promoting its remediation*'.

The issue of waste is now highly significant in the context of site redevelopment, whether a site may or may not contain contaminants. The high cost of disposal and requirements for waste management licensing arising from the Landfill Tax and the application of the European Landfill Directive have major implications for redevelopment of brownfield sites and may dictate the course of the development[22].

Groundwater resources, both in quality and yield, can be put at risk by a variety of human activities. These range from specific point sources of potential pollution such as landfills, chemical works and petrol stations, to diffuse pollution from agricultural land use practices and atmospheric fall-out spreading over large areas. These activities are subject to legislative control to varying extents and also are regulated through guidance, codes of practice and best practice recommendations.

Any work on contaminated land must take into account the requirements of guidance from the Health and Safety Executive on the protection of workers and the general public during the development of contaminated land. There are additional constraints in legislation on waste disposal, pollution control and the Health and Safety at Work Act.

2.5 Codes and standards

The Building Regulations Approved Documents call upon widely recognised documents, including codes and standards published by BSI and guidance published by BRE. Foundation and geotechnical design should comply with the provisions of the appropriate British Standard codes of practice[23]: BS 8004: 1986 *Code of practice for foundations* is particularly important. However, a suite of European geotechnical design codes and standards will, in due course, replace much of the collection of national design documents.

Structural Eurocodes[24] are currently being published; the geotechnical Eurocode is expected to lead to the adoption of a different approach to foundation design from that in BS 8004. The relevant Eurocode for geotechnical engineering, Eurocode 7: *Geotechnical design*, is intended to be applied to the geotechnical aspects of the design of buildings and civil engineering works and is concerned with the requirements for strength, stability, serviceability and durability of structures. Eurocode 7 has two parts; Part 1: *General rules*[25] and Part 2: *Ground investigation and testing*[26]. There are separate European standards for the execution of geotechnical works[27] and for soil testing procedures[28].

Part 1 of Eurocode 7 is a general document covering a broad range of design situations likely to be encountered in normal practice. It is not clear precisely how much of the existing BSI code and standard material will need to be retained, when Eurocode 7 becomes a full European standard, to preserve confidence in the adequacy of design[29].

The Eurocode has introduced new methods of performing geotechnical design calculations. Traditional British geotechnical design has featured the use of a global or 'lumped' factor of safety to ensure that the resistance offered by a foundation, say, is sufficiently greater than the forces attempting to disturb it. The partial factor method separates the global factor of safety into some of its component parts. For example, a partial factor is applied both to the load and to the material strength. In the global factor method, values of applied load and element resistance are assumed to be approximately average values or, more probably, moderately conservative values, and a single factor is used to reduce the loads, or increase the resistance, by an amount derived from experience of the past behaviour of similar designs.

Partial factors deal explicitly with uncertainty in unconnected parameter values and separately for different limit states, whereas the global factor approach embodies all uncertainty into one large factor and cannot explicitly cater for the different limit states. Much experience of calibrating design methods with observations of the performance of structures and analysis of failures has been embodied in the global values that have been used for so long and there is understandable reluctance to abandon a tried and tested method of design until sufficient experience of the use of the partial factor design method has been gained. However, a great deal of work has been undertaken to calibrate the new design methods in the Eurocodes to previous practice to ensure that new designs will be sufficiently safe.

Part 2 of Eurocode 7 gives guidance on the planning of ground investigations, sampling and groundwater measurements. Guidance is also given for the planning of laboratory and field tests and the derivation of ground properties that are used in the geotechnical design of foundations for buildings and other civil engineering works.

2.6 Insurance and litigation

Despite the safeguards built into both the planning process and the Building Regulations, ground and foundation problems continue to occur and structures deform and crack. The cause of foundation movement is most commonly associated with changes in the ground, rather than with the process of constructing a building. Several circumstances have combined to give prominence to the subject of foundation damage in low-rise housing:

❒ Some years ago, insurance companies added subsidence and heave to most domestic household building policies.

❒ Some very dry summers have caused substantial movements of house foundations on shrinkable clays.

❒ Most low-rise housing in the UK is owned by the occupiers, and the house represents their major asset; a foundation problem can seriously affect the value of this asset.

❒ Our society is increasingly litigious.

In 1971, insurance companies added subsidence and later heave to most domestic household building policies. Typical wording in a policy might be: '....*damage to buildings caused by subsidence and/or heave of the site on which the buildings stands and/or landslip.....*'. Cover is limited to subsidence and heave of the site and excludes damage caused by movements within the building itself. It was assumed that claims would be relatively few because few buildings are so severely damaged by subsidence or heave of the ground that major re-building is required. However, insurance policies did not specify how much movement is required for a valid subsidence or heave claim and many owners of low-rise buildings have made claims for relatively minor instability and consequent damage. Figure 2.1 shows how subsidence insurance claims grew between 1975 and 2000.

The emergence of subsidence as a major insurance risk in the late 1970s heightened general awareness of the risks of ground movements to buildings of all types and sizes. For example, the owners of large and prestigious buildings have become more concerned about the impact of nearby tunnel and excavation works, often commissioning monitoring to acquire evidence of any movement to their building arising from the works.

REGULATIONS FOR FOUNDATIONS AND GEOTECHNICS

Insurance considerations increasingly dictate the needs for ground study and foundation design, for low-rise development. NHBC[30] specifies ground and foundation requirements over and above those stipulated in the Building Regulations, as a condition of compliance with their warranty schemes. The NHBC Standards give the technical requirements, performance standards and guidance for the design and construction of dwellings acceptable to NHBC. Alternative standards of performance are acceptable only if, in the opinion of NHBC, the technical requirements are met and the standard achieved is not lower than the performance standard.

The law applying to building problems associated with the ground is complex, with its dependence on precedent. The law develops through specific cases and in the matter of building subsidence has changed substantially over the years[31]. As a result, the law is imposing itself upon the thinking of building professionals. Building owners, occupiers and users are now ready to seek redress for perceived loss, either in income from a late construction completion or through the cost of remedial works to a damaged building. There has been a great increase in litigation as owners and insurers have sought to recover losses from local authorities, building designers, constructors and others. Complex legal issues involving professional indemnity and liabilities of designers, surveyors, contractors, developers, landlords and owners are beyond the scope of this book[32].

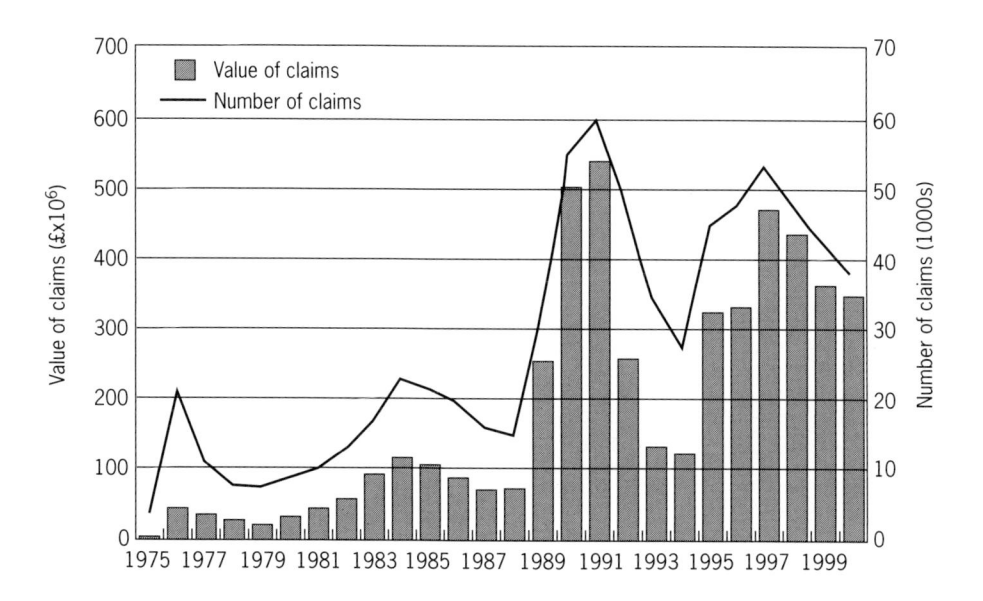

Figure 2.1 Growth of subsidence insurance claims – *after Driscoll and Chown, 2001*

Notes

1. On page 1 of Coduto (1998), where it is attributed to Handy (1995).

2. Collapsing houses were not uncommon in eighteenth century London. Picard (2000) comments on page 42 that *'Ruinous houses in the old parts of London might be good for a few nights' shelter before they fell down – a frequent event in the eighteenth century'*.

3. Hammurabi ruled Babylon for 43 years around 1700 BC and was one of its greatest kings. He revised earlier legal codes and developed the famous Code of Hammurabi, one of the first law codes in history. The code contained nearly 300 legal provisions and covered such matters as divorce, adoption, false accusation, theft, murder, manslaughter, military service, witchcraft, legal problems of trade enterprises, land and business regulations, loans, and debts.

4. This law is quoted by Gadd (1951), p40, who states that there were 282 laws in Hammurabi's code and comments that some of them seem very harsh!

5. The Great Fire of London began in Pudding Lane on 2 September 1666 and ended on 6 September, after destroying most of the City of London. Houses were built mainly of wood and there was no effective firefighting system. About 13,000 houses, 89 parish churches and St. Paul's Cathedral were destroyed. The loss of property has been estimated at £10 m (about £2 billion in today's money). Fire insurance had not been introduced. There were few fatalities but about 200,000 people were made homeless. New buildings of brick and stone replaced the wooden houses which had been destroyed.

6. The term *brownfield* has been widely used in recent years, but it has no universally recognised definition. In planning terms, it is commonly understood as the opposite of *greenfield*, which means land that has not been previously developed. Building professionals usually regard brownfield as any land that has been previously developed, and it is in this sense that the term is used in this book. Chapter 8 examines those aspects of geotechnics that are peculiar to brownfield sites.

7. Benjamin Franklin (1706-1790) was an American writer, publisher, scientist, philanthropist and diplomat; this melancholy observation was contained in a letter from him dated 13 November 1789. Franklin raised money to help build America's first city hospital, the Pennsylvania Hospital and helped establish the American Philosophical Society and an academy that became the University of Pennsylvania.

8. *Managing construction for health and safety*, Construction (Design and Management) Regulations 1994 Approved Code of Practice L54 (Health and Safety Commission, 1995); *A guide to managing health and safety in construction* (Health and Safety Commission, 1995).

9. Picard (2000) pp 42-43.

10. Primary legislation governing the planning process is contained in the following Acts of Parliament:
- The Town and Country Planning Act 1990
- The Planning (Listed Buildings and Conservation Areas) Act 1990
- The Planning (Hazardous Substances) Act 1990.

Each of these Acts has been amended by the Planning and Compensation Act 1991. There is subordinate legislation in the form of several Orders.

11. Some minor classes of development, such as small extensions and alterations to dwelling-houses, receive permission in advance through the Town and Country Planning General Development Order of 1988.

REGULATIONS FOR FOUNDATIONS AND GEOTECHNICS

12. The following Planning Policy Guidance Notes (Department of the Environment, Transport and the Regions, 2000a) are of particular relevance to proposals for housing development:
- PPG1: General policy and principles
- PPG3: Housing
- PPG14: Development on unstable land
- PPG 23: Planning and pollution control
- PPG 25: Development and flood risk.

13. The Building Research Board was established in 1920 and continued until 1965; Lord Salisbury was its first chairman. In its early days, the Board defined the objectives of the Building Research Station. The quotation from the 1936 report of the Building Research Board is by Lea (1971) p85.

14. This quotation from the 1964 report of the Building Research Board is by Lea (1971) p87.

15. The Building Act 1984 provides the legal framework for:
- The Building Regulations 2000
- The Building (Approved Inspectors etc.) Regulations 2000
- The Buildings (Local Authority Charges) Regulations 1998.

16. Approved Documents are revised from time to time. The following revisions are referred to in this book: A – 2004, C – 2004, H – 2002.

17. Approved Document A, Section 2E: *foundations of plain concrete*, gives guidance on the design of strip foundations, including the minimum width of footings. It is concerned with the provision of adequate width and strength to limit to acceptable levels the settlement of foundations on ground that is compressible and laterally variable in its compressibility. Section 2E also contains guidance on the minimum depth of strip foundations. Except where founded on rock, a minimum depth of 0.45 m is specified to avoid frost action. On shrinkable clays, a minimum depth of 0.75 m is required but greater depths will often be needed to ensure that ground movements do not impair the stability of the building. Section 1 makes reference to BS 8004: 1986 *Code of practice for foundations* and warns of the need to research conditions of ground instability from such features as geological faults, landslides and disused mines.

18. There are two types of building control service: that provided by the local authority and that provided by approved inspectors.

19. Contaminated land can be defined as land that contains substances that, when present in sufficient quantities or concentrations, are likely to cause harm, directly or indirectly, to man, to the natural environment or to the built environment.

20. Planning authorities are required to consult statutory consultees over particular types of planning application to ensure that material planning considerations, such as the actual or possible presence of contamination, are properly addressed. Where measures are required to deal with contamination, planning authorities can attach conditions to the planning permission or use planning agreements under Section 106 of the Town and Country Planning Act 1990. Harris et al (1998) have described the key elements and main developments in environmental legislation and policy as they affect the remedial treatment of contaminated land.

21. After five years of consultation and review, these powers came into force on 1 April 2000.

22. The EU Landfill Directive is being implemented in stages through the Landfill (England and Wales) Regulations 2002. The implications of the legislation are described in *The practical guide to Waste Management Law* (Hawkins and Shaw, 2004).

23. The principal BSI documents that are relevant to ground engineering are listed in the References.

24. The structural Eurocodes are European harmonised design codes which are being implemented in stages. Firstly, a trial document is published by national standards bodies with an accompanying National Application Document that specifies how the required level of safety should be attained in the design process and any other national peculiarities. After a process of further harmonisation in which problems encountered during the trial period are resolved, the Eurocode is fully implemented, at which time the corresponding national code should be withdrawn from use.

25. Eurocode 7: Geotechnical design – Part 1: General rules was published by BSI in 1995 as a trial code, DD ENV 1997-1:1995. Following extensive revision of the ENV, a draft of the Eurocode, prEN 1997-1: 2003, was issued in 2003 for public comment prior to publication as a Euronorm (EN). Commentaries on the EN are in preparation, see Frank *et al*(2004).

26. Eurocode 7: *Geotechnical design* – Part 2: *Ground investigation and testing* was published by BSI in 2000 as two trial codes, DD ENV 1997-2: 2000 Part 2: *Design assisted by laboratory testing* and DD ENV 1997-3: 2000 Part 3: *Design assisted by field testing*. In the revision, Parts 2 and 3 of the ENV have been combined and a draft of the Euronorm, prEN 1997-2: 2003, *Geotechnical design – Part 2: Ground investigation and testing*, was issued in 2003 for public comment prior to publication as an EN.

27. CEN technical committee TC288 is preparing standard specifications for the execution of geotechnical works. These standards are listed in the References. The standards for bored piles, displacement piles and micropiles will be of particular interest for building foundations.

28. CEN technical committee TC341 is preparing standard specifications for drilling and sampling, groundwater measurements, laboratory and field tests on soils and rocks, tests on structures and parts of structures. The standards that have been published, or are close to publication, are listed in the References.

29. An account of the progress of the European geotechnical design codes and standards has been given by Driscoll (2004). It is suggested that the European codes and standards may not fully replace the national geotechnical documents until 2008 or later.

30. NHBC (National House Building Council) has published standards for the United Kingdom's house-building industry; they have evolved over time in response to the need to minimise construction defects and ensure compliance with legislation. The standards provide the technical requirements, performance standards and guidance for the design and construction of dwellings which are acceptable to NHBC. These standards are more prescriptive than the Building Regulations and the foundation stability performance required by NHBC appears to be more stringent than that required to ensure the health and safety of people in and around the buildings. Other insurers, such as Zurich, have prepared similar guidelines.

31. The emergence of law relating to latent damage from ground problems resulted in the Latent Damage Act of 1986.

32. A fuller understanding of the legal issues relevant to ground problems can be obtained from Eaglestone and Apted (1988).

CHAPTER 3: TYPES OF GROUND

Ground materials can be categorised in four significant ways:
❐ *The two broad categories of soil and rock.*
❐ *Soils composed of mineral grains should be distinguished from soils containing a significant quantity of organic matter.*
❐ *There is a fundamental difference between the behaviour of coarse soils, such as sand and gravel, and fine soils, such as clay.*
❐ *The distinction between ground consisting of natural soil and rock formed by geological processes and ground made by man, usually described as 'fill' or 'made ground', which can be composed of either natural soil and rock or waste products.*

The following simple classification of ground types has been adopted based on these basic distinctions: coarse soil, fine soil, organic soil, rock and fill. This chapter contains basic information about ground conditions. Most of the problems with the ground are associated with natural soil and fill rather than with rock, so simple parameters are introduced to describe soil composition. Classification tests can be used to identify the general type of soil and to indicate the category to which it belongs. Laboratory tests for the measurement of engineering properties are in Chapter 4, which deals with soil behaviour; site investigation and field tests are in Chapter 5, which deals with ground assessment.

3.1 Introduction

House foundations were of concern long before the science of geotechnics was developed. A well-known parable in the New Testament contrasts the unhappy experience of building a house on sand with the rewarding experience of building a house on rock, with the failure of the house built on sand attributed to rain and floods[1]. Parables link spiritual realities to the everyday experiences of ordinary people; the likely fates of houses built on these two types of ground were clearly known two thousand years ago.

The division of ground materials into the categories of *soil* and *rock* seems a simple distinction but in practice there can be difficulties because civil engineers and geologists have different definitions for these terms. Geologists are concerned with the mode of origin and the structure of a deposit, whereas engineers are more interested in such aspects of its behaviour as how easily it can be excavated, its resistance to erosion and its deformation under load. In this book, the terms soil and rock are generally used in their engineering sense, in which soil can be defined as a natural aggregate of mineral particles that can be separated by gentle mechanical means, such as agitation in water, in contrast to rock, which is a hard and rigid deposit formed from a natural aggregate of minerals[2] bonded together by strong and permanent cohesive forces. Engineers regard most superficial deposits as soil, whereas soil scientists, geographers and geologists use the term in an agricultural sense and describe the superficial deposits underlying the thin layer of agricultural soil as subsoil or drift[3]. Another important distinction is between ground consisting of natural soils and rocks formed by geological processes and ground that has been made by man, which is usually described as fill or made ground[4]. Based on the two basic distinctions above, the following simple classification of ground types has been adopted:

- coarse soil;
- fine soil;
- organic soil;
- rock;
- fill.

While the first four types refer to naturally occurring materials, fill can be composed of either natural soil and rock, or waste products. Most of the problems with the ground are associated with natural soil and fill rather than rock and it is appropriate therefore to first consider soil composition prior to examining the five types of ground.

Properties which are simple to observe and measure enable soil to be classified in ways that are relevant to their load-carrying characteristics. Reference is made to many of these classification tests in this chapter. Some of the more important laboratory deformation and strength tests are described in Chapter 4, which deals with soil behaviour. Chapter 5 on ground assessment includes some important field tests. BS 1377: 1990 *Methods of test for soils for civil engineering purposes*[5], specifies standard procedures for carrying out most commonly used soil tests and European test standards are being prepared[6].

3.2 Soil formation

Soil materials are derived from the physical and chemical weathering of rocks. Soils that remain at the location of the weathering are called residual soils and retain some elements of the parent rock. Other soil types have been transported from the location of the weathering process by various means, summarised in Box 3.1.

Where a soil deposit has only experienced vertical compressive strain under the effect of its weight, it is said to be normally consolidated. After deposition and consolidation, many soil deposits have been subjected to erosion. This means that at any given point in the deposit, the current vertical stress, or overburden pressure, is smaller than the vertical stress experienced in the past. This effect is particularly important for clayey deposits where a heavily overconsolidated soil will be much stiffer

Box 3.1 Origins of soil deposits
Alluvial soil, which is sometimes referred to as alluvium, has been transported to its location by rivers and streams. This may have resulted in a soil deposit which contains layers of different soil types. Marine soil has been deposited in a marine environment. Glacial soil has been transported and deposited by glaciers; it can be heterogeneous with a wide range of particle size, quite unlike soil formed by deposition in water. Soil deposited by wind is usually loose with a relatively uniform grading.

TYPES OF GROUND

and stronger than the same soil in a normally consolidated condition. For clays the distinction between normally consolidated deposits and overconsolidated deposits is critical to an understanding of the response of the soil to loads (Box 3.2).

Glacial soil has been transported and deposited by glaciers and ice sheets. Glacial till is a heterogeneous soil with a wide variety of particle sizes. It can contain rock debris and a clay matrix. It may have been consolidated under a considerable thickness of ice, which has subsequently been removed, and so is often overconsolidated and stiff.

Soil deposited by wind is usually loose and has a relatively uniform grading. Loess deposits are composed of silt sized particles and, although not common in the UK, cover wide areas of the United States, China and Central Europe[7].

Alluvial soil is deposited from still or slow-moving water found typically in lakes, swamps and river flood plains. Deposition under marine conditions, except near river mouths, usually leads to a homogenous deposit because the clay particles floc together to form larger particles which settle out uniformly with the silt and, in some cases, fine sand. Many of the clays in south-east England, such as London Clay and Gault Clay, are marine deposits.

Box 3.2 Overconsolidation
The overconsolidation ratio (OCR) is defined as the ratio of the maximum past vertical effective stress to the current vertical effective stress (effective stress is defined in Section 4.3). Where overconsolidation has been caused by the erosion of overlying strata, the maximum OCR will be close to the ground surface and will decrease with depth. However, near the ground surface overconsolidation is often due to desiccation and capillarity, so that OCR may be larger than would be expected from the geological history of the deposit. Changes in groundwater level can also cause overconsolidation. By definition, a normally consolidated soil has OCR = 1. OCR has considerable significance for soil behaviour. However, it can be determined only when both the current vertical effective stress and the maximum past vertical effective stress (preconsolidation pressure) are known. The former can be readily evaluated, but the latter presents some problems. Overconsolidation is of particular significance for clay soil and there are procedures based on laboratory oedometer tests for determining the preconsolidation pressure (Section 4.5).

Varved or laminated clay consists of alternating thin layers of silt and clay which have been deposited in lakes fed by ice. The thin layers typically may be only a few millimetres thick. Varved clay may be highly permeable in a lateral direction and, where exposed, can erode rapidly because of the presence of the silt layers.

A clay deposit may have very thin laminations of coarse silt or sand which may be no more than a grain or two thick; they are often difficult to detect by eye with the soil at its natural water content. On partial drying, the coarser seams dry first to a light colour and can then be detected. These thin laminations can lead to a much higher horizontal permeability of the soil mass than would be otherwise expected and can transmit pore water pressures (Section 4.4) which influence the behaviour of the soil. In stiff overconsolidated clay these laminations can create planes of weakness.

A residual soil is the in-situ product of weathering of rock and minerals; it may retain some of the characteristics of the parent rock. Quite often it contains pieces of the parent rock, which may give trouble during site investigation, in the sinking of piles and in excavations for foundations. In humid tropical climates, weathering is rapid and may extend to great depths. Residual soil is occasionally encountered in the UK, eg the china clay deposits in Cornwall. The engineering properties of a residual soil are partly governed by the high bonding stresses which existed in the parent rock, so that the soil will usually be fairly stiff and behave as a strongly overconsolidated clay. Occasionally, the soil may become soft where residual ground occupies low-lying waterlogged areas. The soil may vary, therefore, from a soft to a hard material intermediate between soil and rock.

3.3 Soil composition – solid particles

Natural soils and fills are particulate systems made up of individual solid particles ranging in size from a few microns for clay particles, through silts, sands and gravels, up to boulders with dimensions of over one metre. Many clay-sized particles exist as tiny, flat plates. Larger particles tend to be more nearly spherical and, depending on their composition and origin, the particles may be rounded or may contain sharp edges. The solid particles are usually minerals but may include organic matter.

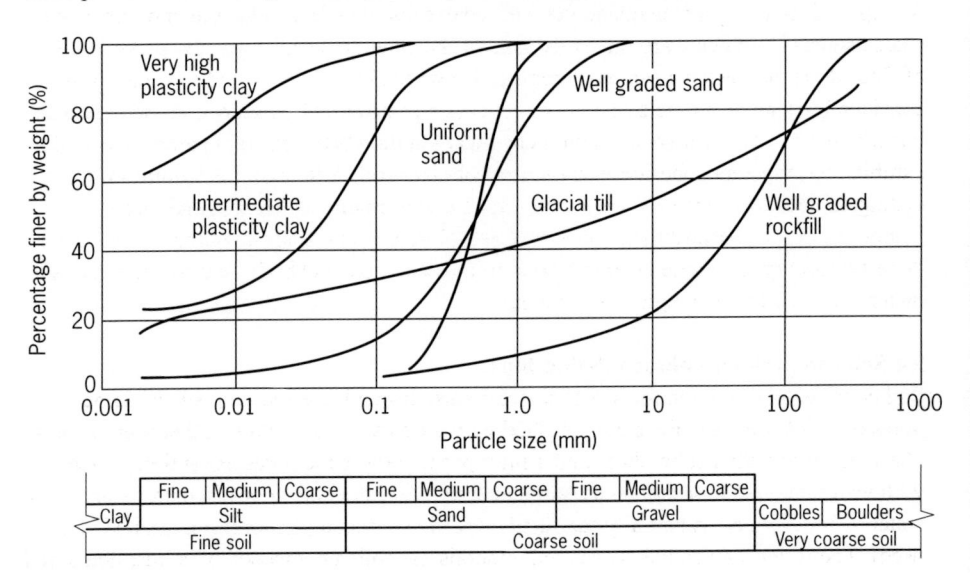

Figure 3.1 Typical particle size distributions

TYPES OF GROUND

Particle size has a strong effect on soil behaviour and provides a useful method of classification (Table 3.1). The distribution of particle sizes is expressed as the percentage by mass smaller than a given particle diameter and can be plotted in the form of a particle size distribution or grading curve, using a logarithmic scale for particle size because the range of particle sizes is very large. Some examples of grading curves for common types of soil are shown in Figure 3.1. It is seen that soils rarely fall completely within one of the categories listed in Table 3.1.

Table 3.1 Particle size classification[8]

Soil type		Particle size – mm
Boulder		>200
Cobble		60 to 200
Gravel	Coarse	20 to 60
	Medium	6 to 20
	Fine	2 to 6
Sand	Coarse	0.6 to 2
	Medium	0.2 to 0.6
	Fine	0.06 to 0.2
Silt	Coarse	0.02 to 0.06
	Medium	0.006 to 0.02
	Fine	0.002 to 0.006
Clay		< 0.002

A helpful distinction can be drawn between coarse soil and fine soil. Sand and gravel are coarse soils, silt and clay are fine soils. One of the consequences of the low permeability of fine soil is that, when a sample is removed from the ground, it is held together for a considerable length of time by an effect generated principally by suctions in the pore water but which may also be associated with chemical bonding between solid particles. Soil that exhibits this apparent cohesion is known as cohesive soil. Although it is helpful to distinguish between soil deposits composed of free-draining coarse soil, and those composed of low permeability, cohesive fine soil, there is no sharp dividing line between these two soil types. Some heterogeneous soil deposits include both very coarse granular material and soil with a high clay content.

3.4 Soil composition - phase relationships

Soil consists of a skeleton formed from solid particles with the spaces (usually termed pores or voids) between the particles filled with air and/or water. Soil is, therefore, a multi-phase material with a solid phase and a fluid phase. Where the pores are entirely filled with water, the soil is described as saturated and where the pores contain both air and water, it is described as partially saturated or partly saturated (the term unsaturated is also used). The water may contain dissolved minerals. As soil is composed of solids, liquid and gas, it can be regarded as a three-phase material:

- the solid phase is usually mineral but may include organic matter;
- the liquid phase is usually predominantly water;
- the gas phase is usually predominantly air.

Some special terms are required to describe the relative proportions of solid particles, water and air. Figure 3.2 presents a diagrammatic representation of the phases in a typical element of natural soil or fill: 3.2(a) shows the phases as they would actually exist and 3.2(b) shows them separated to facilitate the identification of the volumes and weights of the individual phases.

The weight of water expressed as a fraction or percentage of the weight of the solid particles is the gravimetric water content[9], sometimes termed moisture content, w. This is the usual way of expressing water content in soil mechanics, because the measurement is made by weighing the sample before and after drying at 105-110°C. The water content is therefore the moisture lost during drying rather than the total moisture content, as some water, which is chemically bonded to the soil grains, may not be removed in the standard drying process[10]. Water contents for saturated soils typically range from 30 to 80%. In terms of the symbols used in Figure 3.2, water content can be expressed as follows:

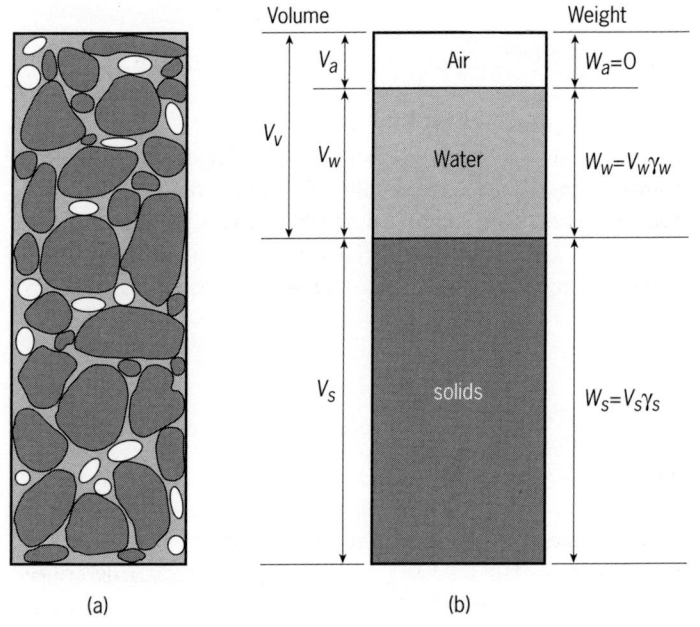

Figure 3.2 Composition of soil

TYPES OF GROUND

$$w = \frac{V_w \rho_w}{V_s \rho_s} = \frac{V_w \gamma_w}{V_s \gamma_s} \qquad (3.1)$$

where:
ρ_w is the density of water (1.0 Mg/m³)
ρ_s is the particle density[11];
the corresponding unit weights are γ_w and γ_s. The density of the solid particles, which is now termed particle density, was previously commonly known as specific gravity. Particle density is not normally used directly for classification purposes but is used in the calculation and interpretation of other test results. For many natural soils and rocks, ρ_s is in the range 2.6 to 2.7 Mg/m³, but is much smaller for organic materials.

The concept of volumetric water content, w_v, which is the volume of water removed by drying as a percentage of the total volume occupied by the soil and water, is used to assist in the determination of shrinking and swelling clay behaviour.

The density of a soil depends on the nature of its formation and on its subsequent stress history. These parameters are commonly used to describe properties related to density:
- void ratio (e)
- porosity (n)
- dry density (ρ_d)
- bulk density (ρ)
- dry unit weight (γ_d)
- bulk unit weight (γ).

The void ratio is the ratio of the volume of the pores to that occupied by the solid particles and is expressed as a decimal number. Porosity is the ratio of the pore volume to the total volume occupied by the soil and is normally multiplied by 100 to give a percentage. Typical values of void ratio for sand are in the range 0.5 to 0.8; for clay they can vary from 1.2 to 8. Using the symbols shown in Figure 3.2, void ratio and porosity can be expressed as follows:

$$e = V_v / V_s \qquad (3.2)$$

$$n = [V_v / (V_s + V_v)]\, 100\ (\%) \qquad (3.3)$$
$$ = [e/(1+e)]\, 100\ (\%)$$

The mass of the solid particles contained in a unit volume of soil is known as the dry density, ρ_d, and is usually measured in Mg/m³. The overall density of the soil is described by the bulk density ρ.

These parameters are all closely linked as the following relationships demonstrate:

$$\rho_d = \rho_s [1 - (n/100)] \quad (3.4)$$

$$\rho = \rho_s \frac{(1+w)}{(1+e)} \quad (3.5)$$

$$\rho_d = \frac{\rho}{1+w} \quad (3.6)$$

Both mass density and unit weight are commonly used. In practice, these terms are not always applied with precision and the expression *density* is often used when *unit weight* is meant. The weight of the solid particles contained in a unit volume of soil is known as the dry unit weight γ_d and is usually measured in kN/m³. The overall unit weight of the soil is described by the bulk unit weight γ. Where the soil is below the water-table, the submerged unit weight γ' is the effective unit weight.

$$\gamma_d = \rho_d g \quad (3.7)$$
$$\gamma' = \gamma - \gamma_w \quad (3.8)$$

where:
    g is the acceleration due to gravity
    γ_w is the unit weight of water
    γ_s is the unit weight of solid particles.

The degree of saturation S_r is the ratio of the volume of water to the volume of pores and is usually expressed as a percentage:

$$S_r = (V_w/V_v)\,100\;(\%)$$
$$= (w\rho_s)/(e\rho_w)\,100\;(\%) \quad (3.9)$$

$S_r = 100\%$ indicates a fully saturated soil; a value smaller than 100% indicates a partially saturated soil. For loose, partially saturated, natural soil or fill, the volume of air voids, V_a, expressed as a percentage of the total volume is a useful parameter:

$$V_a = \left\{1 - \frac{\rho_d}{\rho_w}\left[\frac{\rho_w}{\rho_s} + w\right]\right\} 100\;(\%) \quad (3.10)$$

Note: in equation 3.4, porosity n is expressed as a percentage whereas void ratio *e* and water content *w* are expressed as decimal quantities in equations 3.5, 3.6, 3.9 and 3.10.

3.5 Coarse soil
Coarse soil can be described as coarse-grained, granular or cohesionless. Sand and gravel are the principal coarse soils, although larger materials such as cobbles and boulders are also included. Coarse soil is normally classified according to the size of the constituent

particles and the shape of the particle size distribution curve. Soil which contains a wide range of particle sizes is described as well graded; soil which contains a large proportion within a narrow range of size is described as uniformly graded or poorly graded. The coefficient of uniformity is defined by:

$$C_U = \frac{D_{60}}{D_{10}} \qquad (3.11)$$

where:
D_{60} is the particle size such that 60% of the particle size distribution by mass is finer than D_{60}
D_{10} is the particle size such that 10% of the particle size distribution by mass is finer than D_{10}

Grading can be important in specifying coarse fill as it influences how well the particles pack together; the poorer the grading the greater the opportunity for the particles to adopt a loose configuration. However, a well-graded soil is more susceptible to segregation during placement as a fill.

The particle size distribution is determined by passing the soil sample through a series of sieves with diminishing apertures and measuring the weights of material retained on each sieve. The particle size is, therefore, the size of sieve through which the particle will pass. Dry sieving is suitable only for sand and gravel that do not contain any clay; wet sieving is normally required to separate the finer particles and prevent them from sticking to the coarse fraction[12].

Boulders and cobbles are typically bulky, hard rock fragments; they may be rounded or angular. Gravel is usually formed of quartz or resistant pieces of the parent rock. Although sand is generally formed of quartz grains, in some parts of the world sand deposits are made up entirely of calcareous grains, which are weaker than quartz grains. Gravel is more resistant to erosion than sand. Well-graded sands and gravels are generally less permeable and more stable than those with uniform particle size. Fine uniform sand approaches the characteristics of silt.

The engineering behaviour of coarse soil is determined largely by the mineralogy, shape and size of the grains, the density of packing of the grains and the stress level in the soil. The mineralogy can be determined by simple mineralogical tests and the shape of the particles by inspection. The shape of sand and gravel particles reflects the mode of transportation prior to deposition, so fluvial gravel is likely to be rounded or sub-angular from abrasion during rapid transportation.

Particle shape is categorised as: angular, sub-angular, sub-rounded, rounded, and well-rounded. It is usually sufficient to describe the particles as rounded, sub-angular or angular, although other shapes such as flaky or rod-shaped may be used. A parameter which can be used to describe the degree of packing is the *density index* I_D; this is sometimes known as relative density. Density index provides a useful way of relating the in-situ density of a granular fill to the limiting conditions of maximum density ρ_{dmax} and minimum density ρ_{dmin}:

$$I_D = \frac{(\rho_d - \rho_{dmin})}{(\rho_{dmax} - \rho_{dmin})} \frac{\rho_{dmax}}{\rho_d} \qquad (3.12)$$

Density index I_D is an important parameter which strongly influences compressibility, shear strength and vulnerability to liquefaction; to determine it requires knowledge of three densities, all of which are difficult to determine accurately[13]. Measuring the in-situ density of a coarse soil can be particularly difficult. An alternative approach is to estimate density index from empirical correlations with the results of in-situ tests (Section 5.7), such as the standard penetration test (SPT) or cone penetration test (CPT). However, the resistance to penetration measured in these tests is influenced by the stress level and grain size distribution as well as density index. A classification for coarse soil based on density index is in Table 3.2[14].

Table 3.2 Characteristic densities

Classification	Density index I_D
Very loose	0 – 0.15
Loose	0.15 – 0.35
Medium	0.35 – 0.65
Dense	0.65 – 0.85
Very dense	0.85 – 1.0

3.6 Fine soil

Fine soil, sometimes described as fine-grained or cohesive, includes silt and clay; this can be misleading since there are major differences between plastic fines, consisting of clay minerals[15], and non-plastic fines, typically minerals such as quartz or calcite. The characteristics of fine soil depend upon many factors; these include: the nature of the parent material, the climatic and chemical environment during breakdown of the parent rock and, during deposition of transported soil, the subsequent geological history, groundwater conditions and erosional conditions.

Silt usually consists of bulky shaped particles of primary minerals (minerals, such as quartz or rock flour, that have not been affected by weathering), whereas clay is composed primarily of flat or elongated particles of secondary minerals. Regardless of water content, silt exhibits little or no plastic behaviour and the strength when air dried is low. It is relatively impervious, difficult to compact, highly susceptible to frost heave, and easily erodible. Soil erosion is described in Box 4.3.

TYPES OF GROUND

The surface area per unit mass of flat, plate-like clay particles is vastly greater than that of coarse soils and clay particles carry electrical charges which attract a thin film or layer of adsorbed water onto the particle surface. There are two basic types of particle orientation for clay minerals: flocculated (edge-to-face) and dispersed (face to face). The properties of clay are influenced not only by the size and shape of the particles, but also by their mineral composition (the type of clay mineral) and the chemical environment. The particles that give clay its characteristic properties consist of three minerals; kaolinite, illite and montmorillonite. In practice, the clay-sized soil fraction may include particles that are not made up of clay minerals (for example, very fine fragments of rock known as rock flour) but soils containing only a small proportion of clay minerals can exhibit the characteristic behaviour of clays.

A saturated clay can have a wide range of water contents depending on the stresses acting on it and those to which it has been subjected in the past. The cohesion or cohesive strength of a clay increases with decreasing water content. Low permeability makes it difficult to compact when wet and impossible to drain by ordinary means. When compacted, it is resistant to erosion and is not susceptible to frost heave but is subject to expansion and shrinkage with changes in moisture. Many cohesive soils exhibit sensitivity; this means that their strength in a remoulded or highly disturbed condition is less than in an undisturbed condition at the same water content[16].

Within a certain range of water contents, clay exhibits plastic behaviour and can be classified by measuring two characteristic water contents, the plastic limit w_P and the liquid limit w_L; these are termed the Atterberg limits[17] or consistency limits and they define the range of water contents over which the soil exhibits plastic properties. The liquid limit test[18] determines the water content at which the soil starts to loose its plastic properties and exhibits 'liquid' behaviour; the plastic limit is the water content corresponding to the lower bound of plastic behaviour below which the soil loses its intact behaviour and begins to break up into discrete pieces[19]. Before performing the liquid and plastic limit tests, coarse material should be removed by passing the soil through a 425 μm sieve and test results should always state the proportion of the sample retained on the sieve. There is also a shrinkage limit which represents a lower limit of volume change.

The degree of plasticity of a soil is assessed from its plasticity index I_p, which may also be used to assess the susceptibility of a clay soil to shrinkage and swelling, described in Section 4.8. The plasticity index, like water content, is usually expressed as a percentage.

$$I_p = w_L - w_P \qquad (3.13)$$

Some aspects of the behaviour of clay soils may be determined from the value of the natural water content relative to the liquid and plastic limits. The liquidity index I_L is defined as:

$$I_L = \frac{(w - w_P)}{I_P} \tag{3.14}$$

The water content used to calculate I_L should be determined on the same soil fraction that is used to determine w_P and w_L (the fraction passing the 425 μm sieve). The index ranges from 1.0 when the soil is at the liquid state to 0 when the soil ceases to behave as a coherent mass. I_L is analogous to the density index of coarse soils. The consistency index I_C is sometimes used in preference to the liquidity index:

$$I_C = \frac{(w_L - w)}{I_P} = 1 - I_L \tag{3.15}$$

Typical plastic and liquid limits for some natural clay deposits are in Table 3.3. Two values are quoted for each deposit to illustrate the large variations within soils from the same geological deposit. The liquid and plastic limits may vary significantly with relatively small changes in location and depth within a particular deposit.

Table 3.3 Typical liquid and plastic limits for clay soils

Deposit	w_L %	w_P %	I_P %	Classification
London clay	53	26	27	CH
London clay	71	28	43	CV
Gault clay	63	21	42	CH
Gault clay	75	28	47	CV
Weald clay	49	26	23	CI
Weald clay	62	27	35	CH
Glacial till	28	17	11	CL
Glacial till	46	19	27	CI

Fine soil can be classified on the basis of the Atterberg limits using the plasticity chart[20] shown in Figure 3.3. The classification of each soil listed in Table 3.3 is given in the last column. All the soils plot above the A-line of the chart and so are classified as clay, hence the initial letter of their classification is 'C'. The clays range from very high plasticity CV to low plasticity CL.

In determining the particle size distribution, fine material is separated by sedimentation[21], meaning that the particles are dispersed in a fluid and allowed to settle under the influence of gravity; the particle size is the diameter of the equivalent sphere of similar density

TYPES OF GROUND

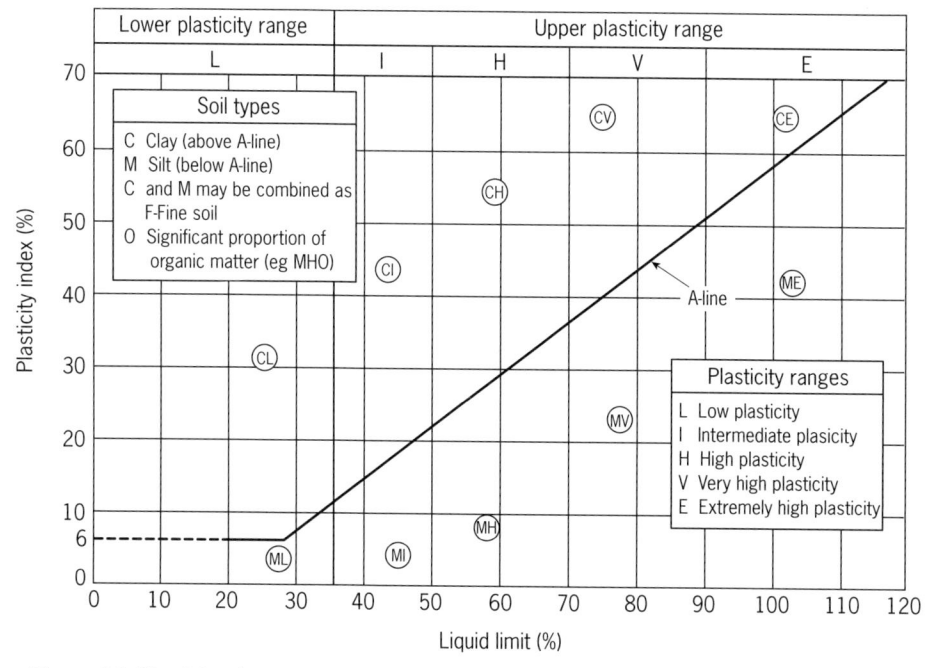

Figure 3.3 Plasticity chart

material that would fall at the same rate. The fines content of a soil F_c (percentage of particles smaller than 0.06 mm) is important since there is a critical magnitude which controls whether the soil behaves as a coarse soil or as a fine soil, although the distinction between coarse and fine soil behaviour is not solely a function of particle size. Coarse soil has been defined as having over about 65% sand and gravel sizes, and fine soil as having over about 35% silt and clay sizes[22]. However, for many aspects of soil behaviour the critical fines value will be smaller than 35% and might typically be about 15%.

The degree of over-consolidation of a saturated clay soil has great influence on its behaviour. Soft clay will be normally consolidated or, more probably, lightly overconsolidated near the surface as a result of groundwater movements, influence of climate or vegetation, and the effects of delayed or secondary consolidation. Most soft deposits will not have suffered much erosion, whereas stiff clay is likely to have been preconsolidated under a considerable depth of overburden, perhaps 200 m or more, which has been subsequently partly or wholly eroded away. Some common characteristics of these two states of clay soil are in Table 3.4. (The two states can be related to undrained shear strength c_u as shown in Table 4.3.)

Table 3.4 Geotechnical features of clay soil

Soft to firm clay $c_u < 75$ kPa	*Stiff clay* $c_u > 75$ kPa
Foundation design normally governed by settlement where foundation width is more than 1.5 m, or where excavation penetrates surface crust of stiffer soil. Consolidation settlement continues over long period.	Foundation design governed by settlement or bearing capacity. Most of settlement usually occurs during loading or within few weeks of loading.
Excavations exposing steep face will usually collapse quickly; if not, probably remain stable over long period. Heave in excavation occurs immediately.	Excavations exposing a steep face usually remain stable for a few weeks or months. May collapse after long period, sometimes suddenly. Heave in excavation may continue over long period.
Usually gains strength over long period. Staged construction makes use of this. Immediate stability is likely to govern slope design and either undrained or drained analysis may be appropriate.	May lose strength over long period. Long-term stability likely to govern slope design and drained analysis should be used. Undrained analysis may be appropriate for short term behaviour if effects of fissuring are taken into account.
Fissuring not usually strongly developed, but laminations, varves or organic inclusions may govern soil behaviour, particularly by influence on permeability.	Fissuring often strongly developed and may form surfaces of weakness which govern behaviour to a large extent. Laminations may influence behaviour.
Shrinkage problems sometimes important but swelling is not.	Shrinkage and swelling problems may be particularly important.

3.7 Organic soil

Organic soil is generally understood to be soil that contains a significant amount of organic material which has been derived from plants or animals and which is still in the process of decomposition[23]. The organic matter may decay causing voids or chemical changes which alter the properties of the soil. Because of its state, organic soil has a distinctive colour, smell and texture and can often be recognised by its dark colour and the characteristic smell of decaying vegetation. Organic content is determined by measuring the quantity of potassium dichromate solution needed to oxidise a small sample of the soil; loss on ignition can also be measured[24]. With some soils mass, loss on ignition is related to organic content; in others, factors unrelated to organic content can account for much of the mass loss on ignition.

The identification of organic soil is important because it is usually much more compressible and much weaker than inorganic soil. Failure to locate these deposits can result in foundation performance which is much inferior to predicted behaviour.

The presence of even a small proportion of organic matter increases the compressibility of the soil and organic soil may be found in inorganic deposits of silt and clay. Thin organic seams have an influence on the permeability of the deposit and may influence the strength of the mass where an organic seam corresponds to a plane of high shear stress. Organic matter may also occur as pockets of material or as fine material dispersed through the clay. In these situations the organic soil may be difficult to detect.

Peat is a decomposed organic material and is often fibrous. Where organic material is predominant, the soil is usually described as peat. Most peat is extremely compressible and structures built over peat may settle by large amounts over long periods of time. The shear strength is variable and sometimes surprisingly high, although large shear strains may be necessary before the full strength is developed. The permeability can often be quite high. Peat can be distinguished from other organic soils by its low density.

3.8 Rock

Rock is an assembly of naturally occurring minerals bonded together to form material which is much stronger and stiffer than soil. It is not surprising, therefore, that in most building projects a rock foundation should present no particular difficulties. Consequently only a brief indication of rock types and behaviour is included[25]. Engineers commonly distinguish between hard rock which needs blasting, and soft rock which can be excavated by normal earthmoving equipment. Geologists are more concerned with origins and divide rocks into three categories: igneous, sedimentary and metamorphic.

Igneous rock comes directly from the earth's mantle as magma: molten rock which is forced up through the crust before cooling and solidifying. There are two primary types of igneous rock. Extrusive rock comes to the surface as lava and forms as the lava flow cools rapidly and solidifies; intrusive rock is formed below the ground surface where it cools slowly. Because it cools more quickly, extrusive rock tends to have a much finer grain structure. Large masses of intrusive rock, such as that underlying Dartmoor, are known as batholiths. Thinner deposits that are formed as the lava forces its way into joints or bedding planes are known as sills, pipes, or dykes, depending on whether the formation is predominantly horizontal, vertical or inclined. Igneous rock is composed predominantly of three minerals: quartz, feldspars and biotite. Depending on its mineral composition, the rock may be acid, intermediate or basic. Acid rock tends to be lighter in colour than basic rock. Some examples of igneous rock are listed in Table 3.5.

Table 3.5 Some common igneous rocks

	Acid	*Basic*
Coarse-grained	Granite	Gabbro
Fine-grained	Rhyolite	Basalt

Sedimentary rock is formed from mineral and rock particles under high pressure and temperature, and possibly chemical action. Clastic rocks, which include sandstone and mudstone, form when deep soil deposits become hardened by pressure from overlying strata and by cementation and, because of their mode of deposition, many clastic rocks are layered. Carbonate rocks, which include limestone and chalk, are formed where organic materials accumulate and are indurated[26]. Carbonate rocks can be dissolved by long exposure to water, especially if it contains a mild solution of carbonic acid. In limestone and chalk, the action of acidic groundwater can produce such solution features as sinkholes; they are described in Section 7.7.

The constituent particles of sedimentary rocks may be primary minerals, such as silica which forms sandstone, or calcium carbonate which forms chalk and limestone. Alternatively, they may be secondary minerals, which form mudstone and shales. Sedimentary rock may also be organic (for example lignite or coal).

Metamorphic rock is derived from sedimentary or igneous rock by the action of extremely high pressure and temperature that can be generated deep in the ground by geological events. Four principal types of metamorphic rock are gneiss, marble, schist and slate.

Intact rock (rock with no significant fractures) is strong compared with soil; its strength is normally expressed as unconfined compressive strength. Its mechanical behaviour can be described by the stress-strain curve under uniaxial compression. The Franklin point load test[27] provides a quick measurement of the strength of unprepared rock core samples, either in the field or in the laboratory. The apparatus consists of a small loading frame fitted with a hand-operated hydraulic ram. The rock core to be tested is held between standardised pointed platens and loaded to failure. The point load strength index Is is a measure of the rock's tensile strength and is given by:

$$I_s = \frac{P}{D^2} \qquad (3.16)$$

where:
> P is the force required to break the specimen
> D is the distance between the platen contact points.

In practice, the strength and deformation characteristics of a rock mass are strongly influenced by the discontinuities within the rock mass and often have little resemblance to the properties of the intact rock. The discontinuities can be thought of as weak links within the rock mass. The orientation, spacing, aperture and filling of the discontinues are likely to have a major effect on rock behaviour.

TYPES OF GROUND

Where there has been little separation or shear movement along a discontinuity, it is referred to as a joint. The orientation of the joint is defined by two angles: the strike is the compass direction of a horizontal line drawn on the plane of the joint, and the dip is the angle between the plane and the horizontal measured in a direction perpendicular to the strike. Most rock contains sets of joints that have similar dips and strikes, so the joint sets form a regular pattern in the rock.

Joints are produced by a variety of processes, including cooling of igneous rock, folding due to tectonic forces, and stress relief near the sides of valleys. Joints may be clean, filled with soft soil or filled with a cemented material, such as quartz or calcite. The strength characteristics of the joint may, therefore, be determined either by the properties of the rock or the infill material.

The mechanical properties of near-surface rock are altered by weathering; this results from a variety of factors including abrasion by wind, water and ice, differential thermal expansion of the constituent minerals, frost expansion of water in crevices and chemical action. Reaction with carbon dioxide is particularly important as it converts the primary rock minerals to secondary clay minerals.

3.9 Fill

The terms fill and made ground describe ground that has been formed by material deposited by man, in contrast to natural soil which has its origin in geological processes. Made ground may be composed of natural soil and rock or may be formed from industrial, chemical, mining, dredging, building, commercial and domestic wastes. Fill commonly encountered in the UK includes opencast mining backfill, colliery spoil, pulverised-fuel ash (pfa), industrial and chemical wastes including iron and steel slags, building and demolition wastes and domestic refuse. Many clay and gravel pits, quarries and disused docks have been infilled. Hydraulic fill is also encountered. The term landfill is often used to describe domestic refuse. Filled sites are frequently used for building developments.

The development of filled sites can present a wide variety of problems, mostly associated with either of these aspects of behaviour:
❐ the ability of the fill to support the building or other form of construction without excessive settlement (it is this aspect of fill behaviour that is of concern in this section);
❐ the presence in the fill of materials which could be hazardous to health or harmful to the environment or the building (this is discussed in Chapter 8).

It is important to distinguish between sites where filling has yet to take place so earthmoving can be controlled and supervised, and sites which have already been filled with little or no control being exercised. This basic distinction is between *engineered fill* and *non-engineered fill*; Figures 3.4 and 3.5 show examples.

Figure 3.4 Placing an engineered rockfill

Figure 3.5 Non-engineered fill – old refuse

Fill that has been placed to an appropriate specification under controlled conditions for subsequent building development can be described as engineered fill (Box 3.3). Engineering design focuses on the specification and control of filling.

Where non-engineered fill (Box 3.4) is used as foundation material problems may be experienced and considerable caution is essential[28]. The particular difficulties encountered when building on fill are described in Section 7.4. Ground treatment may be necessary prior to building development[29]; this is discussed in Chapter 9.

While the hazards associated with the looseness and heterogeneity of the fill and the presence of biodegradable material have long been recognised, settlement was usually considered to be associated with creep mechanisms that had the reassuring quality of a settlement rate which decreased with the time that had elapsed since placement of the fill. It was assumed that the passage of time would alleviate the problem and that building developments could safely proceed provided that the fill had been in place for a sufficient period. The identification of *collapse compression* on wetting has more recently been identified[30] as a major hazard for building on fills. Many partially saturated soils undergo a reduction in volume when inundated with water and this is a particular hazard for partially saturated fill that has been poorly compacted. This compression can occur without any change in external loading if the fill has been placed in a sufficiently loose and/or dry condition (Section 4.9).

3.10 Groundwater

The presence of water in soil voids and rock fissures has a major impact on the engineering behaviour of the ground. Indeed, the difficulties with soils are generally due to the water contained in their voids[31] and not to the soils themselves. A high water-table can present difficulties with foundation excavations and reduce the bearing resistance of the

Box 3.3 Engineered fill
Engineered fill is selected, placed and compacted to an appropriate specification so that it will exhibit the required engineering behaviour. The terms 'controlled fill' and 'structural fill' are sometimes used. Where an engineered fill is to be used as a foundation material, the adoption of a suitable specification for material, appropriate earthmoving plant and layer depth, together with adequate supervision of placement and compaction, should ensure that there are reasonably uniform properties throughout the fill deposit and that the required foundation performance is achieved.

Box 3.4 Non-engineered fill
Non-engineered fill has arisen as a by-product of human activity, usually involving the disposal of waste materials. It has not been placed with a subsequent engineering application in view. Little control may have been exercised during placement so consequently there is the possibility, in some cases probability, of extreme variability. This makes it very difficult to characterise the engineering properties and predict behaviour. Many foundation problems are associated with developments on heterogeneous, loosely placed, non-engineered fills.

ground. The presence of chemicals in groundwater, such as acids and sulfates, can cause damage to foundation concrete if it is not of an appropriate quality.

The principal source of groundwater is precipitation. While much of this water may be lost as surface run-off or by evaporation and transpiration to the atmosphere, some of the water enters the ground and moves downwards by infiltration through the continuous soil voids. The relationships between the processes of infiltration, run-off, evaporation and transpiration are influenced by climate, topography, vegetation and geology.

The term subsurface water encompasses all underground water. In understanding the behaviour of subsurface water, the water-table, which is sometimes known as the groundwater level or the phreatic surface, is a key concept. The water-table is a theoretical surface that is approximated by the level to which water fills open borings which penetrate only a short distance into the saturated zone. The subsurface water below the water-table is in the phreatic zone and the subsurface water above the water-table is in the vadose zone. Technically, only the water in the phreatic zone is *groundwater*, but the term is sometimes used to describe all subsurface water.

The groundwater in the phreatic zone is subjected to a positive pressure as a result of the weight of the overlying water. While the soil in the phreatic zone is often considered to be saturated with the voids full of water, this may not be strictly the case since there may be some discontinuous and occluded air bubbles in the groundwater.

It might be thought that the soil above the water-table would be dry but this is most unlikely. The water in the soil voids has a negative pressure and is held in place by capillary action; the soil in the lower part of the capillary fringe may be saturated. The behaviour of partially saturated soil close to the ground surface is particularly important for the shallow foundations of low-rise buildings.

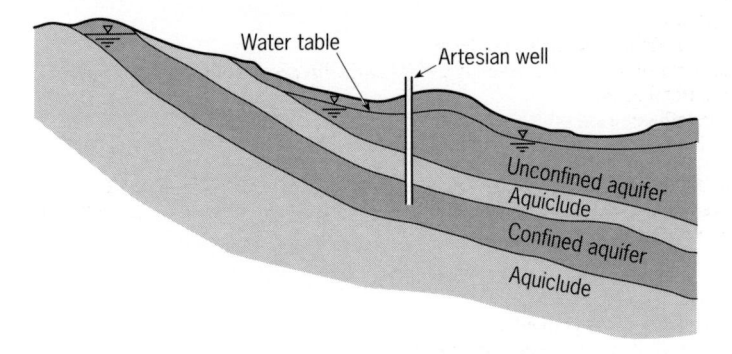

Figure 3.6 Ground profile showing groundwater conditions

TYPES OF GROUND

Groundwater that is in direct contact vertically with the atmosphere through permeable ground with continuous open spaces, such as pores and voids in soils and fissures in rock, is termed *unconfined*. *Confined* water is separated from the atmosphere by impermeable ground. Confined water is sometimes described as artesian when the pressure in the groundwater corresponds to a head of water which is above ground level (Figure 3.6). The first unconfined water to be encountered may be at a *perched water-table*; the perched water is isolated from groundwater at deeper levels by an impermeable layer.

Notes

1. In the Gospel according to Matthew, chapter 7 verses 24–27, the contrast is between a house built on rock and a house built on sand; in the Gospel according to Luke, chapter 6 verses 47–49, the contrast is between a deep foundation on rock and a house built on earth without a foundation.

2. A mineral is a naturally occurring chemical element or compound formed by a geological process and rock is an aggregate of such minerals. Common rock-forming minerals include these silicates: quartz, feldspar, mica and chlorite. There are also oxides, carbonates, sulfates and chlorides.

3. At an early stage in the history of geology, the term *drift* came to be applied to superficial deposits, whereas the underlying bedrock was regarded as the *solid* base on which the drift deposits were laid down. Two different styles of geological map, drift and solid, have been prepared (Chapter 5).

4. *Made ground* refers to ground that has been formed by human activity rather than by natural geological processes. The material of which made ground is composed is described as *fill*. In practice, the terms made ground and fill are often used interchangeably. Fill denotes a large quantity of material which has been deposited by some form of human agency over a wide area and has raised the level of the site or a substantial part of it. A limited amount of fill placed within the foundations of a single building unit is referred to as *hardcore* and the particular problems that can arise with hardcore are considered in Chapter 12.

5. BS 1377: 1990 *Methods of test for soils for civil engineering purposes* specifies standard procedures for carrying out most commonly used soil tests. Parts 1 to 8 cover laboratory tests; Part 9 covers in-situ tests.

Part Title
1 General requirements and sample preparation
2 Classification tests
3 Chemical and electro-chemical tests
4 Compaction-related tests
5 Compressibility, permeability and durability tests
6 Consolidation and permeability tests in hydraulic cells and with pore pressure measurement
7 Shear strength tests (total stress)
8 Shear strength tests (effective stress)
9 In-situ tests

6. European test standards for soil are in preparation under the CEN Technical Committee TC 341 on *Geotechnical investigation and testing*. Technical Specifications are being prepared for the following laboratory tests as CEN ISO/TS 17892.

Part Test
1 Determination of water content
2 Determination of density of fine grained soil
3 Determination of particle density – pycnometer method
4 Determination of particle size distribution

5   Incremental loading oedometer test
6   Fall cone test
7   Unconfined compression test
8   Unconsolidated undrained triaxial test
9   Consolidated triaxial compression tests on water saturated soils
10 Direct shear tests
11 Determination of permeability by constant and falling head
12 Determination of Atterberg limits

7. Soil derived from wind-blown dust is referred to as *loess*. Where the loess has been reworked by river action, it is known as *brickearth*. Loess usually contains a small amount of binder consisting of calcareous material or iron oxide; this allows the soil in its natural dry state to stand at very high vertical slopes. The binder breaks down under the action of water, leaving an extremely unstable loose, fine cohesionless soil. Increasing the water content can cause a substantial reduction in volume. The uniformity and small size of the soil particles, together with the lack of cohesion, make loess an extremely difficult soil to compact and the degree of compaction which can be achieved is very sensitive to the water content.

8. The particle size distribution in Table 3.1 follows BS 5930: 1999 *Code of practice for site investigations*. A slightly different classification is given in the European Standard BS EN ISO 14688-1:2002 *Geotechnical investigation and testing – identification and classification of soil –* Part 1 *Identification and description*. For example, Table 13 in BS 5930 has the silt/sand boundary at 0.06 mm but in Table 1 of BS EN ISO 14688-1:2002 it is put at 0.063 mm. The differences should not have any practical significance.

9. Gravimetric water content can be expressed in terms of mass or weight.

10. A standard test procedure for the determination of water content by oven-drying is specified in BS 1377: 1990 *Methods of test for soils for civil engineering purposes* Part 2, clause 3.2.

11. The measurement of particle density is based on measuring the dry weight of a sample of the soil and the weight of the same sample plus water in a container of known volume. Full saturation is effected by applying a vacuum to the soil-water mixture for several hours. Standard test procedures for the determination of particle density are given in BS 1377: 1990 Part 2, clause 8.

12. Standard test procedures determining the particle size distribution of coarse soil are in BS 1377:1990 Part 2; the wet sieving method is in clause 9.2 and the dry sieving method in clause 9.3. The standard states that the dry sieving method shall not be used unless it has been shown that it gives the same results as the wet sieving method for the type of material under test.

13 Standard test procedures for determining maximum and minimum densities of coarse soil are specified in BS 1377:1990 Part 4, clause 4, together with the derivation of density index. Maximum density is usually determined using an electrical vibrating hammer. In-situ density tests are specified in of BS 1377 part 9, clause 2.

14. A number of different classifications of density can be found in the technical literature, usually based on SPT results. Table 3.2 follows Skempton (1986).

15. Clay minerals comprise a large group of hydrous aluminium silicates with characteristic sheet structures. The more important clay minerals are kaolinite, illite and montmorillonite. For a given weight of soil, the surface area of these particles is far greater than for a granular soil, especially for montmorillonite whose particles are the smallest by far. The liquid and plastic limits of clay minerals show great variations. The plasticity indices of montmorillonites are much larger than those of illites and kaolinites.

16. Highly sensitive clays, known as *quick clays*, are found in areas of Eastern Canada and parts of Scandinavia, particularly Norway. This type of clay is produced by leaching of salt from soil

TYPES OF GROUND

originally laid down under marine conditions. The strength of a quick clay is dramatically reduced by mechanical disturbance. In its natural, undisturbed state the shear strength of the Norwegian quick clay is typically between 50 and 100 kPa. The drop in strength may be a hundredfold and, after considerable disturbance, the soil may become liquid and flow as mud.

17. The liquid and plastic limits are called the Atterberg limits after Albert Atterberg (1846 – 1916), the Swedish soil scientist who developed the tests. In the 1930s, Karl Terzaghi and Arthur Casagrande adapted the tests for civil engineering purposes. BS 1377: 1990 *Methods of test for soils for civil engineering purposes* Part 2 includes standard test procedures for determining the liquid limit (clause 4) and the plastic limit (clause 5.3), and the derivation of the plasticity index and the liquidity index (clause 5.4).

18. Two alternative test procedures for determining the liquid limit are in BS 1377:1990 Part 2. The cone penetrometer is the definitive method for the liquid limit (clause 4.3), but the method using the Casagrande apparatus is also included (clause 4.5). With the cone penetrometer, the liquid limit is defined as the water content at which the cone penetrates 20 mm; penetrations are measured at water contents slightly above and below the liquid limit and the water content corresponding to 20 mm is found by interpolation. With the Casagrande cup, the liquid limit is defined as the water content at which 25 blows cause 13 mm of closure of the groove formed at the base of the cup.

19. A test procedure for determining the plastic limit is included in clause 5.3 of BS 1377: 1990 Part 2. The plastic limit test identifies the lower bound of plastic behaviour by measuring the water content at which the soil can no longer be moulded without breaking up. The test is performed by repeatedly rolling a sample of soil on a glass plate to form a 3 mm diameter thread; each time the thread is rolled out the water content is decreased slightly; the plastic limit is defined as the water content at which it is no longer possible to form the thread.

20. The plasticity chart was developed by Arthur Casagrande. Figure 3.3 shows the modified form in which it is found in BS 5930: 1999.

21. When dispersed in water, particles of different sizes settle at different rates. After shaking to evenly distribute the soil, the density of the upper part of a suspension can be measured at intervals, either by sampling with a pipette and weighing the dried residue, or by using a hydrometer. The maximum particle size remaining in the upper part of the suspension at any given time can be calculated using Stokes Law and the percentage of material within a particular size range can be determined. The time intervals are chosen to measure the proportions of coarse, medium and fine silt and clay. Standard test procedures for determining the particle size distribution of fine soil are specified in BS 1377: 1990 *Methods of test for soils for civil engineering purposes* Part 2; sedimentation by the pipette method is specified in clause 9.4 and sedimentation by the hydrometer method in clause 9.5.

22. This classification is in BS 5930: 1999 *Code of practice for site investigations*. This is not a specification but takes the form of guidance and recommendations.

23. Strictly speaking, any material that contains carbon is 'organic'. However, the term organic soil has a more restricted meaning and is used to describe soils that contain a significant amount of organic material which has been derived from plants or animals. Sand containing calcium carbonate is an example of a soil containing carbon which would not be considered to be an organic soil.

24. BS 1377: 1990 Part 3 includes standard test procedures for the determining organic matter content using Walkley and Black's method (clause 3) and of mass loss on ignition (clause 4).

25. There is extensive literature on rock mechanics and engineering geology if more detailed information on rock behaviour is required. A helpful introduction to engineering rock mechanics is Hudson and Harrison (1997).

26. Induration is the hardening of a rock by the action of heat, pressure or cementation.

27. Franklin point load tests should preferably be carried out by applying the load across a rock core diameter; the results of tests on irregular lumps require a suitable correction. Point load strength index is dependent on core diameter, with large diameter cores giving lower values; all values should be corrected to a reference diameter of 50 mm (Broch and Franklin, 1972).

28. In Section 2E of Approved Document A of the Building Regulations it is stated that there should not be non-engineered fill or wide variation in ground conditions within the loaded area.

29. BRE has carried out an extensive programme of research into the problems of building on fill and the applicability of ground treatment methods (BRE Report BR 424).

30. Although the full significance of collapse compression as a major hazard for building on fills has only relatively recently been recognised by geotechnical engineers, the phenomenon has been recognised in practice for many years. Collapse compression was identified by Proctor in 1933 as a potential hazard in embankment dam fill. Proctor warned that *'It may now be seen that it is possible to compact a soil so firm and hard as to appear entirely suitable for a dam and for this same soil to become very soft and unstable when percolating water saturates it'*. This statement highlights one of the barriers to identifying the hazard of collapse compression; the soil that has been compacted to a firm and hard condition appears much superior to the very soft soil. Proctor developed a procedure involving field and laboratory controls to ensure that clay fills were placed at the correct moisture content to reduce settlement or softening when the embankment dam fill was subsequently saturated.

31. The illuminating comment by Karl Terzaghi that *'in engineering practice, difficulties with soils are almost exclusively due not to the soils themselves, but to the water contained in their voids. On a planet without any water there would be no need for soil mechanics'* was made when he presented the 45th James Forrest Lecture at the Institution of Civil Engineers on 2 May 1939.

CHAPTER 4: SOIL BEHAVIOUR

The rational design of safe and economic foundations and other geotechnical works requires an adequate knowledge of how soils perform in practice. However, the properties of soils and rocks can vary enormously across a site and may be far from constant over time. Furthermore, the ground is more difficult to access and test than structural materials. The greatest uncertainty for the designer of buildings usually is associated with the behaviour of the ground and foundation deficiencies are often more expensive to rectify than structural problems. This chapter examines basic soil mechanics principles with a major emphasis on soil behaviour as it affects low-rise buildings and describes laboratory tests for determining engineering properties. It is usually the behaviour of partially saturated soil under low applied stress at shallow depth that controls the foundation behaviour of low-rise buildings; the influence of atmospheric conditions on the soil near the ground surface is particularly significant.

4.1 Introduction

While a great deal is known about the materials, such as masonry and reinforced concrete, from which the building will be fabricated, much less is known about the behaviour of the ground. The quality of the fabrication processes can be predicted, the performance can be closely observed and the properties of the masonry and concrete will be reasonably uniform and constant throughout the structure and with time. The same cannot be said about the ground and determining its behaviour can be much more difficult; yet the rational design of safe and economic foundations requires an adequate knowledge of how soils perform in practice.

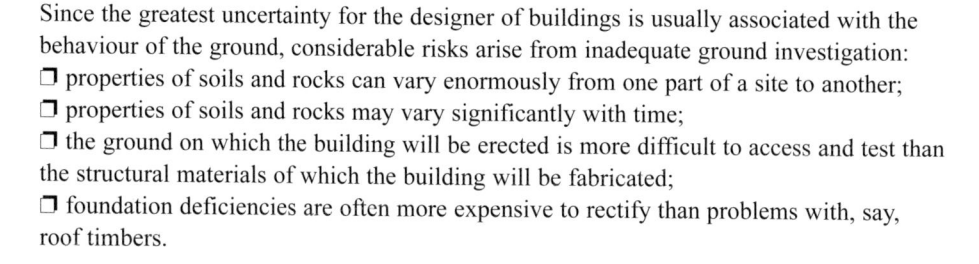

Since the greatest uncertainty for the designer of buildings is usually associated with the behaviour of the ground, considerable risks arise from inadequate ground investigation:
❏ properties of soils and rocks can vary enormously from one part of a site to another;
❏ properties of soils and rocks may vary significantly with time;
❏ the ground on which the building will be erected is more difficult to access and test than the structural materials of which the building will be fabricated;
❏ foundation deficiencies are often more expensive to rectify than problems with, say, roof timbers.

The principles of soil behaviour as they apply to foundations and to retaining walls, embankments and deep excavations are fully described in many standard soil mechanics text books[1] so no attempt is made here to cover the entire subject. Basic soil mechanics principles are described principally with a view to foundation design, construction and performance; the main emphasis is on soil behaviour as it affects low-rise buildings.

The historical development of soil mechanics was primarily related to large-scale civil engineering applications where the behaviour of saturated soil at considerable depth below ground level, under relatively high applied stress, is of major interest. With low-rise buildings, it is usually the behaviour of partially saturated soil under low applied stress at shallow depth that controls foundation behaviour. This means that important features of the behaviour of shallow foundations are not well covered in many accounts of basic soil mechanics. The influence of climate, topography and vegetation on the soil near the ground surface is particularly significant for shallow foundations.

Soil is a multi-phase material (Section 3.4), and loading and deformation affect the soil phases differently. When a soil is loaded, it deforms primarily by a rearrangement of the particles; crushing of particles can also be important in some soils. Individual soil particles are not strongly bonded to each other and the interactions between the particles, as they slide over each other, govern the deformation and strength properties of the soil. The pore fluid exerts a critical influence on the behaviour of the soil and accounts for most of the differences between fine soil (such as clay and silt) and coarse soil (such as sand and gravel). It is necessary to introduce some distinctive concepts and procedures so that the interactions between the phases are properly represented.

Soil mechanics developed by considering deformation and strength as separate subjects. While there have been many attempts to provide a conceptual framework within which both aspects of soil behaviour can integrated[2], for the purposes of this book it is simpler to examine deformation and strength separately. Before doing this, the effect of the flow of water through soil and the significance of pore water pressure will be considered. Most of the chapter describes the behaviour of saturated soil, but towards the end some of the distinguishing characteristics of the behaviour of partially saturated soil are noted. Finally, there is a section on the compaction of partially saturated soil.

4.2 Stress and strain

Stress is the intensity of loading and is measured as load per unit area. Normal stress σ is the load per unit area on a plane normal to the direction of the load and shear stress τ is the load per unit area on a plane parallel to the direction of the load. The principal stresses σ_1, σ_2 and σ_3 are the normal stresses on planes of zero shear stress. In soil mechanics, normal stress is almost always compressive since soil cannot sustain any appreciable tensile stress and so the sign convention is to take compressive stress as positive.

Strain is the intensity of deformation and is measured as the ratio of the change in a dimension to the original dimension. Normal strain ε is the displacement divided by the original length in the direction of the displacement. Vertical strain ε_v is of particular importance in calculating the settlement of foundations.

When a building is erected, the weight of the building is transmitted through the foundations into the underlying ground. Soil is a particulate material and the loads are transmitted through the soil at particle contact points. The stresses at these contact points are very high, whereas elsewhere they are much smaller. There will be a similar situation in rocks due the presence of such discontinuities as joints and fissures. Deformation calculations based on the actual particulate behaviour of the ground would be immensely complex and, consequently, the simplifying assumption is made that the soil can be analysed as a continuum[3]. This means that the stresses calculated for the hypothetical continuum do not exist at any point within the actual particulate soil, but the simplifying assumption works reasonably well provided that the dimensions of the problem under analysis are large compared with the particle size of the soil or the joint size of the rock.

When a soil is loaded it deforms. If the soil reverts to its original size and shape when the load is removed, the soil has undergone elastic (recoverable) deformation. If some of the deformation produced by an applied load remains when the load is removed, the soil has undergone plastic (irrecoverable) deformation. While soil deformations are rarely fully elastic, except at very low strains, elastic analysis[4] is often used to assess the vertical stresses in the ground due to foundation loading. The distribution of vertical stress beneath a foundation is not particularly sensitive to the stress-strain relationship of the ground and the simplest approach is to assume that the soil has linear elastic stress-strain properties in which stress and strain are related through Hooke's law[5].

In calculating the settlement of a foundation due to the deformation of the soil under the weight of the building, the crucial factor is the stress-strain relationship of the soil. This can be determined in an oedometer test (Section 4.5).

4.3 Effective stress and pore water pressure

The principle of effective stress states that the deformation and strength behaviour of a soil is controlled by the effective stress. This most important principle in soil mechanics is described in Box 4.1. The principle of effective stress was at the heart of the development of soil mechanics in the twentieth century[6].

The interaction between the soil skeleton and the pore water is the key to understanding the behaviour of soil. The effective stress σ' is basically the stress carried by the soil particles. The total stress σ and the pore water pressure u_w are quantities that can be measured, whereas the effective stress is a deduced quantity and for a saturated soil:

> **Box 4.1 The principle of effective stress**
> The principle of effective stress states that all the measurable effects of a change of stress, such as compression, distortion and change of shear strength, are due to changes in effective stress. For a saturated soil, the effective stress is the total stress minus the pore water pressure; an understanding of the deformation and strength behaviour of a soil therefore requires knowledge of both the total stress and the pore water pressure.

$$\sigma' = \sigma - u_w \quad (4.1)$$

The vertical effective stress σ'_v within a soil deposit depends on the weight of the overlying soil, the pore water pressure u_w and any applied external load. A simple situation is illustrated in Figure 4.1 in which there is a groundwater level or water-table at depth h_w below ground surface and no flow of water through the ground. The top of the water-table is under atmospheric pressure. Point A, at depth z_a below the ground surface, is below the water-table (ie $z_a > h_w$) and there is a positive pore water pressure u_w, such that:

$$u_w = \gamma_w(z_a - h_w) \quad (4.2)$$

where
γ_w is the unit weight of water.

The vertical total stress σ_v acting at point A is:

$$\sigma_v = \gamma z_a \quad (4.3)$$

where:
γ is the bulk unit weight of the soil.

Therefore, the effective vertical stress in the soil is:

$$\sigma'_v = \sigma_v - u_w = (\gamma - \gamma_w) z_a + \gamma_w h_w \quad (4.4)$$

In practice, the bulk unit weight of the soil will vary with depth and, in particular, will be smaller nearer the ground surface where the soil is not saturated.

While it is relatively simple to evaluate the vertical effective stress, it is quite difficult to assess the full three-dimensional stress state in a soil deposit because the horizontal or lateral effective stress σ'_h depends on the method of deposition, the stress history, and other factors. The coefficient of earth pressure at rest K_0 relates the horizontal effective stress to the vertical effective stress for the situation in which there are no lateral strains:

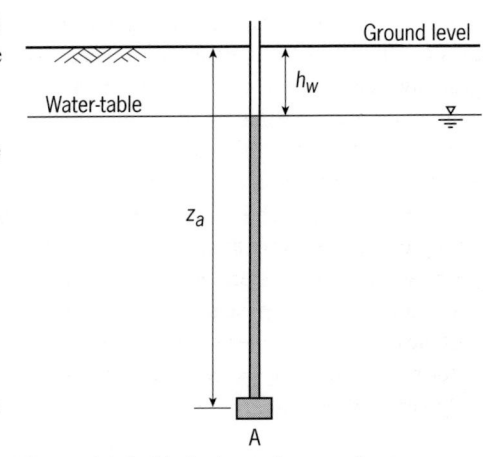

Figure 4.1 Stable hydrostatic groundwater conditions

$$K_0 = \frac{\sigma'_h}{\sigma'_v} \qquad (4.5)$$

Since K_0 is the ratio of horizontal effective stress to vertical effective stress, to calculate the horizontal total stress σ_h the pore pressure has to be added:

$$\sigma_h = K_0 \sigma'_v + u_w \qquad (4.6)$$

In a normally consolidated soil (Section 3.2) K_0 can be estimated from the following simplified relationship:

$$K_0 = 1 - \sin \phi' \qquad (4.7)$$

where:
ϕ' is the angle of shear resistance described in Section 4.6.

More complex relationships have been suggested for overconsolidated soils[7].

In many soils K_0 is likely to range from about 0.4 to 1.5 but in heavily overconsolidated clay larger values of K_0 could be operative; in theory it can range up to the full passive value. Box 4.2 describes limiting values of earth pressure.

4.4 Groundwater flow

Many of the problems which geotechnical engineering has to address are concerned with the flow of water through the ground. Catastrophic failures in geotechnical engineering have often resulted from instability of soil masses due to groundwater flow. For example, if a large excavation is required in ground with a high water-table, the rate of flow of water into the excavation is crucial to the design of a dewatering scheme and to the stability of the excavation. The ability of water to flow through soil, termed permeability, is dependent on the size and distribution of the pore spaces between the solid particles and on how these alter as the soil deforms. Permeability is reduced where the soil is only partially saturated.

Water will flow through the pores of a soil if

> **Box 4.2 Lateral earth pressure**
> Limiting values of lateral earth pressure can be related to the earth pressure acting on a smooth vertical retaining wall. Active earth pressure is the minimum earth pressure, which is realised when the smooth wall moves away from the soil; passive earth pressure is the maximum earth pressure, which is experienced when the wall moves into the soil. Much more movement is required to mobilise the passive pressure than the active earth pressure. For a smooth vertical wall, the active earth pressure coefficient K_a and the passive earth pressure coefficient K_p are functions of the angle of shear resistance ϕ':
> $K_a = (1 - \sin \phi')/(1 + \sin \phi')$
> $K_p = (1 + \sin \phi')/(1 - \sin \phi')$

there is a hydraulic gradient. To define hydraulic gradient we must introduce the concept of head[8]. Consider the two points A and B in Figure 4.2. There is a horizontal ground surface and the points are at depths z_a and z_b respectively below ground level. The water levels in standpipes installed at the two points are at depths h_a and h_b respectively below ground level.

The pore water pressure at A is:
$$u_{wa} = \gamma_w (z_a - h_a)$$
the pore water pressure at B is:
$$u_{wb} = \gamma_w (z_b - h_b).$$

The pore pressure at B is greater than at A, but it is clear that water will flow from A towards B and not from B towards A! This is because it is the difference in hydraulic head between the two points (the difference in the water level in the two standpipes) that is the sole force driving the flow of water from A to B and the head at A is greater than the head at B. Hydraulic gradient i is defined as the loss of head per unit distance along a flow-line $(h_b - h_a)/L_{ab}$. The same units should be used for both head and distance, so that i is a pseudo-dimensionless quantity. Where the ground surface is not level, head has to be calculated with respect to some other arbitrary horizontal datum line.

For one-dimensional flow, it is normally assumed that the velocity of the pore water is proportional to the hydraulic gradient[9]. This linear relationship between hydraulic

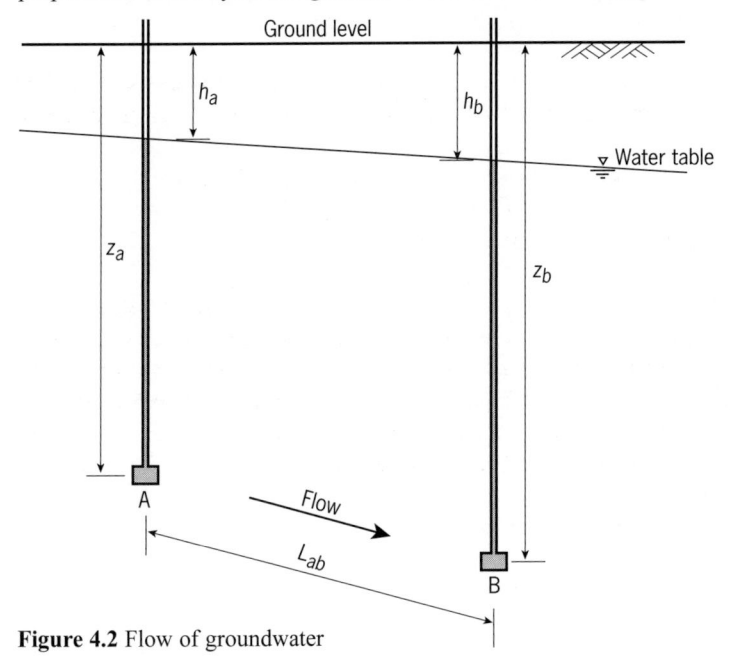

Figure 4.2 Flow of groundwater

gradient and rate of flow is referred to as Darcy's Law and the coefficient of proportionality is known as the coefficient of permeability, or hydraulic conductivity, of the soil, k. The rate of flow q through a sample of soil with a cross-sectional area A is:

$$q = vA = kiA \qquad (4.8)$$

$$v = nv_s \qquad (4.9)$$

where:
>v_s is the actual seepage velocity
>v the approach velocity (the velocity of water required to supply the soil sample)
>n is the soil porosity, expressed as a decimal.

Ranges of the coefficient of permeability for different soil types are in Table 4.1. No other soil parameter exhibits such a large variation between coarse and fine soils.

Table 4.1 Typical values of the coefficient of permeability

Soil type	k (m/s)
Clean gravel	$> 10^{-2}$
Clean coarse sand	$10^{-4} - 10^{-2}$
Fine sand	$10^{-6} - 10^{-4}$
Silty sand	$10^{-7} - 10^{-5}$
Silt	$10^{-9} - 10^{-6}$
Homogeneous clay	$< 10^{-9}$

Other important effects and processes are influenced by water movement and permeability:
- rate of dissipation of excess pore water pressure and associated primary consolidation of clay (Section 4.5);
- collapse compression of loose, unsaturated, natural soil and fill (Section 4.9);
- liquefaction of saturated, coarse soil and fill (Section 7.3);
- loss of ground due to erosion of fine particles from permeable soil.

As water flows through soil, it exerts a frictional drag on the soil particles resulting in head losses. This seepage force must be taken into account when calculating, for example, the stability of slopes. If the flow of water is upwards, the force can exceed the weight of the soil so that there is no contact between particles (effective stress is reduced to zero). For coarse soil, the strength is effectively reduced to zero and the soil is described as *quick* (as in quicksand). Quick conditions can occur wherever the upward seepage force exceeds the submerged unit weight of the soil:

$$i > \frac{\gamma'}{\gamma_w} \qquad (4.10)$$

A principal difference between coarse and fine soil is that the behaviour of fine soil is strongly time-dependent, largely because of its low permeability. Settlement takes place over a period varying from days to decades; soil strength may increase or decrease owing to slowly changing pore pressures over long periods. A steep embankment or cutting in stiff overconsolidated clay, which has shown no signs of instability, may fail suddenly many decades after construction because the suctions induced in the clay during construction have slowly dissipated. This causes an increase in water content and a reduction in strength, to the point of failure. Sand will not stand on a vertical face unless there is some cementing of grains or suction associated with a low water content but this source of strength may be only temporary. A consequence of the low permeability of fine soil is that, when a sample is removed from the ground, it is held together for a considerable length of time by an effect generated principally by suctions in the pore water but which may also be associated with chemical bonding between solid particles.

High permeability is an important engineering feature of coarse soil; during embankment construction, it can be utilised in drainage blankets placed under and within clay fill embankments to give a much increased rate of dissipation of excess pore pressure developed due to the weight of overlying soil. Similarly, vertical sand or band drains are often used to ensure rapid dissipation of excess pore pressure in soft clay soil underlying embankments.

The presence of a deposit of coarse soil under a water-retaining structure can lead to excessive leakage and, more importantly under excessive water pressure, there may be a loss of material due to piping with local collapse of the ground and partial collapse of the structure it supports. Uncontrolled seepage can cause the migration of fine particles. Where water flows out of a fill, for example, local instability may occur at the exit point. Where water is flowing through a fill that is susceptible to internal erosion, the process can be prevented or controlled by protecting the fill with filters that halt the loss of fine material. Erosion of soil is described in Box 4.3.

Permeability parameters for fine soil can be derived from laboratory oedometer tests (Section 4.5). The permeability of coarse soil may be determined in either variable or constant head tests, which can be performed in either the field or the laboratory[11]. Because of the difficulty in recovering undisturbed coarse samples, laboratory measurements are normally made on recompacted samples, which inevitably introduces changes in porosity and particle orientation. The variable head test is performed by recording the change with time of the head in a standpipe; it is appropriate for relatively permeable soil and should be used only where the test specimen is saturated. In a constant head laboratory test, the volume of water which flows through the permeameter cell is measured in a specified time period.

Box 4.3 Soil erosion
Soil erosion can occur at the surface of the ground or at depth within the ground:
- the action of water and wind can erode the surface of slopes in many types of soil;
- the flow of water through the ground can cause internal erosion by the removal of soil particles, usually in suspension;

A helpful distinction can be made between two forms of internal erosion:
- localised erosion;
- more general erosion which may be associated with internal instability of the soil.

Internal erosion is a particular hazard in the fill and foundations of embankments which retain water and in the foundations of other types of water retaining structure. Embankment dams usually contain filters and drainage elements to control seepage and prevent the development of internal erosion.

The term *piping* is usually applied to a localised process that starts at the exit point of seepage and in which a continuous passage or 'pipe' is developed in the soil by backward erosion. When the pipe approaches the source of water there is a sudden breakthrough and a rush of water through the pipe, which greatly accelerates internal erosion. The hydraulic gradient at the point where the water flows out of the ground is critical but is difficult to predict as it depends on localised weaknesses in the soil. Fine sand and silt are most susceptible to piping.

4.5 Deformation and compressibility

Buildings apply loads to the ground. These loads produce increases in effective stress that induce strains in the soil and the ground deforms. Where the building is founded on rock, ground deformations should be very small but where a building is founded on soft clay or peat, large ground movements may occur even with the relatively light loading of a two-storey house; these deformations may damage the building.

The most commonly encountered ground deformation behaviour of foundations is settlement and this often results from the soil undergoing predominantly one-dimensional compression, sometimes called confined compression. In this type of response to loading, vertical strain occurs with little lateral movement of the soil so that only the behaviour of soil under laterally confined conditions needs to be considered. Young's modulus E and Poisson's ratio υ are the parameters which are normally used to describe elastic behaviour. However, these parameters describe behaviour in uniaxial compression with no lateral restraint, which has no direct relevance to the stress conditions that occur in soil under a building foundation. The constrained modulus D and the coefficient of earth pressure at rest K_o are of direct application to confined compression and, for an elastic material, they are simply related to Young's modulus and Poisson's ratio[12]. Significant lateral soil movement can be expected when a foundation load produces stress conditions in the soil that are close to failure.

While the deformation of a coarse soil under confined compression exhibits many of the same characteristics as a fine soil, the magnitude of the reduction in volume under a given increase in stress can be an order of magnitude smaller. The most fundamental difference between the deformation behaviour of fine soil and coarse soil is related to the time-dependency of the response to loading. In coarse soil, the water can be forced out of the pores quite rapidly and the application of load produces an immediate compression[13].

Consolidation
When a clay soil is loaded, the particles are forced closer together and water is expelled. Because the water is bound to the fine clay particles and the very small pores have limited permeability, this change in water content occurs slowly. Consequently, the initial effect of applying a load to a saturated clay is simply to increase the pore water pressure. Water is practically incompressible under the range of stresses commonly encountered[14], so the instantaneous increase in pore pressure under saturated conditions is equal to the applied loading and there is no immediate increase in the effective stress between particles.

The flow of water out of the affected area is controlled by the hydraulic gradient generated by the excess pore pressure and the drainage condition at the soil boundaries. The excess pore pressure may dissipate over a period of months or even years in some situations. As the water slowly flows away from the loaded area, the applied load is transferred to the soil skeleton and the soil compresses, a process known as consolidation. When the load is removed, the clay tries to absorb water in order to swell but it may not recover to its original volume. A saturated clay can exist under a range of water contents depending on the stresses that are acting on it and those to which it has been previously subjected.

Oedometer compression test
Consolidation parameters for fine soil are normally measured in a laboratory oedometer test[15] although other types of laboratory tests[16] are sometimes used. The conventional oedometer consolidation cell is illustrated in Figure 4.3. The test specimen[17], typically 76 mm diameter and 19 mm high, is contained in a steel ring, which prevents lateral movement, and is loaded vertically through a lever system by weights on a hanger. The vertical compression of the soil specimen is measured using a dial gauge. Porous discs placed in contact with the top and bottom faces of the specimen ensure that it can drain freely from both ends.

The cell is flooded with water when the first load is applied. If the soil swells, further loads in the standard sequence are applied immediately until swelling ceases. However, this procedure does not allow the swelling pressure of desiccated clays to be evaluated accurately. Where this parameter is of interest, weights are carefully added to the hanger to just maintain the dial gauge reading at its original value. A standard oedometer test consists of a series of load increments with each increment approximately doubling the vertical stress (Box 4.4). After each load is applied, the dial gauge is read at prescribed intervals, which enable the compression of the specimen to be plotted against the square root of time.

SOIL BEHAVIOUR

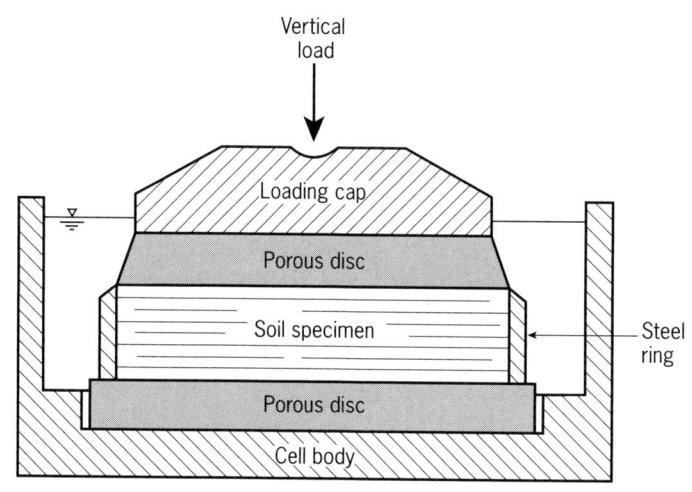

Figure 4.3 Laboratory oedometer consoldation cell

There are two alternative methods of presenting the test results: a strain plot and a void ratio plot. The strain plot is a direct representation of the data obtained in the laboratory test. To compute the void ratio at the various stages of the test, it is necessary to relate void ratio e to vertical strain ε_v:

$$e = (1 - \varepsilon_v)(1 + e_o) - 1 \qquad (4.11)$$

The initial void ratio can be computed from the initial dry density ρ_d and the particle density ρ_s:

$$e_o = (\rho_s / \rho_d) - 1 \qquad (4.12)$$

Figure 4.4 shows some typical oedometer test results plotted as the variation of ε_v and e, with the applied vertical stress plotted on a logarithmic scale. Each load increment is represented by a single point corresponding to the void ratio at the end of that load stage.

Box 4.4 Application of load in an oedometer test

In a standard oedometer test, a series of load increments apply vertical stresses to the specimen of, say, 25, 50, 100, 200, 400, 800 and 1600 kPa, each increment approximately doubling the vertical stress. When a load is applied, the excess pore pressure Δu is increased immediately throughout the soil and subsequently dissipates with time. Each load is held constant until movement ceases, which is normally within 24 hours. After each load is applied, the dial gauge is read at standard intervals of 0, 15 and 30 seconds, 1, 2, 4, 8, 15, 30 and 60 minutes, 2, 4, 8 and 24 hours the compression of the specimen can then be plotted against the square root of time.

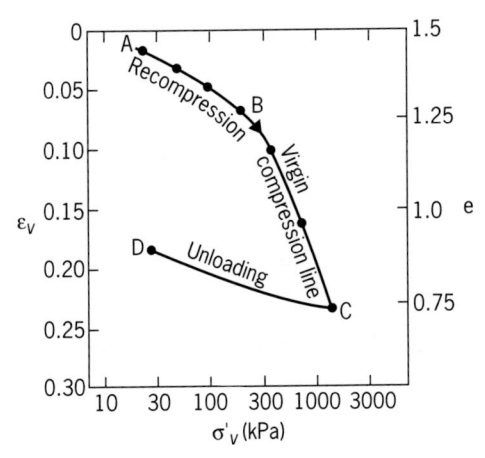

Figure 4.4 Vertical strain and void ratio as a function of vertical effective stress in a clay soil

At higher stresses, the plot approaches a straight line which is known as the virgin compression line or normal compression line. Figure 4.4 also shows the changes in void ratio as the sample swells during unloading, using a similar procedure to that adopted for loading. The response of the soil during unloading is much stiffer than during the loading since the soil has experienced plastic deformation.

The compressibility of a high plasticity clay[18] is shown in Figure 4.5. A-B forms the recompression curve and σ'_p is the preconsolidation pressure, which is the maximum effective stress to which the sample has been subjected in the past. The compression curve can be used to estimate the preconsolidation pressure. B-C is the virgin compression line which is characterised by largely irreversible strains. Unloading from C to D is largely reversible and this is shown by reloading D-E-C. On further loading C-F the virgin compression line is rejoined.

Coefficient of volume compressibility
The stress-strain behaviour of a clay soil under confined compression is normally expressed in terms of the coefficient of volume compressibility m_v which for a load increment of $\Delta\sigma_v$ can be expressed in terms of either the vertical strain or, alternatively, the void ratio at the start of loading e_o and the change in void ratio Δe produced by the loading:

$$m_v = \frac{\Delta\varepsilon_v}{\Delta\sigma'_v}$$
$$= \frac{\Delta e}{[\Delta\sigma'_v(1+e_o)]} \qquad (4.13)$$

In the study of the deformation of materials, it is customary to use a modulus which is in terms of the ratio of stress to strain; soil mechanics is unusual in having a parameter which is the ratio of strain to stress. The constrained modulus D, which is the inverse of m_v, has much to commend it but is used much less frequently. Figure 4.4 shows that the value of m_v depends on the stress level. Consequently, an oedometer test should be carried out incrementally over a wide range of stress to ensure that the appropriate value is available. The value of m_v which is normally quoted is that corresponding to a stress 100 kPa greater than the estimated vertical effective stress which was acting on the specimen when it was in the ground.

Compression index

The results of an oedometer test can also be described by the compression index C_c, which is the slope of the compression curve on a plot of e versus log σ'_v:

$$C_c = \frac{\Delta e}{\log(\sigma'_{v2}/\sigma'_{v1})} \qquad (4.14)$$

where increasing the vertical stress from σ'_{v1} to σ'_{v2} has reduced the void ratio by Δe. Although C_c appears to be a more complex parameter than m_v, it has one major advantage. The virgin compression line is approximately straight so C_c, which is the slope of that line, has a nearly constant value over a wide range of stress, whereas m_v is strongly stress dependent.

Compressibility classification

A clay soil will have consolidated as the result of subsequent deposition above it during its geological history. Because lateral strain is prevented by the adjacent soil, which is equally loaded, the boundary conditions are essentially those which exist in the oedometer. Many clay soils have preconsolidation pressures that are far greater than their current overburden pressure and are described as heavily overconsolidated (Section 3.2). Even normally consolidated soil may have a surface crust that is slightly overconsolidated as a result of fluctuations in groundwater level or desiccation.

A classification of soil based on compressibility is in Table 4.2. Heavily overconsolidated clays, such as London clay, Gault clay and Oxford clay, are likely to have a very low compressibility. A heavily compacted gravel fill may also be in this category. Some glacial tills have a low compressibility, as may compacted sandstone rockfill. Weathered London clay may have medium compressibility. A normally consolidated alluvial clay is likely to have high

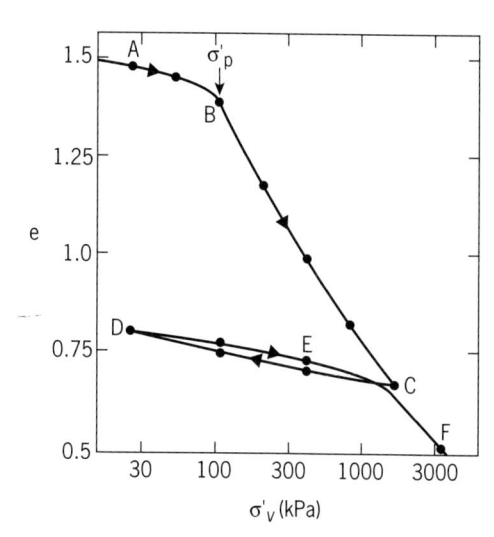

Figure 4.5 Compression of a high plasticity clay: loading, unloading and reloading – after Lancellotta, 1993

compressibility, as is old domestic refuse. Very organic alluvial clays and peats may have a very high compressibility, as may recently placed domestic refuse.

Table 4.2 Compressibility classification

Qualitative description of compressibility	Coefficient of volume compressibility m_v m²/MN	Constrained modulus D MPa	Effect of $\Delta\sigma_v = 100$ kPa on 2 m depth of soil ε_v %	Settlement mm
Very low	< 0.05	> 20	< 0.5%	< 10
Low	0.05 – 0.10	10 – 20	0.5 – 1%	10 – 20
Medium	0.10 – 0.30	3.3 – 10	1 – 3%	20 – 60
High	0.30 – 1.50	0.67 – 3.3	3 – 15%	60 – 300
Very high	> 1.50	< 0.67	> 15%	> 300

Rate of consolidation

The test results for each load increment in an oedometer test can be used to derive the coefficient of consolidation c_v which is needed to calculate the timescale associated with settlement. The coefficient of consolidation[20] is related to the permeability of the soil, k:

$$c_v = \frac{k}{m_v \gamma_w} \quad (4.15)$$

It is convenient to describe the progress of the consolidation in terms of a non-dimensional time factor T, which can then be related to the proportion of the total compression U that has taken place. Different assumptions about such matters as the dissipation of pore pressure give somewhat different relationships between T and U as shown by the hatched area in Figure 4.6. The elapsed time, t, is related to the time factor, T, by the coefficient of consolidation, c_v, and the length of the drainage path, h:

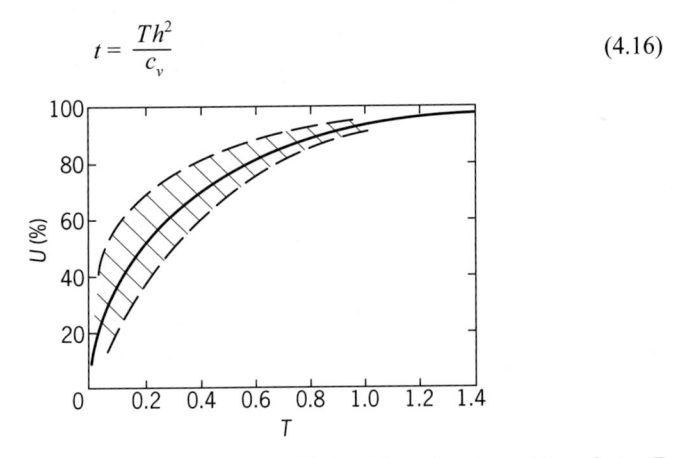

$$t = \frac{Th^2}{c_v} \quad (4.16)$$

Figure 4.6 Percentage consolidation U as a function of time factor T

SOIL BEHAVIOUR

Secondary compression

Some further compression will occur under conditions of constant effective stress when all the excess pore pressure has dissipated. This is known as secondary compression. The secondary compression of saturated clay exhibits a linear relationship between compression and the logarithm of time that has elapsed since the load was applied and can be described by the coefficient of secondary compression in terms of either vertical strain or void ratio. Equation 4.17 shows C_α expressed in terms of vertical strain:

$$C_\alpha = \frac{\Delta\varepsilon_v}{\log(t_2/t_1)} \quad (4.17)$$

where:

vertical compression $\Delta\varepsilon_v$ has occurred between times t_2 and t_1 after the load was applied.

4.6 Shear strength

The safety of a building or a geotechnical structure is dependent on the strength of the ground. If the soil on which a building is founded fails, the building is likely to collapse. The strength of soil can, therefore, be of crucial importance. Only the static loading of the ground is considered here. Some consideration of the effect of dynamic loading on loose coarse soil[21] is in Section 7.3.

The strength of a material, which is the maximum stress it can sustain before failure, can be expressed in different ways. The strength of steel can be described in terms of a tensile strength and the strength of concrete as an unconfined compressive strength, but these concepts are of limited relevance to soil where the individual particles are only lightly bonded together. Nevertheless, a soil can exhibit considerable strength, supporting the weight of a high-rise building or an earth dam, by virtue of the confining stress. A mass of sand or clay is able to support vertical loads because of the horizontal stress exerted by the surrounding soil.

The low strength of coarse soil under low confining stress leads to low bearing capacity for narrow surface footings. The bearing resistance is further reduced if the water-table is close to the foundation because buoyancy forces reduce the effective stress in the ground. The increasing strength of the soil under increasing confining stress can be exploited by wide foundations which create a confined stress environment under the middle of the foundation, or by placing the foundations at a substantial depth below the surface (1.0 m or more) where overburden stress increases the strength. Medium to high density sand and gravel deposits provide excellent bearing layers for deep foundations and piles. Foundation design is described in Chapter 6.

Direct shear test

The direct shear box[22] was one of the earliest methods of measuring the failure condition in soil (Figure 4.7). The shear box is usually square in cross-section and typically 20 mm high with an area of 36 or 100 cm². The soil specimen is placed in the box, which is split at its mid-height to allow the top half to slide horizontally with respect to the base. A vertical stress is applied to the soil and it is sheared by displacing the bottom half of the box while the upper half is restrained. The shear force is plotted as a function of shear displacement; usually the change in thickness of the specimen is also recorded. Coarse and fine soils can be tested in the shear box but the specimen is not sealed so porous stones should be provided to allow free drainage of the top and bottom surfaces of the soil.

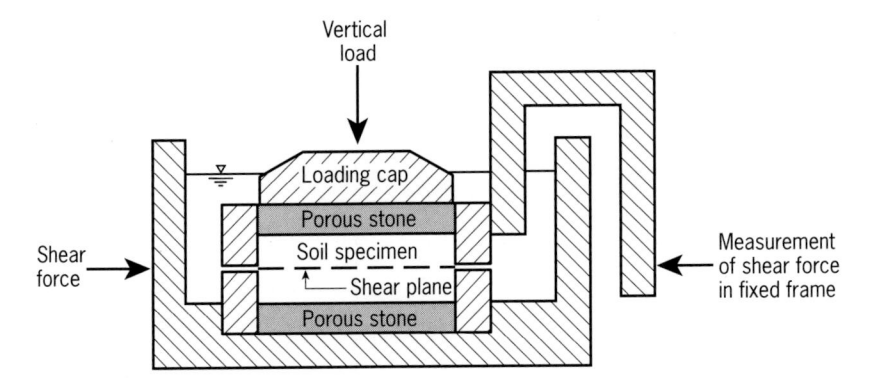

Figure 4.7 Laboratory shear box

The direct shear test has the advantage of being simple and relatively inexpensive but it has some limitations and there are some difficulties in interpreting the test results[23]. Modified versions of the direct shear test, such as the reversing shear box and ring shear apparatus, are used to measure residual strength parameters; these are of relevance to calculating the stability of a slope containing existing slip surfaces.

Triaxial compression test

The triaxial test is now the preferred method of measuring strength. A vertical cross-section through a triaxial cell is shown in Figure 4.8. The cell is pressurised so that the cylindrical specimen of soil enclosed in a rubber membrane is subjected to a confining pressure. The test specimen is loaded axially by means of a piston passing through the top of the cell and the load is measured by either a proving ring mounted externally between the loading frame and the top of the piston or, preferably, by a load cell mounted inside the cell between the piston and the soil specimen. In a standard triaxial compression test, the piston moves downwards at a constant rate.

SOIL BEHAVIOUR

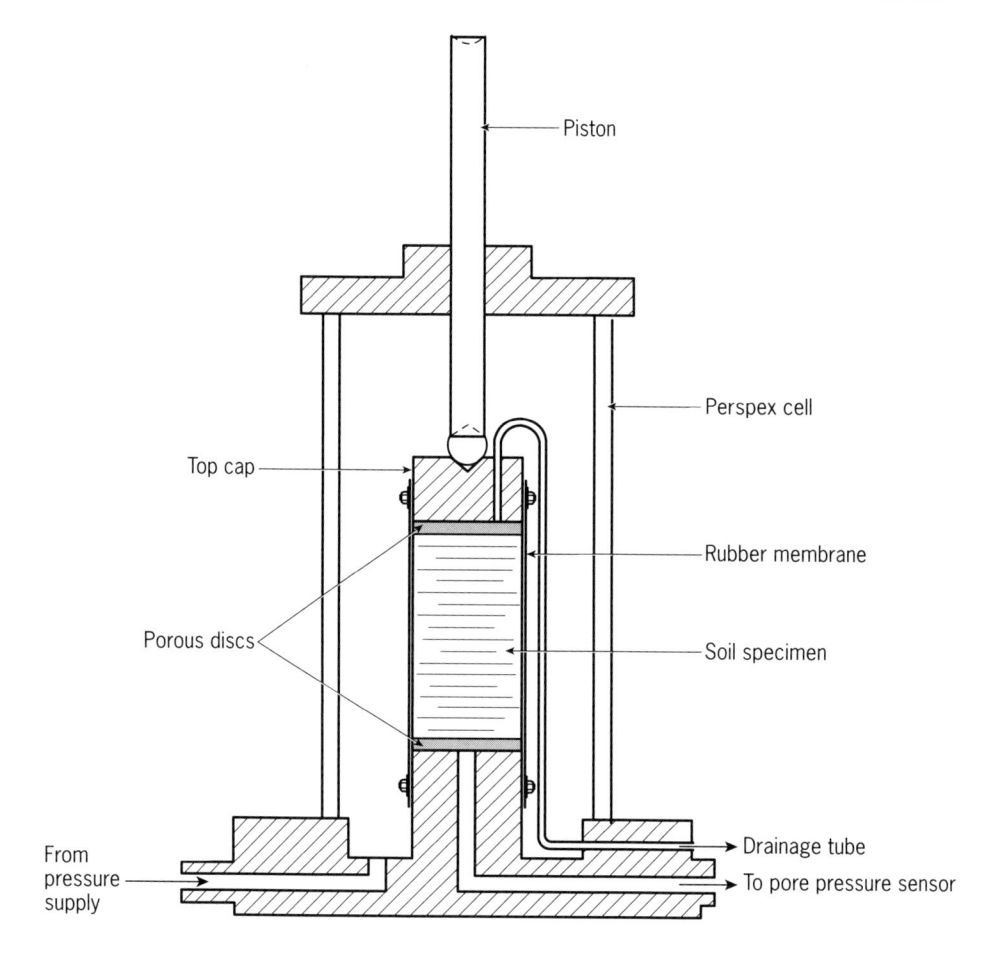

Figure 4.8 Laboratory triaxial test

The cylindrical specimen is contained within an impermeable membrane so that the drainage from the soil can be controlled. With no drainage permitted, the soil can be tested under undrained conditions while measuring the pore water pressure u_w. Hence the vertical and radial effective stresses, σ'_1 and σ'_3, can be determined[24]. While the shear box constrains the soil to fail along a predetermined horizontal plane, the triaxial test is more like field conditions where failure occurs along the plane where the ratio of shear stress to normal stress is highest.

The test specimens have standard diameters of 38 mm or 102 mm and a length of twice the diameter. Because of the influence of soil fabric on the strength of clay, the size of the specimen can have a significant effect on the shear strength; larger specimens contain more discontinuities and will therefore yield lower strengths; this should be more representative of field conditions.

The stresses acting on the triaxial test specimen can be expressed in terms of two stress path parameters s' and t' which are related to the minor and major principal effective stresses as follows:

$$s' = 0.5\,(\sigma'_1 + \sigma'_3) \tag{4.18}$$
$$t' = 0.5\,(\sigma'_1 - \sigma'_3) \tag{4.19}$$

The soil may be consolidated and then loaded to failure under either undrained or drained conditions. The three common forms of test for clay soils are:
❐ unconsolidated undrained triaxial compression test, without pore water pressure measurement[25];
❐ consolidated undrained triaxial compression test with pore water pressure measurement[26];
❐ consolidated drained triaxial compression test, with volume change measurement[27].

Effective stress shear strength parameters
The effective stress shear strength parameters may be derived from consolidated undrained triaxial compression tests with pore water pressure measurement or from consolidated drained triaxial compression tests. The failure condition for each triaxial test can be represented by a Mohr circle and by performing a series of triaxial tests at different consolidation pressures, the effective stress shear strength parameters c' and ϕ' are determined as shown in Figure 4.9. The Mohr-Coulomb failure criterion is expressed in terms of effective stress as a linear relationship[28]:

$$\tau_f = c' + \sigma' \tan \phi' \tag{4.20}$$

where:
    τ_f is the shear strength
    σ' is the normal effective stress on the failure plane
    c' is the cohesion intercept (the apparent shear strength at zero effective stress)
    ϕ' is the angle of shear resistance.

For many soils, the cohesion intercept is zero. While drained tests can be used for fine soil, undrained tests with pore pressure measurement are normally preferred because they can be performed more quickly[29].

Depending on the particle size distribution, the shape of particles, the density of packing of

SOIL BEHAVIOUR

the particles and the state of confining stress in the ground, a soil undergoing shear may try to expand or to contract. Whether the volume change is dilatant or contractant controls how its resistance develops as shearing progresses. For example, after reaching a peak value (τ_{max} in Figure 4.10) the soil resistance reduces as shearing progresses because the soil is dilating. This is sometimes known as strain softening. Eventually, at a large strain, the soil reaches a constant volume at the critical void ratio e_{crit}, though straining is still occurring. At extremely high strain levels, for example on the failure surface in a landslide, the strength of a clay soil may fall to its lowest possible level, the residual strength. A soil may therefore have three strengths: peak or maximum[30], constant volume ϕ'_{cv} and residual ϕ'_r, depending on how much deformation has occurred.

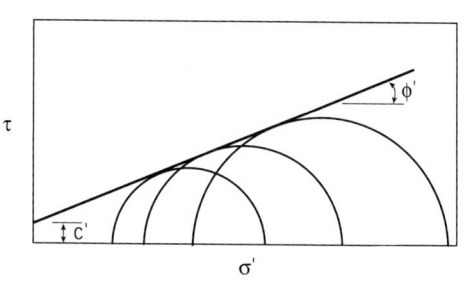

Figure 4.9 Mohrs circles and failure envelope derived from triaxial tests at different consolidation pressures

For sands ϕ'_{cv} is typically in the range 32° to 37° and the maximum value for a dense sand sheared at a relatively low stress could be 10° greater[31]. Values of ϕ' for low plasticity clays may be only moderately lower, but medium and high plasticity clays[32] are likely to have much smaller values and in some cases ϕ'_r may be as low as 10°.

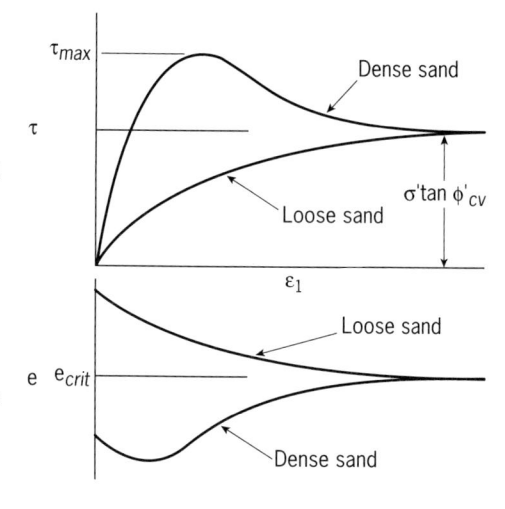

Figure 4.10 Dilatancy of sand

Undrained shear strength

Undrained behaviour is of practical importance for saturated, low permeability clay when applied loads change more rapidly than pore pressure dissipates. When soil is sheared without allowing any drainage, the measured strength is known as the undrained shear strength, c_u.

$$c_u = 0.5(\sigma_1 - \sigma_3) \qquad (4.21)$$

This strength may be appropriate for design in situations where loading is occurring quickly or there is no effective drainage. Undrained shear strength is a function of water content and can be estimated in the field (Table 4.3).

Table 4.3 Consistency of clays[33]

Consistency	Simple field test	c_u kPa
Very soft	Finger easily pushed in up to 25 mm	< 20
Soft	Finger pushed in up to 10 mm	20 – 40
Firm	Thumb makes impression easily	40 – 75
Stiff	Can be indented slightly by thumb	75 – 150
Very stiff	Can be indented by thumb nail	150 – 300
Hard	Can be scratched by thumb nail	> 300

The undrained shear strength c_u of the soil is normally derived from unconsolidated undrained triaxial compression tests on undisturbed specimens. A typical rate of strain for the test is 2 % per minute[34]. The confining pressure is selected to ensure that any gas in the sample is dissolved, but would normally be twice as large as the overburden pressure acting on the soil when it was in the ground. The deviator stress is calculated from the applied axial load, allowing for the change in cross-sectional area of the specimen as it is compressed; c_u is half the peak deviator stress.

Undrained strength can also be measured in consolidated undrained triaxial compression tests in which the specimen has been consolidated back to the estimated in-situ overburden pressure. These tests tend to produce higher strengths than unconsolidated tests[35].

Clay soils can be characterised by a plot showing the variation of undrained shear strength with depth. For a normally consolidated soil, c_u is related to the effective overburden pressure σ'_v and typically the ratio c_u/σ'_v lies in the range from 0.2 to 0.4. With a high water-table, effective stresses will be low, but some desiccation is likely to occur near the ground surface, forming a stiffer crust.

4.7 Partially saturated soil

Most of this chapter has been concerned with the behaviour of saturated soil. The standard procedures for laboratory compressibility and strength tests are for testing saturated soil. However, most low-rise buildings are founded at relatively shallow depths on partially saturated soil above the water-table (Section 3.10). The behaviour of partially saturated soil is much more complex than the behaviour of saturated soil.

Soil above the water-table is not subjected to a positive hydrostatic pore water pressure. Evaporation and transpiration[36] may lower the pore water pressure sufficiently for water in the soil near ground surface to partially drain out of the voids in the soil. This creates a zone, which is part saturated and part partially saturated, within which the pore pressures are negative. This negative pore pressure in the soil above the water-table results in attractive forces between the soil particles. Soil suction can be significant in fine soil and result in an apparent cohesion with enhanced resistance to particle movement. However, small perturbations in ground conditions can destroy the suction.

Suction can be measured by a number of different techniques. In the filter paper test, soil specimens are placed in intimate contact with pieces of filter paper and the whole assembly sealed and left to come to equilibrium (Box 4.5). For a particular porous material, there is a unique relation between the water content and the pore suction; this allows the measured moisture content to be converted directly to a suction.

The principle of effective stress is rightly regarded as a basic element in understanding soil behaviour but, with partially saturated soil, the concept of effective stress is much more complex than for saturated soil. For partially saturated soil, it has proved extremely difficult to get agreement on either the definition of effective stress or a general framework to describe soil behaviour. Recently, much soil mechanics research has been concentrated in this area[38]. Some important features of the behaviour of partially saturated soil are dealt with in the following sections of this chapter.

4.8 Shrinkage and swelling of clay

Some clay soils are liable to a change in volume; they expand and contract under meteorological influences, swelling when wetted and shrinking when dried. The surface layer of these soils is inherently unstable because of this tendency to change volume. Evaporation and transpiration cause shrinkage during summer months; rainfall during winter months, and the removal of vegetation, allow the soil to recover from any moisture deficit and to swell. These processes cause ground movements that can damage buildings on shallow foundations.

Box 4.5 Filter paper test[37]
Samples of the soil are placed in intimate contact with pieces of filter paper; the whole assembly is sealed, eg using cling film, and left to come to equilibrium. The filter paper is then removed and weighed before and after drying in an oven to determine its moisture content. For a particular porous material, there is a unique relation between the moisture content and the pore suction. So, provided the material has been calibrated, the measured moisture content can be converted directly to a suction. The test was developed using U100 (100 mm diameter) samples. These are relatively expensive to obtain but it has been suggested that meaningful measurements can be made on reconstituted samples of high plasticity clay; this reduces significantly the costs of obtaining the samples.

As explained in Chapter 3, clay soils contain a significant proportion of extremely fine particles, smaller than 0.002 mm. In coarse soil, the water simply fills the available voids. Clay minerals form small plates, and an increase in water content forces the plates apart causing the soil to expand; a reduction in water content allows the particles to adopt a denser configuration, causing the soil to shrink.

The volume of a clay soil reduces as a result of an increase in the stress acting on the solid particles of the soil skeleton. This is normally brought about in one of two ways:
❐ an increase in the imposed loading produced, for example, by raising ground level or by applying foundation loads;
❐ a reduction in the pore water pressure.

If the applied load is reduced, or free water is made available to soil with high suction, the soil draws the available moisture back into the soil. Because of its low permeability, swelling and shrinking volume changes occur slowly, often over years. Once the pore water pressure reduces below atmospheric, the water is in a state of tension and imposes a suction on the soil particles[39]. High suction is produced by evaporation and extraction through the roots of transpiring vegetation, a process known as desiccation (Box 4.6); it can also occur due to other causes[40].

The increase in suction in the soil water increases the stress acting on the soil particles, bringing about a reduction in volume. There is a supply of water from surrounding soil and the suctions reach limiting values. The soil is able to achieve a new equilibrium at a lower water content. A soil which is described as desiccated may have a water content that is reduced by only 1 or 2% below its undesiccated condition. Desiccation can reduce the water content of firm shrinkable clay to values below the plastic limit, making the soil much stiffer and giving it a dry and friable appearance. Desiccation often causes the ground to crack near the surface. In highly shrinkable soil, cracks 25 mm wide and 0.75 m deep are not uncommon during dry summer months.

> **Box 4.6 Desiccation**
> Desiccation is often associated with a complete removal of moisture; for example, a desiccator is a device for removing the last traces of free water from crystals. Desiccation to geotechnical engineers is any reduction in the natural water content of a soil caused by evapotranspiration. Desiccation can be defined as drying of the soil resulting from an increase in suction over the normal equilibrium value.

The physical process of desiccation involves the removal by external forces of water from the soil. The process creates suction in the remaining soil water so it goes into a state of tension, bringing the soil particles closer together (shrinkage) and increasing the forces between them (the soil becomes stronger and stiffer). However, factors other than desiccation, such as loading followed by dissipation of positive excess pore water pressure, may also bring the particles together with similar changes in the soil strength and stiffness. It follows, therefore, that the most reliable way of determining desiccation is to measure the suction in the soil water.

The natural water content w and the Atterberg limits w_L and w_P are routinely measured in site investigations and attempts have been made to use them to determine whether or not a soil is desiccated. Water contents of samples are taken from two boreholes, one in the test

SOIL BEHAVIOUR

area, the other a control borehole well away from any desiccating influences. Desiccation can be identified by a comparison between these borehole water content profiles but there can be considerable uncertainties in interpreting the results[41]. Alternatively, a comparison of profiles of suction measurements can be made (Figure 4.11) where both water content and suction measurements give a good indication of the effect of desiccation.

Unlike water content profiles, a single suction profile can provide a reliable indicator of desiccation but it is important to appreciate that a sample of saturated, undesiccated, overconsolidated clay will register a suction value. This is because equal but opposite stresses will be generated in the soil water when the sample is released from the ground stresses which have been confining it.

The volume change potential depends on the proportion of clay particles (clay fraction) and the mineral composition of the clay particles. Clay that is described as shrinkable usually contains a relatively large proportion of montmorillonite, which has the greatest

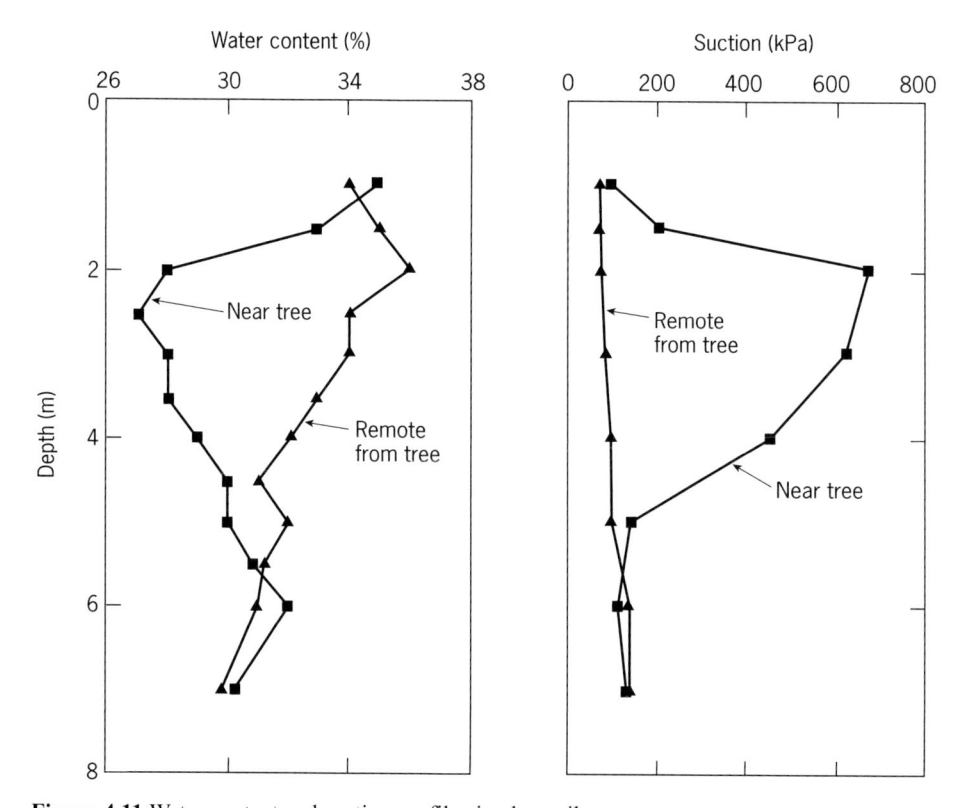

Figure 4.11 Water content and suction profiles in clay soil

potential for absorbing water and hence for changing volume. The plasticity indices of montmorillonites are much larger than those of illites and kaolinites.

Most of the overconsolidated clays that are prevalent throughout south-east England can be classed as shrinkable and their water contents w are close to the plastic limit w_P. Close to the ground surface, the water contents are influenced by evaporation, transpiration and rainfall and, as a consequence, may fluctuate from as little as 15% in dry summer weather to 40% in wet winters. The potential for a clay soil to change volume is normally referred to as its *shrinkage potential* and this is usually assumed to be proportional to the plasticity index ($I_P = w_L - w_P$). A helpful classification is shown in Table 4.4 in terms of a modified plasticity index I'_P which makes allowance for the fact that the plastic and liquid limit tests are carried out on only that fraction of the soil which is finer than 425 µm. This fraction can be expressed as a percentage (% < 425 µm) and I'_P is related to I_P as follows:

$$I'_P = I_P (\% < 425 \text{ µm})/100\% \qquad (4.22)$$

Table 4.4 Clay volume change potential[42]

Modified plasticity index I'_P %	Volume change potential
> 60	Very high
40 - 60	High
20 - 40	Medium
< 20	Low

4.9 Collapse compression on wetting

Most poorly compacted, partially saturated fills and some natural soils undergo a reduction in volume when inundated or submerged for the first time; if this occurs subsequent to construction on the ground, buildings can suffer serious damage. This often represents the most serious hazard for buildings on fill. The phenomenon, which can occur under a wide range of applied stress, is usually termed *collapse settlement* or *collapse compression*. It was originally described as collapse settlement because it was considered to be associated with a collapse of the soil structure. Collapse compression is a widespread phenomenon which can occur without any change in applied total stress.

Collapse compression received little attention in the early years of modern soil mechanics. This may have been because the phenomenon might seem to conflict with the principle of effective stress. For example, the submergence of a fill by a rising ground water-table increases pore pressures so reducing effective stresses. The ground might be expected to heave but in practice saturation often causes settlement due to collapse compression. Despite this lack of attention from those leading the development of soil mechanics as a scientific discipline, the phenomenon was recognised in practice. Collapse compression was used to compact highway fills in the 1920s. In contrast to these highway embankment fills, where collapse compression was used as a method of ground improvement, collapse compression was identified as a potential hazard in embankment dam fills[43]. When collapse compression began to receive more attention, much of the early work was concerned with natural sands and silts[44].

In the UK, the hazard of collapse compression is largely associated with poorly compacted fills. The problems associated with building on fill are described in Section 7.4. Various mechanisms may cause collapse compression on wetting in fills (Box 4.7) but from a practical perspective it is more important to be able to determine collapse potential than to know the precise mechanism which causes it to occur. The effect of inundation on a partially saturated fill can be examined in oedometer tests[45].

Box 4.7 Mechanisms causing collapse compression in fills

Collapse compression can be attributed to three basic processes:
❏ Weakening of interparticle bonds; there may be intergranular bonds within the fill which are weakened or eliminated by an increase in water content. These bonds may be due to capillary forces, silt and clay particles or cementing agents. This may be the predominant cause of collapse compression in sand fill.
❏ Weakening of particles in coarse fill; the parent material from which the fill is formed may lose some strength as its water content increases and approaches saturation. This may be the predominant cause of collapse compression in relatively uniformly graded rockfill.
❏ Weakening or softening of aggregations of particles in fine fill; where a fill is formed of aggregations of fine particles, such as lumps or clods of clay, these aggregations may lose strength as the water content increases. This may be the principal cause of collapse compression in stiff clay fill.

Building developments and road construction in the UK have been increasingly taking place on opencast mining backfills[46] which may be composed of mudstone and sandstone fragments, shale or clay. Collapse compression has been identified as a major hazard; there are two different causes of wetting:
(a) rise in the water-table can trigger settlement at depth within the fill;
(b) downward infiltration of surface water can trigger settlement close to ground level.

Some typical magnitudes of collapse compression for different types of fill are listed in Table 4.5. Where wetting is caused by a rising groundwater level, the situation is usually well defined. Figure 4.12 shows the large settlement that occurred when the ground water level rose 34 m through a mudstone and sandstone opencast backfill during a three-year period from 1974 to 1977. Vertical compressions were locally as large as 2%, but the average compression measured over the full depth of inundated backfill was smaller than 1%[47]. Where wetting is associated with the percolation of water downwards from the

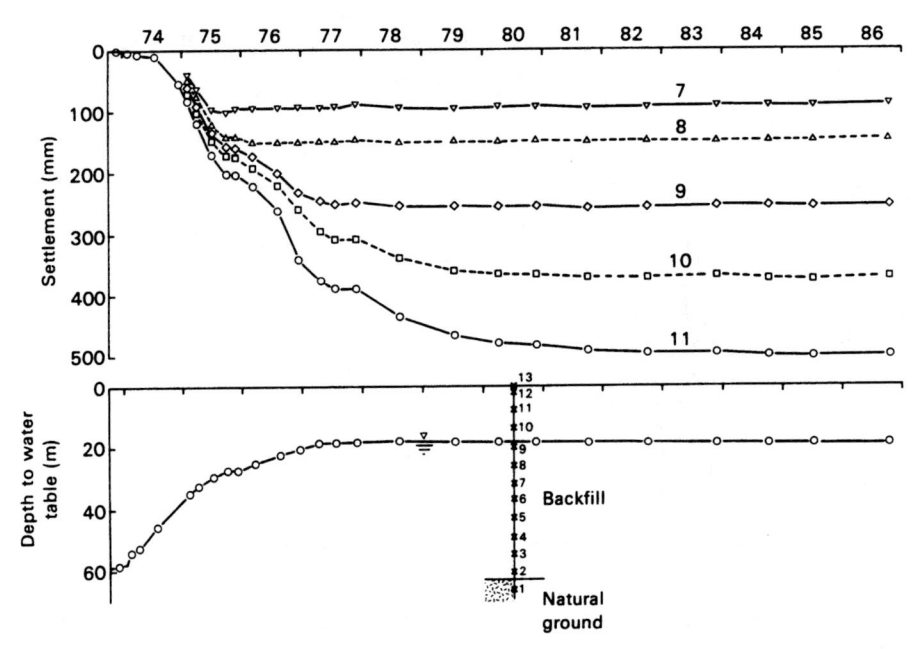

Figure 4.12 Settlement caused by rising groundwater level in uncompacted coarse fill

ground surface, the progress of collapse compression is less predictable. Figure 4.13 shows that after the completion of an inundation test from 1 m deep trenches, the surface settlement of a stiff clay backfill did not reduce to the pre-inundation rate for another six years. About 40% of the total settlement occurred after the addition of water had ceased[48].

Table 4.5 Collapse compression measured in some non-engineered fills

Fill type	Collapse compression %	Cause of wetting
Mudstone/sandstone	2	(a)
Clay/shale fragments	5	(a)
Stiff clay	6	(b)
Stiff clay	3	(b)
Colliery spoil	7	(b)

4.10 Compaction of partially saturated fine soil

The term *compaction* is used to describe processes in which mechanical equipment is used to compress soil into a smaller volume, thus increasing its density. As solid soil particles and water are virtually incompressible, the increase in density is achieved by a reduction of the air voids. The density of a soil has a major influence on its behaviour and compaction normally improves engineering properties because a dense soil generally has superior properties to the same soil in a loose condition.

SOIL BEHAVIOUR

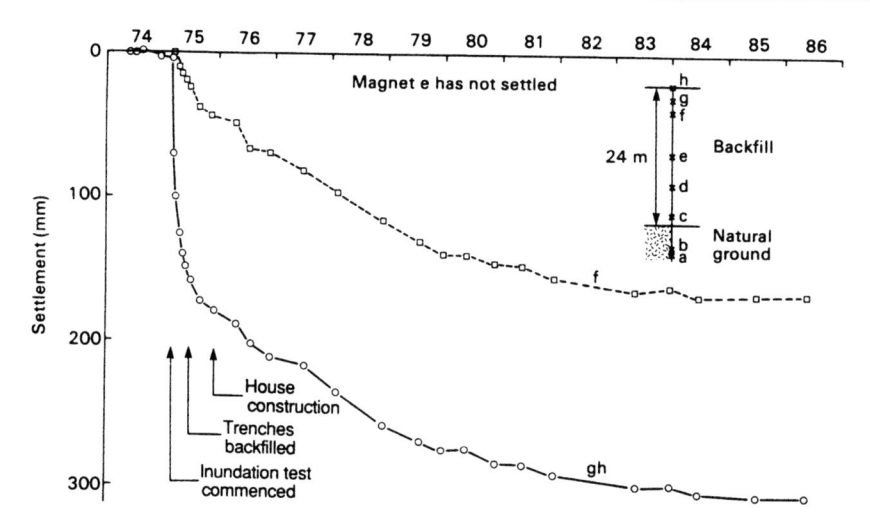

Figure 4.13 Settlement caused by downward percolation of water from trenches in uncompacted clay fill

Natural soils can be compacted in-situ to a considerable depth by dropping heavy weights (dynamic compaction is described in Section 9.3) but, more commonly, compaction is applied by various types of roller to fill material as the fill is placed in relatively thin layers – Section 9.5. Appropriate forms of laboratory testing can be helpful in optimising the compaction process and minimising subsequent compression of the fill. The approaches to this are different for coarse and fine fills. Two types of compaction are commonly used in laboratory tests: multiple blows with a rammer of fixed weight or, for coarse soil, an electrical vibrating hammer. These compaction methods are intended to model processes that are used in the field.

In assessing field behaviour of coarse fill, it is helpful to relate the in-situ degree of compactness to that determined in standard laboratory compaction tests. The degree of packing in coarse fills can be described by the density index, I_D (Section 3.5).

The amount of densification of a clayey soil that can be achieved for a given compactive effort is a function of water content. The results of laboratory compaction tests, in which soil specimens are compacted by multiple blows with a fixed weight rammer, are plotted as the variation of dry density with water content; Figure 4.14 shows two compactive efforts. Such plots are a simple and useful way of representing the condition of a partially saturated fill; they can be used to identify the two parameters of prime interest: the maximum dry density and the corresponding optimum water content w_{opt}. The results can be compared to theoretical densities for various levels of saturation (for example 0%, 5% and 10 % air voids).

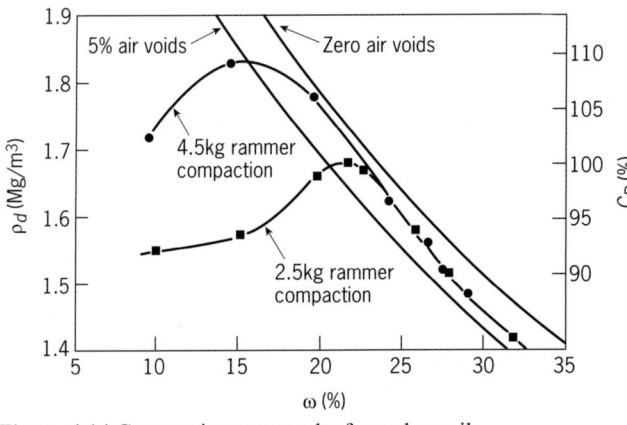

Figure 4.14 Compaction test results for a clay soil

The percentage ratio of the in-situ dry density to the maximum dry density achieved with a specified degree of compaction in a standard laboratory compaction test is termed the relative compaction (C_R). When quoting a value of relative compaction, the type of compaction must be specified, and this is usually, although not always, the standard 2.5 kg rammer Proctor compaction test[49]. In Figure 4.14, C_R is based on this test. The concept that in field operations the fill should be placed at its optimum moisture content ignores the major difficulty in relating the compactive effort applied in the field to that applied in a laboratory test.

Some fundamental features of the compaction of clayey soils are:
❒ The density to which a particular compactive effort will bring a clay fill depends on the water content of the fill.
❒ For a given compactive effort there is a maximum dry density which is achieved at the optimum water content w_{opt}. The expressions maximum dry density and optimum water content refer to a specified compaction procedure and can be misleading if taken out of the context of that procedure.
❒ At a water content significantly dry of optimum, the specified compaction procedure will result in a fill with large air voids.
❒ At a water content significantly wet of optimum, the specified compaction procedure will produce a fill with a minimum air voids, typically between 2% and 4%.

Notes
1. Many soil mechanics text books have been written for civil engineering students; Coduto (1998) and Budhu (2000) are recent examples. A good introduction to the subject is given by Lancelotta (1995). A comprehensive account from a practical perspective can be found in Lambe and Whitman (1979). Hudson and Harrison (1997) provide a helpful introduction to rock mechanics. Blyth and De Freitas (1984) is a highly respected undergraduate text book on engineering geology.

2. Critical state soil mechanics was developed at the University of Cambridge (Schofield and Wroth, 1968). The critical state model is a simplification of real soil behaviour, but it can provide helpful insights into the response of the soil to complex loading conditions, as shown in Atkinson (1993).

3. In continuum mechanics, materials are treated as though they completely fill the region that they occupy without any voids or gaps. In reality, soils and rocks are not like this; soils are formed of

discrete particles and rocks contain fissures and joints.

4. It may seem surprising that analyses which are strictly applicable only to homogeneous, elastic, isotropic materials are appropriate for soils which are often heterogeneous, non-elastic and anisotropic. Furthermore, soil is a particulate material and in using elastic theory it is being analysed as though it is a continuum. However, the distribution of vertical stress beneath a foundation is not particularly sensitive to the stress-strain relationship of the soil and there are considerable practical advantages in using simple isotropic linear-elastic theory since it involves only two parameters, Young's modulus E and Poisson's ratio υ.

5. The law of elasticity discovered by Robert Hooke in 1660 states that, for relatively small deformations, the displacement or size of the deformation is directly proportional to the deforming force or load (*Ut tensio, sic vis*). Thus the applied force F equals a constant k times the displacement or change in length δL:
$$F = k\ \delta L$$
The value of k depends not only on the type of elastic material but also on its dimensions and shape. In school physics, the usual example is a metal wire of cross-sectional area A and length L, where the small increase in its length δL when stretched by an applied force F doubles each time the force is doubled. Hooke's law may be expressed in terms of stress and strain:
$$F/A = E\ (\delta L/L)$$
where:

E is Young's modulus for the wire.

Under elastic conditions the deformed object returns to its original shape and size upon removal of the load. In the general case, a deforming force may be applied to a solid by stretching, compressing, bending, or twisting.

6. Karl Terzaghi (1883-1963) was the first to recognise the significance of effective stress during his research into soil consolidation in the 1920s.

7. It has been suggested that for a soil in an overconsolidated state K_{0oc} can be related to K_{0nc} for the soil in a normally consolidated state as follows:
$$K_{0oc} = K_{0nc}\ (OCR)^{0.5}$$
where:

OCR is the overconsolidation ratio.

8. The total head at a point in a soil mass is considered to have three components; the elevation head, the pressure head and the velocity head. Normally the velocity head can be neglected in seepage problems and the head is considered to be the sum of the elevation head, which is the difference in elevation between the point and a given datum, and the pressure head, which is the difference in elevation between the point and the level to which the fluid rises in a standpipe.

9. Darcy's law is valid for laminar flow under hydraulic gradients of practical interest. The principal soils to which it may not be applicable are clean gravel, where turbulent flow may occur, and clay under low hydraulic gradient, where the flow rate is very small.

10. Excess pore water pressure is the pore water pressure in excess of the equilibrium water pressure. Below the water-table, the equilibrium pore water pressure is related to hydrostatic conditions as described in Section 4.3.

11. A standard laboratory test procedure for determining permeability using a constant head in a permeameter cell is in BS 1377: 1990 *Methods of test for soils for civil engineering purposes* Part 5, clause 5.

12. Young's modulus E describes the ratio of major principal stress to major principal strain, and Poisson's ratio υ describes the ratio of minor to major principal strain in uniaxial compression with no lateral restraint. The constrained modulus D and the coefficient of earth pressure at rest K_o are of

direct application to confined compression, D being the ratio of applied vertical stress to the resulting vertical strain and K_o the ratio of minor to major principal effective stress. For an isotropic, linear elastic material, E and υ are related to D and K_o:

$$(E/D) = \{(1 - K_o)(1 + 2K_o)\}/(1 + K_o)$$
$$= \{(1 - 2\upsilon)(1 + \upsilon)\}/(1 - \upsilon)$$
$$= K_o/(1 + K_o)$$

when $K_o = 0.5$, then $E/D = 0.67$ and $\upsilon = 0.33$.

13. If a load is applied to a saturated coarse soil surrounded by an impermeable barrier, in theory an effect similar to the loading of a clay soil can be produced, since the applied load is carried by the increase in pore water pressure without affecting the effective stresses. Allowing the water to slowly drain through the barrier results in a gradual dissipation of pore pressure and a corresponding increase in effective stress. Only a small flow of water is required to allow this transfer to take place because the soil grains are already in contact with one another. While such conditions might be produced in the laboratory, the drainage conditions in the ground allow the water to flow and consequently coarse soil compresses instantaneously and does not consolidate over a lengthy period, as does clay.

14. Although the water in a saturated soil is essentially incompressible, when highly compressible air bubbles appear the water becomes compressible.

15. Part 5 of BS 1377:1990 specifies compressibility, permeability and durability tests and describes the determination of one-dimensional compression parameters.

16. The oedometer measures one-dimensional consolidation parameters (no lateral strain occurs during consolidation). This boundary condition models what happens to a soil layer as it consolidates as the result of subsequent deposition and under the middle of a building. In many other situations, lateral strains occur during consolidation and this can be modelled more closely in the triaxial apparatus. In this test, the soil is compressed under isotropic effective stress, the volume change is monitored by measuring the volume of water expelled from the soil and the pore pressure measured at the base of the specimen. The volume change occurs three-dimensionally and the test tends to show a greater compressibility than that measured in the oedometer. BS 1377:1990 Part 6 specifies consolidation and permeability tests in hydraulic cells and with measurement of pore pressure.

17. A *sample* is a portion of soil or rock recovered from the ground using some sampling technique whereas a specimen is a part of a soil or rock sample which is used in a laboratory test. In practice laboratory test specimens are often referred to as samples.

18. Lancelotta (1995) describes the compression of this high plasticity clay ($w_L = 64\%$, $w_P = 27\%$).

19. There are several procedures for determining the preconsolidation pressure from the results of an oedometer test. The most commonly used method, which is that proposed by Casagrande (1936), can be found in many soil mechanics text books (eg Lancellotta, 1995).

20. The coefficient of consolidation c_v can be derived from two curve fitting methods; the compression data is plotted against either square root time or logarithm of time as described in BS 1377 Part 5: 1990 clause 3.6.3. The two methods usually show reasonable agreement.

21. Sensitivity to vibration is a characteristic of loose sand; not only may it settle markedly under vibratory loading, such as under machine foundations, it may also suffer liquefaction, a state which occurs when the pore water pressure in the sand equals the total stress so that the effective stress is zero. This state can develop during earthquakes in saturated sand layers.

22. Procedures for direct shear tests are specified in clause 5 of Part 7 of BS 1377: 1990.

23. There are several difficulties interpreting the direct shear test, including the non-uniformity of the stresses acting on the shear plane due to the restraint applied by the ends of the box. Another

problem is that the soil is constrained to fail along a predetermined horizontal plane.

24. In a triaxial compression test, the vertical effective stress is the major principal effective stress σ'_1 and the radial effective stress is the minor principal effective stress σ'_3.

25. A test procedure for an unconsolidated undrained triaxial compression test without pore water pressure measurement is specified in clause 8 of Part 7 of BS 1377: 1990.

26. A test procedure for a consolidated undrained triaxial compression test with pore water pressure measurement is specified in clause 7 of Part 8 of BS 1377: 1990.

27. A test procedure for a consolidated drained triaxial compression test with volume change measurement is specified in clause 8 of Part 8 of BS 1377: 1990.

28. For many soils, the relationship between shear strength and normal stress is actually curved, particularly at low stresses. However, for practical purposes in most cases a linear relationship can adequately represent soil behaviour.

29. A drained triaxial test should be performed sufficiently slowly to avoid the development of any significant pore water pressure in the soil; for clay this may be very slow. The testing rate should be related to the coefficient of consolidation which can be calculated from the volume change measurements made during the consolidation stage. The shear stage of a drained test can be expected to take about ten times longer than that of an undrained test with pore pressure measurements. For a 100 mm diameter sample, it may require one month to bring the soil to failure.

30. Failure in a triaxial compression test can be defined as the maximum deviator stress ($\sigma'_1 - \sigma'_3$) or the maximum effective stress ratio (σ'_1/σ'_3). In a drained test, the maximum deviator stress and the maximum stress ratio will occur at the same strain, but this may not be the case in an undrained test.

31. See, for example, Bolton (1986).

32. Very low residual shear strength can be correlated with the clay fraction (Skempton, 1964) and the plasticity index (Lupini *et al*, 1981).

33. Based on Table 13 in BS 5930:1999.

34. BS 1377: 1990 Part 7 clause 8.1.2 states only that failure is normally produced within a period of five to 15 minutes.

35. Undrained shear strength can be measured in consolidated undrained triaxial compression tests in which the sample has been consolidated back to the estimated in-situ overburden pressure. The tests are performed at a much slower strain rate than unconsolidated tests, to allow the pore pressures to equalise; a typical strain rate is 2% per hour. These tests tend to produce higher strengths than unconsolidated tests because the consolidation phase produces a sample with a smaller water content than existed in the ground; this effect is counterbalanced to some extent by the strain rate being much slower than that used in the unconsolidated tests. For soft soils, the consolidation may compensate for strength lost as a result of disturbance during the sampling operation.

36. Evaporation is a process in which a liquid changes into a vapour, the rate of change being greater at higher temperatures. In the hydrological cycle, evaporation involves the conversion to water vapour and the return to the atmosphere of the rain and snow that reached the ground. Transpiration is the extraction of moisture by vegetation. The combined process of evaporation and transpiration is sometimes termed evapotranspiration.

37. The filter paper test has been described by Chandler and Gutierrez (1986) and Chandler *et al* (1992).

38. A comprehensive text book on the behaviour of partially saturated soils has been written by Fredlund and Rahardjo (1993). The proceedings of a conference held in Paris also provide a wealth

of information (Alonso and Delage, 1995).

39. The suction referred to here is sometimes known as the matrix suction. The total suction is the sum of the matrix suction and the solute or osmotic suction which derives from soil water chemistry.

40. Substantial suction has also been detected beneath the sites of furnaces, where the high temperatures have driven out the soil water.

41. If the plasticity of the soil varies appreciably between the two boreholes, difference in water content may simply reflect the different void ratios of the soil, not desiccation. With the uncertainty of such comparisons and the expense and impracticality of drilling comparison boreholes, efforts have been made to find alternative means of determining desiccation. An approximate rule-of-thumb indicating a potential for significant desiccation in highly shrinkable clay when $w < 0.4\ w_L$ has been widely used but it has been found to be unreliable, especially for soils of lower plasticity.

42. Table 1 of BRE Digest 240 *Low-rise buildings on shrinkable clay soils*. The classification applies only to overconsolidated clays. A normally consolidated clay may have a considerably greater volume change potential than is indicated by this classification.

43. Proctor (1933) developed a procedure involving field and laboratory controls to ensure that cohesive fills were placed at the correct water content to reduce settlement or softening when the embankment dam fill was subsequently saturated. These procedures for the placement and compaction of fill were introduced in a series of four articles published in *Engineering News Record*.

44. Clevenger (1956) studied the characteristics of loess as a foundation material in the USA; Jennings and Knight (1957) described large sudden settlements of sandy soils in Southern Africa where movements were associated with the presence of water in the ground (eg from broken drains); Pilyugin (1967) carried out a major field infiltration trial in loess in the Northern Caucasus.

45. Double-oedometer tests are carried out on pairs of identical samples (Jennings and Knight, 1957). One sample is loaded in the as-compacted condition and inundated with the maximum load applied to the sample; the other is soaked prior to loading. The difference in vertical strain at each loading increment is a measure of the vulnerability to collapse compression under that applied stress. The double oedometer test procedure has been one of the most common approaches to evaluating collapse potential and the test procedure gives a measure of collapse potential over a wide range of stress.

46. In 1973, BRE began an investigation of opencast backfills which at that time were usually placed without systematic compaction to determine whether they were susceptible to collapse compression on wetting. This work identified collapse compression as a major hazard and determined its magnitude in a variety of field situations (BRE Report 424 *Building on fill: geotechnical aspects*).

47. BRE carried out a major programme of settlement monitoring at Horsley restored opencast coal mining site (Charles *et al*,1993).

48. BRE carried out a major programme of settlement monitoring at Corby restored opencast ironstone mining site (Burford and Charles, 1991).

49. Modern procedures for the placing and compacting fill using simple laboratory and field tests were introduced by Proctor (1933) for use in earthfill dam construction. The laboratory testing carried out by Proctor demonstrated the fundamental features of the compaction of clay soils. BS 1377: 1990 Part 4 specifies two rammer weights for use in compaction tests; 2.5 kg and 4.5 kg. The volume of the compaction mould is 1000 cm^3. Using the 2.5 kg rammer, the soil is compacted into the mould in three layers using a drop height of 300 mm; with the 4.5 kg rammer, the soil is compacted into the mould in five layers with a drop height of 450 mm. The rammer tests are performed on material passing a 20 mm sieve. The vibrating hammer compaction test, which is also specified in BS 1377, uses an electrical vibrating hammer with a tamper diameter of 145 mm and a mould 152 mm diameter. The maximum particle diameter for the vibrating hammer test is 37.5 mm.

CHAPTER 5: GROUND ASSESSMENT

The ground is a major area of risk for a building project and it is important to determine the nature and behaviour of those aspects of a site that could have a significant effect on the project. Ground investigation involves acquiring all types of relevant information about the ground, particularly identifying adverse ground conditions that may cause problems. The main purpose is to control and reduce ground-related risks; the investigation should provide the engineering parameters needed to produce a practical and economic foundation design and, where appropriate, a ground improvement strategy. This chapter presents the component parts of ground investigation, including procedures needed to obtain the required information in cost-effective ways. It describes techniques used to obtain samples from the ground and to determine the characteristic properties of the ground, including in-situ measurement of these properties. For low-rise developments, emphasis should be put on desk studies, aerial photograph interpretation and careful design of fieldwork and testing. It is normally necessary to carry out some soil exploration and classification. Where possible, this is done by excavating trial pits, so that the soil can be examined in situ, and by simple tests to determine, for example, particle size distribution, plastic and liquid limits and chemical composition.

5.1 Introduction

In any building or construction project, the ground is a major area of risk; an inadequate investigation of the ground will do little to reduce this risk. However, an appropriate and adequate site investigation together with, where necessary, appropriate ground treatment (Chapter 9), can greatly reduce those risks[1]. Building development should not begin until sufficient information about the site has been collected to identify and assess those features that will affect the support given to the buildings by the ground on which they will stand. The distinction between site investigation and ground investigation is explained in Box 5.1; ground investigation is more narrowly defined than site investigation.

The objective of a site investigation is to ensure economic design and construction by reducing to an acceptable level the uncertainties and risks that the ground poses to the

> **Box 5.1 Site investigation and ground investigation**
> A distinction is made between site investigation and ground investigation. Site investigation is the more comprehensive concept and involves acquiring all relevant types of information concerning the site; this could include site access, meteorological and environmental conditions as well as the ground conditions. The more narrowly defined ground investigation should provide a description of ground conditions relevant to the proposed works and establish a basis for the assessment of geotechnical parameters relevant for all construction stages. It is important to identify adverse ground conditions that might cause problems, such as the presence of fill material, shrinkable clay or unstable slopes. The ground investigation should provide the geotechnical parameters needed to produce a practical and economic foundation design and, where required, a ground improvement strategy to allow the site to be successfully developed.

project. In order to fulfil this overall objective, the site investigation should aim to determine the nature and behaviour of all the aspects of a site and its surroundings that could significantly influence or be influenced by a building project. Site investigation should form the first step of the design process and is essential if development work is to be carried out safely, economically and on schedule. This chapter focuses principally on ground investigation since the wider aspects of site investigation are outside the scope of the book. It concentrates principally, but not exclusively, on ground investigation for low-rise buildings, in contrast to much of the published guidance on ground investigation which is primarily concerned with site investigation for civil engineering works[2].

By identifying the probable soil conditions and the distribution of the various types of soil, the likely foundations that are needed for the proposed structures can be established. Without an adequate ground investigation, the hidden dangers lying beneath the ground surface cannot be known. Once a preliminary desk study and walk-over survey have established the likely soil conditions and identified the hazards which may be present at a site, the risks can be avoided or mitigated in a number of ways, including:
❐ carrying out further site work to allow an adequate engineering design to be prepared;
❐ adopting a conservative design;
❐ designing in accordance with general good practice based on previous construction experience in the area.
Figure 5.1 presents a flowchart for managing ground conditions using a multi-staged ground investigation.

Ground investigation should provide information for the engineering design of new works in a cost-effective way[3] in order to:
❐ reduce construction costs through economic foundation and earthworks design;
❐ avoid contractual claims due to unforeseen ground conditions;
❐ eliminate structural defects which would result from unacceptable ground movements;
❐ prevent chemical attack on foundations;
❐ detect health hazards from contaminated land.

Building on previously developed sites, which are often described as 'brownfield' in contrast to 'greenfield', is in the public interest, but there can be particular risks associated with previous land use. Ground-related hazards for the built environment on brownfield sites include poor load-carrying properties of the ground, harmful interaction between building materials and aggressive ground conditions, gas generation from biodegradation of organic matter within the ground, and combustion. Where such hazards are not identified, or are misdiagnosed, remedial measures may be required with significant implications for whole-life building costs. The situation becomes more difficult and complex where there are also hazards for human health and detrimental effects on the natural environment and resources such as groundwater.

GROUND ASSESSMENT

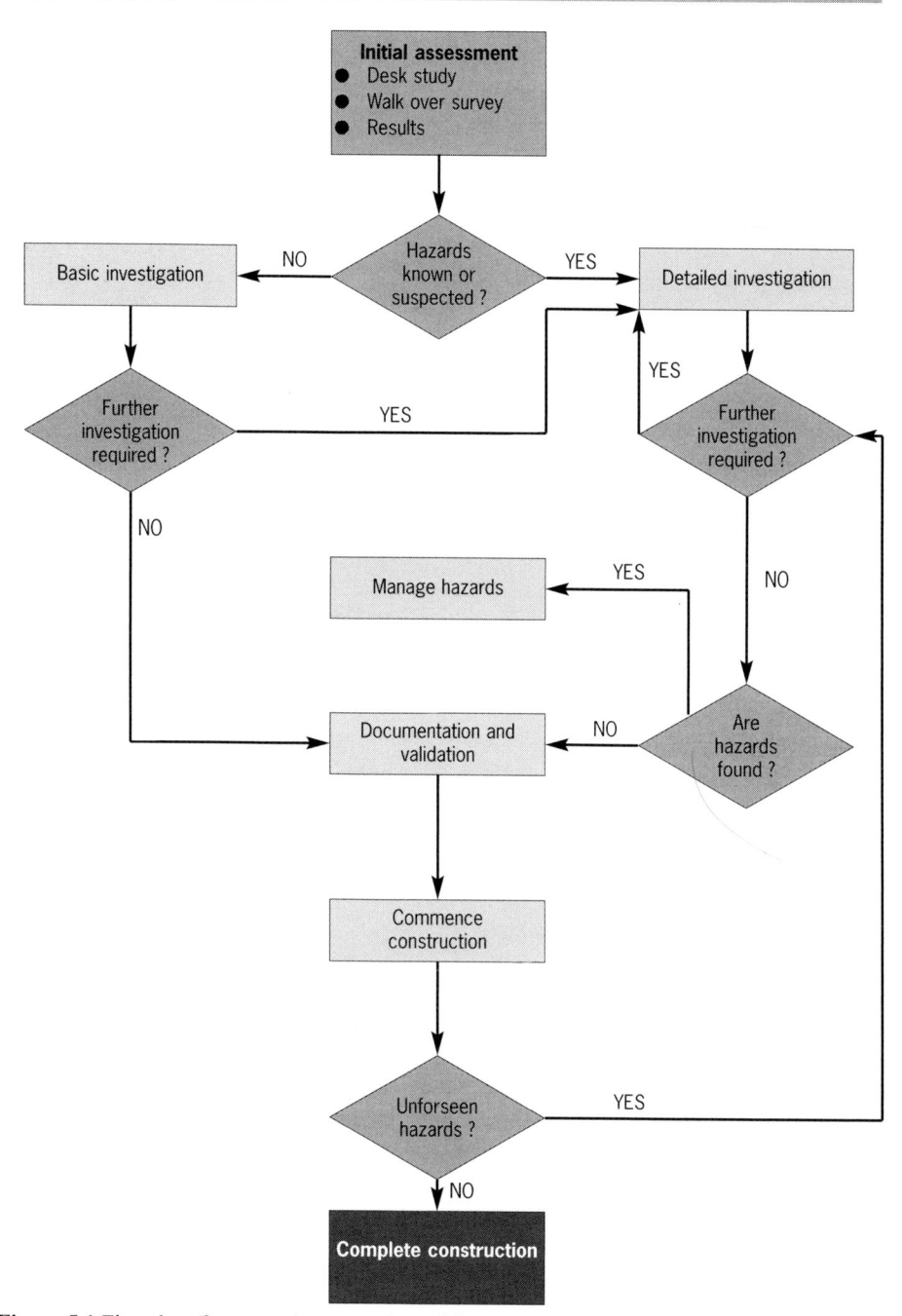

Figure 5.1 Flowchart for managing ground conditions – *after NHBC, 1999*

Building development on brownfield sites is described in Chapter 8 but it should be noted that, wherever possible, there should be an integrated ground assessment strategy, including geotechnical and geoenvironmental risks[4]. Geotechnical hazards are essentially linked to the physical properties of the ground, whereas geoenvironmental hazards are principally concerned with contamination and are linked to the chemical properties of the ground and the groundwater conditions[5]. This chapter is principally concerned with the identification of geotechnical hazards and the assessment of the geotechnical properties ot the ground; the investigation of geoenvironmental hazards on brownfield sites is described in Chapter 8.

The ground investigation should provide the information required to form a conceptual model of the ground conditions[6], which has the following elements:

❏ Three-dimensional stratigraphic model of the ground (geological regime); soil or soil and rock profiles need to be determined, but the boundaries between soil types are not always distinct.

❏ Physical and mechanical properties of the different strata (engineering regime); those geotechnical properties that are relevant to foundation performance may need to be measured.

❏ Modifications caused by previous human activities; an understanding of the history of the site and of the ways in which it has been affected by various types of human activity is required.

❏ Groundwater model (hydrological regime); many ground-related problems are associated with groundwater and an understanding of groundwater conditions is essential.

❏ Soil-structure interaction model; the physical behaviour of the ground under foundation loading and the response of the building to ground movements caused by foundation loading or other phenomena should be evaluated.

❏ Soil contamination model (geoenvironmental regime); the distribution of contaminants and source-pathway-receptor scenarios should be identified (Chapter 8)[7].

The conceptual model of the site is enhanced as information is gathered and the site investigation may be multi-staged. At an early stage in the investigation, the developing ground model should make it easier to identify likely hazards. The later stages of the investigation can concentrate on these hazards and evaluate the risks that they pose. The ground model should assist in assessing the consequences and effects of site activities.

This chapter presents the component parts of a ground investigation; there are essentially two stages:
(a) Collection of available information on the conditions at the site; this is normally achieved by a desk study and a walk-over survey.
(b) Acquiring new information for reassessment of information collected during the desk study and walk-over survey and subsequent detailed engineering design; this is normally

GROUND ASSESSMENT

achieved by direct methods of ground investigation, such as boreholes, trial pits, soil sampling and laboratory testing, and in-situ testing occasionally supplemented by geophysical measurements.

5.2 Ground investigation
A ground investigation may include the following activities:
- desk study;
- walk-over survey;
- direct investigation
  - laboratory measurement of ground properties;
  - in-situ measurement of ground properties;
  - geophysical measurement of ground properties.

It is essential that an adequate and appropriately structured report is prepared which provides a factual account of all the investigatory work and an interpretation of the observations and test results.

Low-rise buildings
Site investigation for low-rise buildings presents particular problems due to the small scale of many developments and the sometimes inadequate geotechnical input. BRE has published a set of Digests on *Site investigation for low-rise building:*

322 Procurement
318 Desk studies
348 The walk-over survey
381 Trial pits
383 Soil description
411 Direct investigations
472 Optimising ground investigation

For low-rise developments, much emphasis should be put on desk studies, aerial photograph interpretation and careful design of fieldwork and testing. Desk studies often give better value for money than drilling deep boreholes and extensive programmes of laboratory testing. However, following a desk study, it is normally necessary to carry out some soil exploration and classification. Where possible, this is done by excavating trial pits, so that the soil can be examined in situ, and by simple tests to determine, for example, particle size distribution, plasticity and chemical composition.

Hazards to be identified
Foundation problems are described in Chapters 6 and 7. The load imposed on the ground by typical low-rise housing is likely to be in the range from 50 to 100 kPa; soft clay soils and peat could pose problems owing to inadequate bearing capacity or excessive settlement due to loading. Problems identified during construction may necessitate design

variations, such as the need for increased depth of footings due to removal of soft spots. Problems may be encountered with such construction processes as piling and dewatering.

When building on clay, consideration should be given to the following:
❐ Shrinkage potential of the clay and the locations of trees and other vegetation: to determine whether special precautions are needed to protect foundations from volume changes in the surface soil (Sections 4.8 and 7.5).
❐ Chemical characteristics of the clay and the water it contains: to determine the possibility of sulphate attack on foundation concrete (Section 6.11).
❐ Magnitude and inclination of slopes on the site: to determine the likelihood of there being pre-existing landslips (Section 7.6).
❐ Shear strength of the clay: to determine the maximum foundation load it can safely sustain by limiting the mobilised strength (Sections 4.6 and 7.2).
❐ Compressibility of the clay: to determine the settlement that will occur under the foundation loading (Sections 4.5 and 6.8).

> **Box 5.2 Separate employment of geotechnical specialist and site investigation contractor**
> Separate employment of a geotechnical specialist and a site investigation contractor is common for large ground investigations. Three or four companies are selected to tender for the field and laboratory work on the basis of previous experience, the skills of their staff and quality of their equipment. The lowest tender price is generally accepted, but the contract is subject to re-measurement as work proceeds.
> Competitive tendering encourages low bids for investigation work, which can reduce the quality of the investigation; this adverse effect can be reduced by careful selection of the tenderers. Since the contractor's engineering skills can be utilised only after the tendering process, the competence of the geotechnical specialist is particularly important. Appropriate supervision is vital. It may be advantageous for the contractor to perform on a dayworks basis those parts of the work which are critical to the success of the investigation or very complex, or where variations may be needed as the work proceeds.

Procurement
A geotechnical specialist[8], experienced in determining all the geotechnical requirements of the project and optimising the ground investigation[9], should be appointed as early as possible during project planning to assess the potential hazards in the ground. Good communication between the developer, structural designers and the geotechnical specialist is essential during the ground investigation.

Two systems of procurement are commonly used[10]. In the system that is widely used on large civil engineering projects (Box 5.2), a geotechnical specialist and a site investigation contractor are separately employed. The independent geotechnical specialist should carry out a thorough desk study and formulate an appropriate plan for an investigation; this will form the basis of the tender for the field and laboratory work. The system commonly used to procure ground investigations for low-rise building developments involves a package-deal contract in which the desk study, planning and execution of field and laboratory work and reporting are carried out by one company (Box 5.3).

5.3 Desk study

An initial desk study has an important role in assessing the suitability of a site for development and is particularly important in the investigation of sites where low-rise building is proposed[11]. Site investigations for low-rise building projects are normally very restricted compared with large civil engineering or high-rise developments, owing to financial constraints. A desk study may indicate problems that cannot readily be recognised by physical investigations and the subsequent ground investigation can be properly focused, with savings on the cost of trial pits and boreholes.

> **Box 5.3 Package deal contract for site investigation**
> A package deal is often suited to low-rise building developments because of the relatively simple contract documentation and flexibility. The desk study, planning and execution of field and laboratory work and reporting are carried out by one company, either for a lump sum or on the basis of measurement of work agreed as the investigation progresses. Up to three ground investigation companies may be invited to tender on the basis of past experience and reputation. They should have sufficient suitably qualified and experienced staff to carry out the investigation, including desk study, aerial photograph interpretation, design and execution of ground investigation and reporting, and to self-supervise drilling and testing. A lump sum contract can be negotiated, which is helpful for financial forecasting, and responsibility for ground investigation is not divided between two parties.

The essential feature of a desk study is that it brings together all available documents relevant to the nature and history of the site in order to address the following questions:
- What is the nature and condition of the ground?
- What are the potential implications of the ground conditions for the structure and, particularly, its foundations?
- What will be the impact of the development on the local environs?

Ground conditions which could have an adverse affect on low-rise buildings or which may place constraints on the type of foundation that can be used need to be identified. As much information as possible should be collected from existing records and archives and much factual and interpretative information is publicly available in the UK. Some of the more common sources of information are summarised in Table 5.1, including Ordnance Survey maps, geological maps[12] and books[13], and aerial photographs[14]. There is considerable variation in the quality and coverage of geological maps across the UK[15] and, needless to say, the availability and scale of geological maps varies greatly in different countries. Civil engineering journals, mining records and reports of previous site investigations can also provide important information.

Table 5.1 Information from maps and aerial photographs in the United Kingdom

Topic	Sources
Slopes and landslips	Slopes can be identified and measured on Ordnance Survey maps (at scales of 1:10 000, 1:10 560 and 1:25 000). Landslipping is shown on some geological maps and aerial photographs should be studied[16].
Water	Springs, ponds, rivers and drainage features can be identified on Ordnance Survey maps. Old maps may show changes in the position of water-courses when compared with more recent maps. Large scale maps may indicate high groundwater by symbols for boggy land.
Vegetation	Aerial photography provides a record of site vegetation. The height of trees or bushes can be estimated from stereo pairs of photographs. Large scale topographical maps will show the positions of hedge lines and woodland.
Made ground	Large scale topographic maps made at different dates allow identification of changes on the site, such as infilling of hollows or old pits, removal of vegetation or demolition of old buildings. Infilled ground also may be identified from aerial photographs.
Geology	The geology of a site can be determined from geological maps. *Drift* (superficial deposits) or *solid & drift* (bedrock & superficial deposits) maps are preferred. Drift maps, especially at large scale, may identify significant thicknesses (greater than 1 to 2 m) of *Head*[17].
Mining	Many materials have been mined and quarried and geological maps can identify layers containing minerals such as coal. Old mine shafts can sometimes be located from old topographical maps. There are many sources of information on old mining activities[18].

Geological records can provide information on the nature of the ground (sand, clay or rock) and the thickness of the different soil or rock strata. Adverse ground conditions associated with particular soil and rock types, or particular characteristics of the site, can be identified. This should help the builder to anticipate ground and groundwater conditions. However, geological maps and records classify soils and rocks according to their presumed age, not according to their nature (whether they are sand, clay or rock). Therefore, a single geological unit may vary in composition, and hence in its behaviour under foundations, from one location to another. This type of variation can usually be discovered from geological records of the area.

Present or past mining activity can be a major threat to the satisfactory performance of low-rise buildings. Geological maps may give some general warning of the existence of minerals beneath the site, but it is often difficult to assess the likely risk to a proposed structure. When coal has been extracted, the Coal Authority[19] will often be able to provide valuable information. In some cases, the help of specialist mining consultants will be required, and special ground investigation boreholes may be needed to investigate possible cavities (Section 7. 7).

GROUND ASSESSMENT

An economic building layout cannot be planned without some knowledge of the surface features, such as hills, slopes, valleys, and streams, on the proposed site. Significant depths of cut necessitate retaining wall (Section 12.5) or slope design; where buildings are to be placed on fill (Section 7.4), suspended floors or special foundation designs, such as piling, may be required. Existing slopes of more than about 7° in a clay subsoil could contain pre-existing slip surfaces, which could be re-activated by increased loading or steepening of the slope during construction (Section 7.6). Level ground in valleys will probably have a high water-table and the subsoil may be soft or loose and unsuitable for simple foundations. Springs, ponds or rivers may also be a sign of a high water-table. The movement of construction plant could be difficult and the problems of controlling groundwater in excavations should be considered.

The locations of previously demolished buildings are of interest because old foundations may have been buried and this can cause additional cost and delay to construction owing to difficulties of demolition or excavation. Old basements may hide voids or uncompacted demolition debris. A factory may have left the ground contaminated by toxic waste, presenting a health hazard to construction workers and a source of chemical attack on new foundations. Identification of areas of fill is particularly important for low-rise buildings with shallow foundations as structures built on top of or across uncompacted or untreated filled ground will be at risk from large differential settlements which are caused by the variation in ground compressibility. These brownfield hazards are considered in Chapter 8.

Local sources of reference material should be consulted; these include: libraries, county archives, local history and natural history societies, planning authorities, universities and polytechnics. Local authorities have extensive experience of building in their area and can comment on general ground conditions, the possibility of flooding, occurrences of structural damage in the area associated with ground movement and previous site usage. Where possible, information should be obtained on the types of foundation commonly used in the area and of any problems encountered. Local authorities also can be consulted about mining and quarrying activity. Libraries and archives may hold records and maps of the site showing the position of old field boundaries, ponds, streams and pits or quarries, some of which may have been removed or infilled.

Public utilities, such as gas, electricity, water and telephone, can give information on the position of their services in and around the site; these must be avoided during ground investigations and may need to be re-routed during the development of the site.

Once the nature of the ground has been discovered from the desk study and the walk-over survey, the problems associated with that type of ground should be determined. Table 5.2 provides a checklist appropriate for low-rise building developments.

5.4 Walk-over survey

The walk-over survey, which is sometimes described as a site reconnaissance or site inspection, is an integral part of the site investigation process and should not be omitted. It is a particularly important part of ground investigation for low-rise developments[20]. The object of the survey is to check and add to the information collected during the desk study. In conjunction with a good desk study, it provides valuable information, which cannot be obtained in any other way. The whole site should be covered carefully on foot, making full use of available maps and photographs, and all the information obtained during the desk study. The survey should not be confined to the site itself, but should include the surrounding area and its building stock.

Some simple tools that are likely to be needed include the following:
- 20 m or 30 m tape to measure the position of features of interest;
- compass to orientate the site map;
- pocket penetrometer or hand vane to assess the strength of clay soil;
- abney level or clinometer to measure the ground slope angle in the area;
- auger, spade and polyethylene bags for taking soil samples;
- camera for visual records.

Table 5.2 Checklist for desk study and walk-over survey of site for low-rise building

Topic   Questions to be addressed

Topography, vegetation and drainage
- Does the site slope? What is the maximum slope angle?
- Is there evidence of landslips on or adjacent to the site or on similar ground nearby?
- Are there springs, ponds or water-courses on or near the site?
- Are, or were there, trees, or hedges growing in the vicinity of proposed construction?

Ground conditions
- What geological strata lie below the site and what problems are associated with this geological setting?
- Is the site covered by alluvium or other soft deposits?
- Is there information on the strength and compressibility of the ground?
- Is shrinkable clay present?
- In these soil conditions, is it likely that groundwater will attack concrete?

Previous use
- Is there evidence of previous building development?
- Has there been mining or quarrying activity in this area?
- Are there coal seams or other mineral resources under the site?
- Have there been changes in ground level (eg by placement of fill)?

Proposed building
- What area will the buildings occupy?
- What foundation loading is expected?
- How sensitive will the structure be to differential foundation movement?
- What soils information is required for the design of likely types of foundation?
- Is specialist geotechnical advice required?

GROUND ASSESSMENT

The geology and other features of special interest noted during the desk study should be marked on a site map prepared at a suitably large scale. Exposures of soil or rock (for example in railway cuttings) should be examined and samples taken. Land underlain by chalk or limestone may contain naturally occurring voids or pipes filled with soft soils, which can collapse or settle under a structure (Section 7.7). This type of ground is often associated with dry valleys and surface hollows or with areas where streams disappear into the ground. Shallow holes can be excavated using an auger to get some indication of the type of soil and to obtain samples (Section 5.5). Where holes are made to examine soil conditions, the presence or absence of groundwater should be noted.

The positions of trees and hedges are of concern when they are close to proposed buildings and when the site is underlain by shrinkable clay (Sections 4.8 and 7.5). The growth of trees on the site may lead to further settlement of shallow foundations, while the removal of trees and hedges may cause heave. The magnitude of these movements will depend on the size, type and location of the trees, the plasticity of the clay and the groundwater level. Dense tree cover may also indicate poor ground conditions; unstable slopes, for example, are sometimes tree covered because the disruption of subsoil drainage by ground movements makes the ground boggy and difficult to cultivate. The location of such features as trees, hedges, pits and exploratory holes can be marked on the site map, and the existence of previously identified features can be confirmed. If the site is on shrinkable clay, a record of the positions and heights of trees, shrubs and hedges is needed, together with, if possible, their species. The absence of any trees, shrubs or hedges indicated on previous records or aerial photographs should be noted.

The positions of springs, ponds and other water should be marked on the site map. The absence of such features which are shown on Ordnance Survey or geological maps and aerial photographs may indicate that ground levels have been raised by placing fill on the site. The position of any infilling being carried out at the time of the site visit should be noted. Areas that may have been previously filled should be identified by comparing the available Ordnance Survey maps and aerial photographs with what can be seen on the site. Signs of mineral extraction in the area may include old mine buildings, derelict or hummocky land, surface depressions, and evidence of infilling or spoil heaps.

Slope angles can be obtained more accurately during the site visit than from Ordnance Survey maps. Slope angles can be interpreted in terms of the types of materials underlying the site. Very flat ground near streams or rivers is probably associated with soft or loose alluvium. Landslips may be indicated by broken or terraced ground, or boggy, poorly drained conditions on hill slopes. Except in very windy conditions, trees normally grow vertically; ground movements may cause them to change inclination. Kinks in hedge lines may also be a sign of past movements.

Valuable information can be obtained by talking to local people and local builders may be able to provide information on the ground problems and typical foundation designs used in the area. Structures in the area should be examined and signs of damage recorded. The location of structures that are marked on maps or aerial photographs, but which no longer exist, should be noted.

Drilling rigs or hydraulic excavators may be needed for detailed ground investigation work and potential access problems should be identified. Record photographs should be taken of the condition of gates and tracks, so that any damage caused can be properly quantified.

Further studies may be necessary, particularly where the walk-over survey has yielded new information and it is thought that more may be gained from existing records, or if the findings from the walk-over survey conflict with those of the desk study. Following this process, a report should be produced summarising the sources used, the information obtained, and the relevance of the findings to the proposed development. The report should explain and evaluate the various risks associated with construction on site. It should outline the nature and extent of ground investigation that may be necessary and assess the need for specialist advice.

5.5 Direct investigation
The previous stages of ground investigation should have given a general indication of the soil strata and site conditions. This information, in conjunction with details of the proposed buildings, should assist in identifying what additional information is required from direct investigations to assess the engineering and chemical properties of the ground. Direct investigations involving trial pits and boreholes are sometimes termed physical investigations or intrusive investigations

The direct investigations should enable values to be obtained for parameters that define the characteristics of the soil sufficiently well for safe and economical foundations to be designed[21]. The type of foundations to be used will depend on the ground conditions, the magnitude of the proposed structural loading, and the sensitivity of the structure to settlement. Spread foundations are typically 1 to 2 m deep and require a minimum undrained shear strength of soil of about 50 kPa to safely support a foundation pressure of 100 kPa (Chapter 6), otherwise it may be appropriate to pile the buildings. Deeper boreholes will be needed during the ground investigation if piles are to be used. The required data can be obtained in a number of ways and only the simpler and more common methods are described in this and the following sections of the chapter. Detailed information on soil test procedures can be found in British Standards[22] and text books[23].

The direct investigation should be designed to ensure an ordered and controlled progression of work, which the exploratory nature of the work requires, and to be cost-

GROUND ASSESSMENT

effective. Since a multi-storey building will need foundations very different from those for a low-rise dwelling, there will be different requirements for the physical investigation. Likely locations of buildings need to be known before a trial pit and borehole ground investigation is carried out, so that exploratory holes can be positioned close to the proposed building positions. Trial pits normally should not be excavated at the actual location of a small building which will have shallow foundations[24]. Consideration should be given to all the information from the previous stages of the ground investigation process, such as faults across the site, strata outcrops, and spring lines. This information should influence both the position of the building and the location of trial pits and boreholes.

It is important to assess the variability of each stratum and the major fabric features, such as fissures and fissure spacing, cementation of particles and particle size. These have a considerable effect on the interpretation of laboratory and in-situ soil tests. For example, the strength along fissures will be less than that of the intact soil and the strength of a specimen of intact clay will be greater than that of a specimen containing fissures.

Trial pits

For an accurate assessment of the soil, a trial pit is often the most suitable method[25]. They can be excavated to depths of about 5 m, either by hand or by using an excavator with a hydraulically operated backhoe. The excavation must be properly supported to allow personnel to safely inspect the undisturbed soil strata[26]. The features in the soil profile that will have the greatest effect on building foundations should be identified. All available information should be recorded to allow proper correlation with similar data from other trial pits on the site, including reduced level, strata depths, visual log, sketch of pit, soil strength or density, soil description, samples taken, ease of excavation, ingress of water, date of excavation, pre-existing slip surfaces and stability of the sides[27].

Boreholes

Exploratory boreholes may be required. Hand augering is the simplest and cheapest boring method. Auger holes are made by turning a light auger into the ground. Posthole and helical augers are commonly used (Figure 5.2). The equipment is light and portable and small

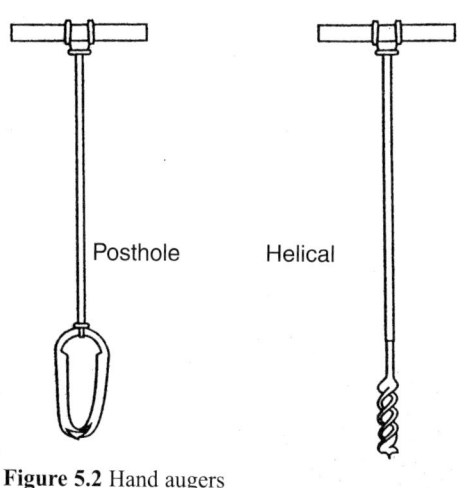

Figure 5.2 Hand augers

undisturbed samples can be obtained. However, the technique is limited in depth and diameter and difficult to progress in hard strata. Common types of hand augers are generally available in diameters up to 200 mm. Provided the soil is not too stiff, does not contain boulders or large gravel and does not collapse, depths in excess of 5 m are easily achieved to obtain an assessment of the soil profile with depth. In clay, 38 mm diameter sample tubes may be driven in to the base of an auger hole to obtain an undisturbed soil sample. In free-draining soils an indication of the water-table level may be obtained.

Deeper exploratory boreholes are generally bored or drilled using one or a combination of three methods:
❒ light cable percussion boring;
❒ power auger boring;
❒ rotary core drilling.
The first two methods are used in soft to stiff soils, the third is normally reserved for stiff soils and rocks. Tthe equipment for light cable percussion boring, which is sometimes termed shell and auger drilling, is the most readily available of these methods.

Boreholes in ground other than rock can be sunk using a light cable percussion rig (Figure 5.3). The rig, which is mounted on two wheels and can be towed by a 4-wheel-drive vehicle, can work in remote and inaccessible areas. The rig comprises a derrick and pulley wheel from which a weighted boring tool attached to the end of a cable is operated using an engine powered winch. The borehole is advanced by successively dropping a clay cutter, sometimes with an added sinker bar, into the ground from a height of some 2 to 3 m. The clay cutter comprises a cylindrical tool with a circular cutting shoe. Boreholes are usually fully cased throughout. When advancing a borehole partly filled with water, the recovered material can be highly disturbed and care needs to be taken in interpretation.

Figure 5.3 Light cable percussion rig

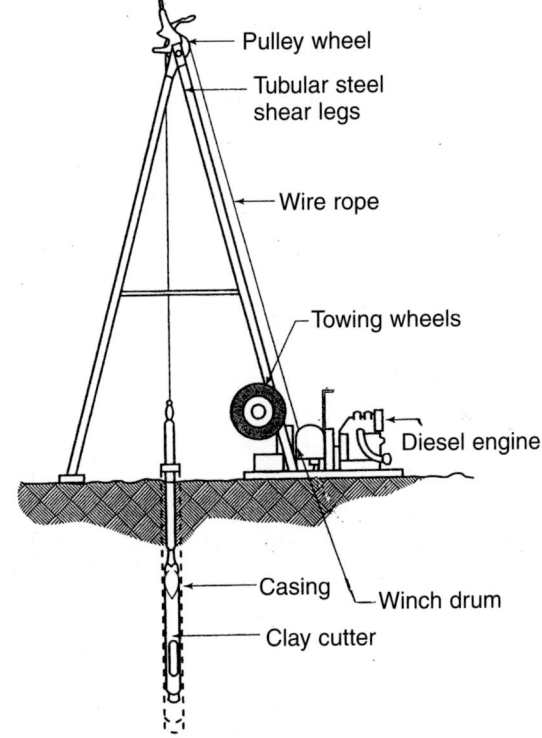

GROUND ASSESSMENT

For example, in boulder clay, the recovered material may be just a gravelly silty sand with its original structure lost. Material below the base of the borehole can be badly disturbed by the displacements caused by the blows from the heavy clay cutter.

Sampling

Soil samples are often described as disturbed or undisturbed (Box 5.4). Disturbed samples of soil may be obtained from different depths for index testing. The samples should be put in plastic bags or containers and fully labelled with site location, trial pit number and depths. A trial pit also allows small-scale in-situ tests to be carried out in the sides and base of the pit. In clay, hand vane and pocket penetrometer tests can be used to obtain a crude provisional assessment of strength. An assessment of the strength of clay or the state of compaction of coarse soils can be obtained by applying simple field tests such as the standard penetration test and the cone penetration test (Section 5.7).

Disturbed soil samples for index tests and identification can be obtained from the boring spoil and put into suitable plastic bags or containers and fully labelled with site, borehole number and depth. Undisturbed samples[28] suitable for soil strength tests require the use of sampling tubes, the most common of which is the 100 mm diameter U100 open drive sampler. This sampling tube is lowered to the bottom of a suitable cleaned borehole and driven into the soil by means of a sliding hammer; it has a detachable cutting shoe.

For higher quality samples in medium to stiff soils, a thin-walled, open-tube sampler can be used in which the wall of the sample tube is much thinner and has a ground cutting edge, thereby reducing soil disturbance. The soil disturbance is further reduced by pushing, rather than hammering, the sampler into the base of the borehole using a ground frame and a series of pulleys or a hydraulic ram. For higher quality samples in soft to medium stiff soils, a piston sampler can be used which also uses a thin walled sampling tube, but incorporates a piston to aid recovery and to stop the ingress of debris whilst the sampler is lowered to the bottom of a borehole.

Window sampling is increasingly being used to obtain soil samples. A window sampler consists of a steel tube with a

> **Box 5.4 Disturbed and undisturbed soil samples**
>
> A disturbed sample is a sample that has been obtained with no attempt to retain the in-situ soil structure. It is sometimes called a bulk sample. Where a bulk sample is taken from below water in a borehole, it may be unrepresentative of the soil deposit.
>
> An undisturbed sample is recovered with the objective of obtaining a sample in which the in-situ soil structure and the in-situ water content have not been modified in any way. Unfortunately, in practice a truly undisturbed sample cannot be obtained and special techniques are required to obtain high quality 'undisturbed' samples. Since the behaviour of the ground in mass is often controlled by the presence of weaknesses and discontinuities, even a high quality undisturbed sample may not be fully representative of the in-situ soil mass.

longitudinal slot along part of its length and is fitted with a shoe with a sharp cutting edge. The tube is advanced into the ground by monotonic thrust, dynamic impact or percussion. After the tube has been removed from the ground, the sample is removed for assessment. The lightweight equipment can be mobilised at locations where access is restricted and the technique permits rapid operation. Windowless samplers are similar to the window sampler, but do not have the longitudinal slot. They are normally fitted with a clear plastics liner to contain and protect the sample. Versatile track-mounted rigs are available which can be used not only for window sampling but also for dynamic probing (Section 5.7) and to obtain U100 samples.

Tubes containing undisturbed soil samples should have the sample ends trimmed and waxed to prevent moisture loss, and should be fully labelled (site name, borehole number, and depth). With U100 sample tubes it is essential to note which end is the top of the sample because, once the cutting shoe is removed, the tubes are the same at both ends. The orientation of thin-walled sample tubes is obvious as the tubes have a cutting edge at the lower end and location holes at the top.

5.6 Laboratory measurement of ground properties
Laboratory tests are carried out on specimens which have been taken from the ground using sampling techniques and which are as representative as possible of that ground. Soil and rock laboratory tests are carried out[29]:
❒ to characterise and establish the variability of the ground;
❒ to obtain design parameters.
In characterising the ground, visual appearance is important and classification tests include particle size distribution, Atterberg limits, organic content, water content and particle density (Chapter 3). The testing performed for low-rise building developments falls largely, although not entirely, within this category. The design of foundations for a high-rise building almost certainly requires some quantification of the strength and deformation properties of the supporting ground. Strength tests, compaction and consolidation tests, seepage and permeability tests are described in Chapter 4. For desiccated clays, suction measurements may be needed (Section 4.8).

Where site-specific design parameters are needed, it may be helpful if the ground investigation initially concentrates on the simple tests required to classify the ground and delineate areas of similar ground conditions and then subsequently focuses on determining the properties of the different areas. Unfortunately, constraints of time and budget often make this approach impractical and it is necessary to rely initially on visual appearance to identify variations in ground conditions.

Classification tests allow a particular soil to be placed into a group, which broadly reflects its anticipated geotechnical properties. The principal classification tests are water content

GROUND ASSESSMENT

(Section 3.4), particle size distribution (Sections 3.5 and 3.6) and Atterberg limits (Section 3.6), supplemented in some circumstances by particle density determinations (Section 3.4).

Variation in the natural water content of clay indicates changes in strength, and is often caused by varying amounts of desiccation or softening. Water content data can, therefore, be helpful in ensuring that the softer samples, which are indicated by their higher water contents, are selected for strength evaluation.

Compaction tests on coarse soil can be used to determine the density index, which is a measure of the degree of packing or compactness (Section 3.5). The compaction of partially saturated fine soil and the significance of the standard Proctor compaction test are described in Section 4.10.

The principal laboratory techniques used to measure soil strength are the various forms of triaxial and direct shear tests described in Section 4.6. Because these tests are relatively time-consuming and expensive, simpler but less reliable tests, such as the laboratory vane test[30], are sometimes used to give an indication of the undrained shear strength of clay. A typical laboratory vane has a height and width of 12.7 mm and is a scaled down version of the instrument used to measure strengths in situ (Section 5.7). Because of the effects of soil fabric and variable boundary conditions, the accuracy of the test is questionable and the results are seldom used for obtaining design parameters. The hand vane and the pocket penetrometer, which are much cruder instruments, are sometimes used on site in the sides of trial pits and in the laboratory on the ends of tube samples; they should be regarded more as an aid to the visual classification of clay soil than as a method of measuring strength for design purposes.

Consolidation parameters for fine soil are usually measured in an oedometer (Section 4.5). The horizontal permeability may be significantly greater than the vertical permeability and this may have important implications for predicting the behaviour of a field situation where the drainage is predominantly horizontal, such as the consolidation of a layer of alluvium under an earth embankment.

5.7 In-situ measurement of ground properties

When specifying a ground investigation programme, in-situ testing should be considered since it has certain advantages over the traditional combination of borings, sampling and laboratory testing:
- continuous or near continuous data;
- speed of operation;
- cost savings.

BRE has published a simple guide to in-situ ground testing in seven parts[31].

The soil can be tested while it is still in the ground, thereby removing the effects of sampling disturbance. In-situ tests can be used to assess the variations in ground conditions across a site; in some circumstances they give a better estimate of soil properties than laboratory tests. A wide range of tests is available and only those that are more commonly adopted for ground investigation in the UK are described:

- standard penetration test;
- dynamic probing;
- cone penetration test;
- vane shear test.

Standard penetration test (SPT)

Despite its lack of sophistication, or perhaps because of it, the standard penetration test continues to be very commonly used. The equipment consists of a 63.5 kg hammer which drops 760 mm onto an anvil and drives, via the drive rods, a thick-walled, 50 mm diameter split spoon sampler into the base of the borehole. The test is performed by counting the number of blows N required to drive the sampler 300 mm after a seating drive of 150 mm[32]. Table 5.3 lists the applications, advantages and disadvantages of the SPT.

Considerable care must be taken when assessing the test results since N values are not free from operator influence and the equipment also varies in different parts of the world. Nevertheless, while appreciating their limitations, some correlations between the blow count N and soil properties can be useful[33].

For clay soil a very approximate correlation with undrained shear strength is given by:

$$c_u = 5N$$

where:
c_u is in kPa.

For normally consolidated natural sand deposits, a correlation between a normalised blow count[34] and density index is given in Figure 5.4.

Table 5.3 Standard penetration test

Application	Standardised blow count N in coarse soil and clay may be used in empirical correlations to assess bearing capacity, settlement, density index, angle of shear resistance, and undrained shear strength.
Advantages	Simple, robust and inexpensive equipment and test procedure. Highly disturbed sample permits identification of the soil. Correlations exist to assess various soil characteristics and performance indicators.
Disadvantages	Test results are sensitive to operator techniques, equipment malfunction, poor boring practice and diameter of borehole. Test is insensitive in loose sands and results in fissured clays may be misleading.

GROUND ASSESSMENT

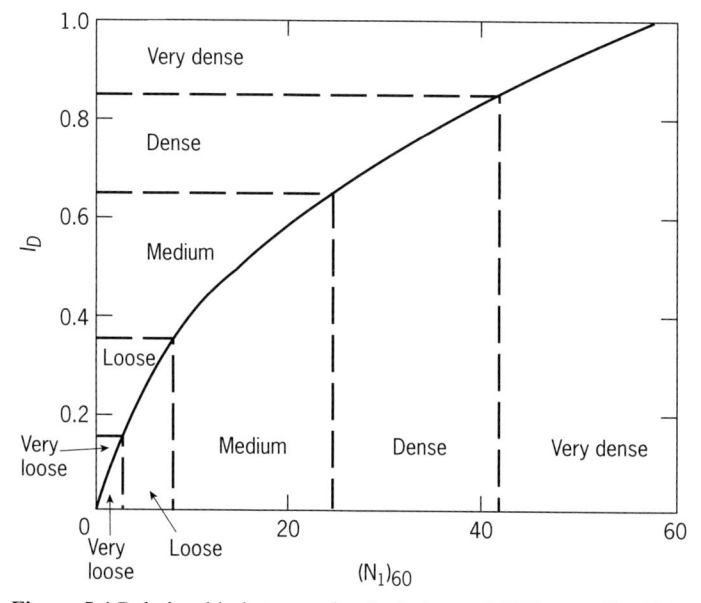

Figure 5.4 Relationship between density index and SPT normalised blow count for sands

Dynamic probing (DP)

Dynamic probing is similar to the SPT, but is generally carried out from the ground surface, not from the base of a borehole. A solid cone is advanced into the ground by a falling weight striking an anvil, which is attached to the cone via extension rods[35]. Several forms of the test are commonly used with different energy delivered per blow, up to that used in the SPT. The number of blows N_{10} used to drive the cone each successive 0.1 m into the ground (or N_{20} to drive it 0.2 m) is recorded to give a near continuous record of the penetration resistance of the soil – see the example in Figure 5.5 for the DPH (heavy dynamic probing) test specification. There is no provision for measuring the force on the cone itself, unlike the cone penetration test (CPT).

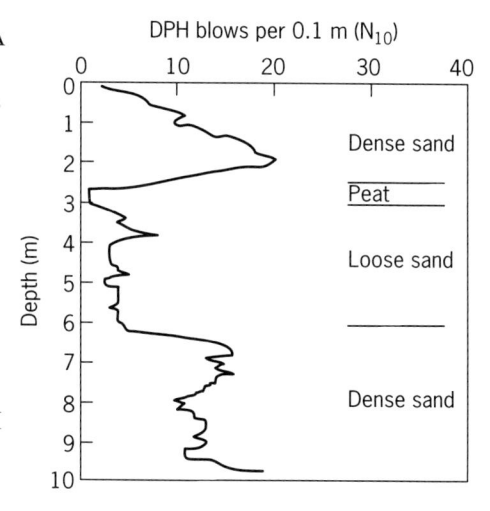

Figure 5.5 Dynamic probing test results

In cohesive soil and in other soils at depth, friction along the extension rods becomes significant. To minimise friction, the rods are rotated at one metre intervals and the torque measured with a torque-measuring wrench. The torque reading is used to correct the blow count for the friction on the rods. The dynamic point resistance[36] q_d can be calculated from the corrected blow count. DP supplements conventional sampling and more complex penetration tests, and is particularly useful in delineating areas where weak soil overlies stronger strata and in quickly assessing the variability of the soil conditions. Table 5.4 summarises applications, advantages and disadvantages of the test.

Table 5.4 Dynamic probing

Application	Penetration resistance can give an indication of density index of coarse soil and undrained shear strength of clay; can assess effect of ground treatment. Test is mainly used as a profiling device.
Advantages	Can be used without a borehole. Equipment is simple, portable and robust. A large number of tests can be carried out in a day. Access is possible to remote and awkward locations.
Disadvantages	Test is crude and correlations with soil properties may be unreliable. Results need to be interpreted in association with a borehole investigation. Results are at less frequent intervals than those from CPT.

The Mackintosh probe is even lighter weight; a 4 kg hammer falling 0.3 m drives a 27 mm diameter cone into the soil. The number of blows required to penetrate each 0.3 m interval is recorded producing a profile of number of blows with depth. This tool can be useful for identifying the depth of soil overlying rock, locating a competent stratum at shallow depth, assessing the extent of a soft layer on a site or extending the cover of more expensive tests by calibration and correlation.

Cone penetration test (CPT)

The electrical cone penetration test has steadily gained in popularity in the UK since the 1970s. The test is carried out using a penetrometer which incorporates a friction sleeve behind the cone and allows separate measurement of cone resistance q_c and sleeve friction f_s (Figure 5.6). The dimensions of the 60° apex angle cone with a cross-sectional area of 10 cm² are specified in the British Standard test procedure[37], although cones of other cross-sectional areas are in use.

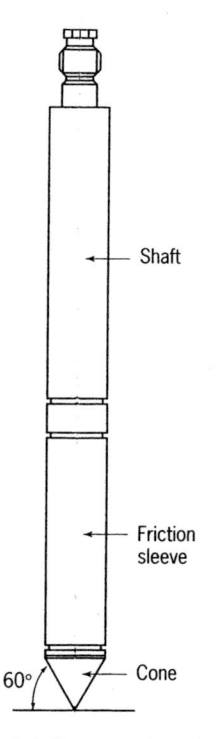

Figure 5.6 Cone penetrometer

GROUND ASSESSMENT

The cone measures friction and cone resistance as it is pushed into the soil at a constant rate of penetration of 20 mm/s on the end of a series of hollow push rods with an outside diameter similar to that of the cone. Piezocones (CPTUs) are additionally fitted with transducers to measure the pore water pressure u generated in the soil during penetration, thereby improving the profiling capability. Figure 5.7 shows some CPT test results through a soil profile in which dense sand overlies stiff clay; Figure 5.8 shows some CPTU test results in hydraulically placed pfa[38]. Table 5.5 summarises the applications, advantages and disadvantages of the CPT[39].

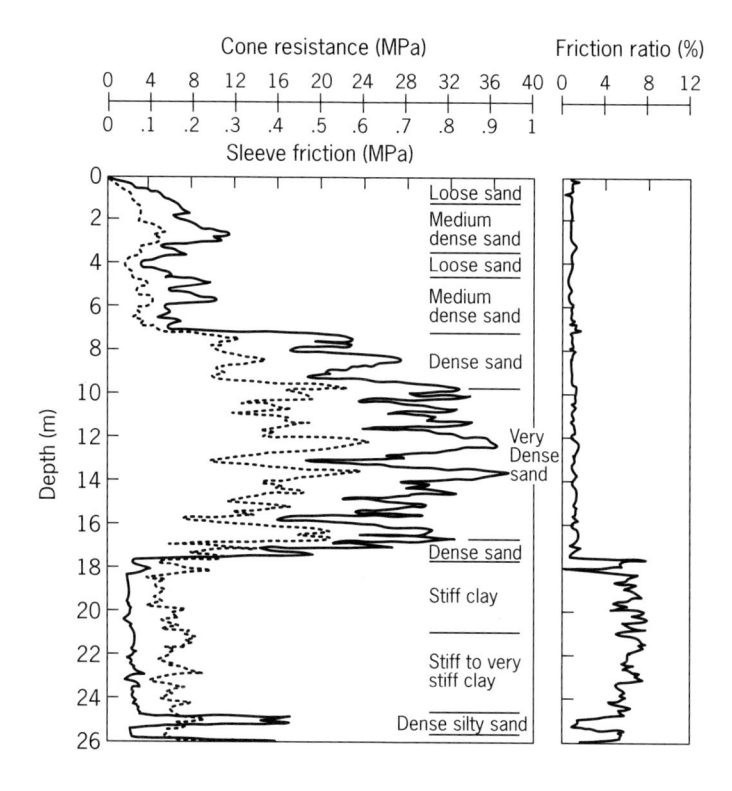

Figure 5.7 CPT results in dense sand overlying stiff clay

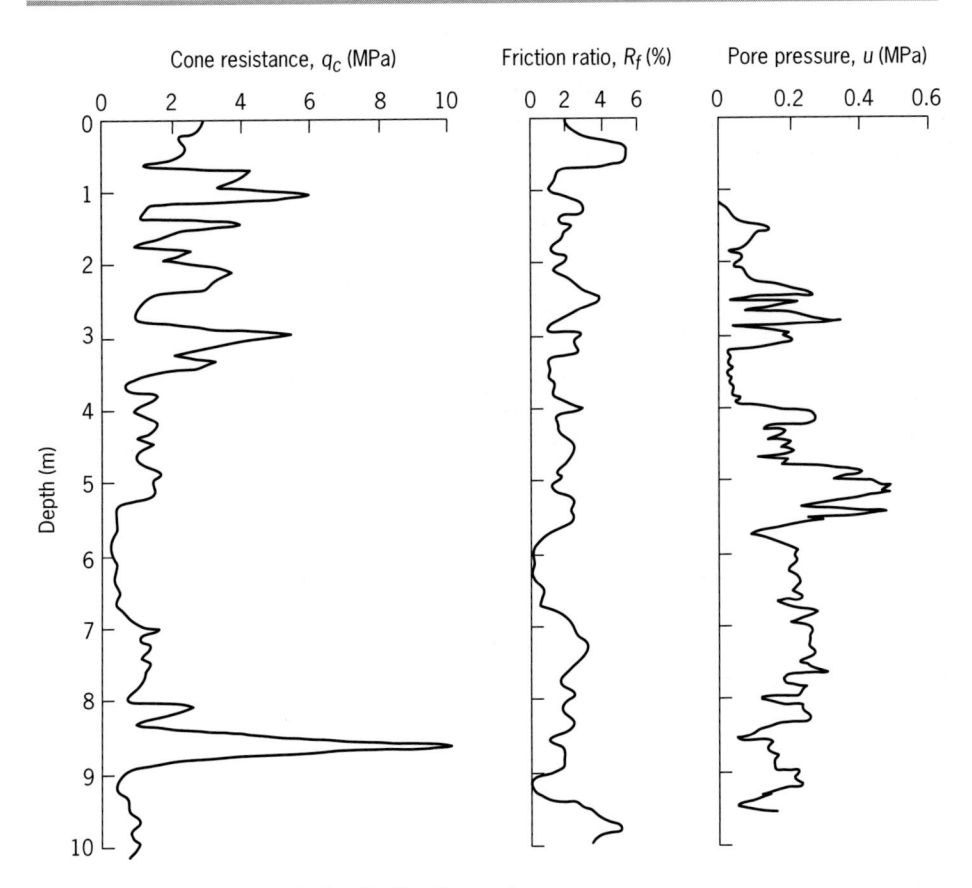

Figure 5.8 CPTU results in hydraulically placed pfa

Table 5.5 Cone penetration test

Application	Cone resistance and sleeve friction measured in silt and sand, also stiff clay and dense gravel with suitably powered machines. Rapid appraisal of stratigraphy and detection of thin layers. Estimates of settlement of foundations on sand and design of driven piles using empirical correlations.
Advantages	Rapid means of obtaining a continuous resistance profile in coarse soil and clay and mapping variations in ground conditions. Other soil parameters can be inferred from results. Well-documented and proven application to foundation solutions using empirical formulae. Cheap in comparison to boring.
Disadvantages	Coarse, dense gravel necessitates large capacity equipment for penetration. Refusal often encountered at cemented layers or boulders. Difference in results using different cones. Boreholes are necessary to confirm soil profiles.

GROUND ASSESSMENT

There have been numerous attempts to correlate the CPT cone resistance q_c with the SPT blow count N. The ratio q_c/N increases as the mean particle size of the soil D_{50} increases, but there is a large scatter in the results from sites where both types of measurement have been made[40] (Figure 5.9).

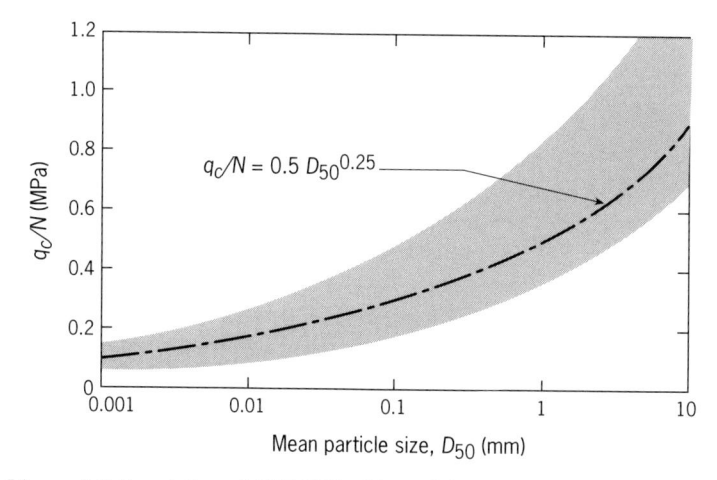

Figure 5.9 Correlation of CPT-SPT with particle size

(Graph shows $q_c/N = 0.5 D_{50}^{0.25}$ with axes q_c/N (MPa) vs Mean particle size, D_{50} (mm))

Vane shear test

The vane test provides a means of measuring the undrained shear strength of soft or firm clay soil in situ. The vane comprises four blades set at right angles to each other in the shape of a cruciform. The British Standard[41] specifies vanes 100 mm or 150 mm long depending on soil strength and with length to width ratios of 2. A vane length of 150 mm is suitable for clays with a shear strength of up to 50 kPa; a vane length of 100 mm is adequate for strengths in the range 50 to 75 kP. There are two different types of apparatus:

❏ With the borehole apparatus, the vane is lowered into the borehole on extension rods and pushed a short distance into the undisturbed soil below the bottom of the borehole; the distance is at least three times the borehole diameter or twice the vane length, whichever is the greater.

❏ With the penetration vane test equipment, the vane and a protecting shoe are jacked or driven into the ground and at the required depth the vane is advanced at least 0.5 m from the protective shoe into the undisturbed soil below the protecting shoe.

Applied torque *T*

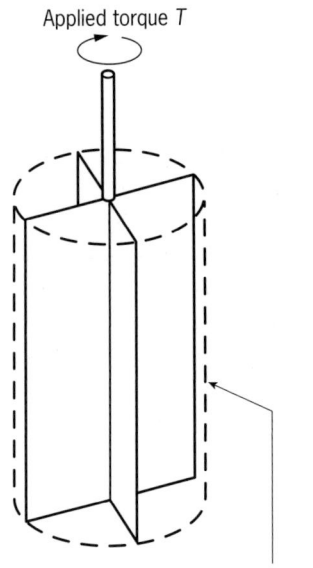

Assumed cylindrical shear surface with shear stress = c_u at maximum torque

Figure 5.10 Shear surface in vane test

A torque-measuring device is connected to the top of the uppermost extension rod. The vane is slowly rotated and the maximum torque required to shear the soil is used to calculate the undrained shear strength. Figure 5.10 shows the cylindrical shear surface created by the rotation of the vane. The speed of rotation is between 6° and 12° per minute. Continued rotation of the vane allows calculation of the remoulded undrained shear strength. Hand-operated versions of the vane shear test are available. These have much smaller blades (20 to 40 mm high) which can be pushed into the side or base of a trial pit, a U100 sample or, by using extension rods, into the bottom of a hand-augered hole. Table 5.6 summarises the advantages and disadvantages of the vane test.

Table 5.6 Vane test

Advantages	Permits rapid and direct in-situ measurement of undrained strengths up to 100 kPa. Remoulded shear strength may also be measured. Can be used direct from the surface, or from the base of a borehole. Small hand-operated versions are available.
Disadvantages	Results affected by silty and sandy pockets, or significant organic content. Anisotropy can lead to unrepresentative strength values. Poor maintenance gives excessive friction between rods and guide tubes, and in bearings. Test should be used in conjunction with laboratory measurements of undrained shear strength and measurement of plasticity index in order to assess the validity of the results.

5.8 Geophysical measurement of ground properties

Another type of in-situ testing involves the use of various principles of physics to determine the soil profile and to measure ground properties[42]. There has been steady growth in the use of geophysical techniques in ground investigation work since this type of testing has major advantages:

GROUND ASSESSMENT

❐ field work is relatively rapid and, with modern data logging facilities and processing software, the results can be presented very quickly;
❐ non-intrusive surveys can be carried out from ground surface;
❐ representative values of some soil parameters can be measured.

A major limitation is that additional forms of ground exploration are needed to enable geophysical measurements to be interpreted with confidence.

A number of geophysical techniques are available, including ground probing radar[43] and electrical resistivity[44], but in many situations seismic methods are the most useful. A seismic wave transmits energy by vibration of soil particles, but the wave velocity is quite distinct from the velocity at which individual particles oscillate. In compression waves, the particles vibrate in the direction of wave propagation; in shear waves the particles vibrate in a direction perpendicular to the direction of wave propagation. Rayleigh waves are distortional stress waves that propagate near the ground surface.

Seismic methods may be used for many purposes. They include: determining the location and depth of boundaries, such as bedrock beneath overburden and the depth of the water-table, identifying localised features including voids and, possibly, measuring soil properties and their spatial variation. Where seismic methods are used to measure soil properties that are needed to calculate the response of the ground to static loads, the measured small-strain dynamic properties have to be related to the static properties of interest for deformation prediction. Shear wave velocity and damping are of particular interest.

In refraction surveys the time is measured for waves to pass from the source to a number of geophones placed on the ground surface at different distances from the source. The analysis of refraction measurements is based on the assumption of an increasing velocity with depth and a relatively consistent soil profile along the survey line. The seismic refraction technique is commonly used to delineate the depth to bedrock.

Measurements of Rayleigh wave velocity are frequently undertaken; two forms of surface wave source are in use:
❐ impact sources, such as a hammer or drop weight which produce a transient pulse;
❐ vibrators, which produce continuous waves.
Where an impact source is used, the data is usually analysed using the spectral analysis of surface waves method[45] (SASW). This is an important method for evaluating the variation with depth of stiffness moduli at small strains. It has also been used to detect underground obstacles and cavities, to evaluate liquefaction potential and to evaluate the effectiveness of ground treatment. Vibrator sources have been widely used with the continuous surface wave system[46] (CSW).

Notes

1. Ground was identified as the construction element about which least is known in *Site investigation in construction* which was prepared in four parts by the Site Investigation Steering Group of the Institution of Civil Engineers (1993). A number of important recommendations were made concerning the planning, procurement and quality management of geotechnical investigations. Clayton (2001) has discussed the management of geotechnical risk. See also Hatem (1998) and Clayton *et al* (1995).

2. Although BS 5930: 1999 *Code of practice for site investigations* covers both civil engineering and building works, much of this code of practice is focused on site investigation for large-scale civil engineering works rather than for small building works.

3. Against a background of increasing litigation and consequent risks to developers, BRE Digest 322 *Site investigation for low-rise building: procurement* recommends that expenditure on ground investigation should be a minimum of 0.2 % of the cost of the project. The majority of this sum should be spent on activities that bring the greatest returns in terms of risk appreciation, reduction in construction costs and increases in the effectiveness of ground investigation. The developer should take an active role in the investigation process, particularly in determining the amount of investigation required to quantify each area of risk. BRE Digest 472 describes ways of optimising ground investigation. The Association of Geotechnical and Geoenvironmental Specialists (1998a and 1998b) has provided guidelines for good practice in site investigation.

4. The Association of Geotechnical and Geoenvironmental Specialists (2000) has published *Guidelines for combined geoenvironmental and geotechnical investigations*.

5. British Standard BS 10175: 2001 *Code of practice for the investigation of potentially contaminated sites* and CIRIA Special Publication SP103 (Harris et al, 1995) give guidance on site investigation and assessment of contaminated sites. Both documents recommend a risk based approach to identify and quantify the hazards that may be present and the nature of the risk they may pose.

6. The development of a conceptual model is strongly advocated in BS 10175: 2001 *Code of practice for the investigation of potentially contaminated sites*; it states that the purpose of a site investigation is to gather the information needed to form such a model in order to be in a position to assess the presence and significance of contamination of land. Geotechnical engineers develop a conceptual model of the physical ground conditions, although they may not always recognise it as such.

7. *Guidance for the safe development of housing on land affected by contamination* has been prepared by the Environment Agency and NHBC (2000).

8. Suitably qualified and experienced geotechnical engineers and engineering geologists can be found with some consulting engineers and with many specialist site investigation contractors, but in some situations it will be preferable if the geotechnical specialist is an independent consultant.

9. BRE Digest 472 *Optimising ground investigation*.

10. BRE Digest 322 *Site investigation for low-rise building: procurement*.

11. BRE Digest 318 *Site investigation for low-rise building: desk studies*.

12. Geological maps are divided into *solid* (those that show the solid materials below the drift deposits), *drift* (those that show alluvium, head and glacial materials), and *solid & drift*. Because they may omit the shallowest deposits, solid maps can give a misleading impression of the surface ground conditions and drift or solid and drift maps should be used wherever possible. From 2004, most newly published maps will be classified as *bedrock* (replacing solid), *superficial deposits* (replacing drift) and *bedrock & superficial deposits* (replacing solid & drift).

GROUND ASSESSMENT

13. After studying the geological map, the types of materials associated with each of the soil and rock types in the area should be identified using the descriptions in the Regional Geology Guides published by the British Geological Survey for the United Kingdom. Each regional guide covers a wide area (eg South-west England), so descriptions are rather general. More specific descriptions are given in the sheet memoirs which relate to the area shown on a 1:63 360 or 1:50 000 map. Information about the British Geological Survey (formerly known as the Institute of Geological Sciences) can be found at *www.bgs.ac.uk*. See also Powell (1998).

14. Sources of aerial photographs are listed in BRE Good Building Guide 39 *Simple foundations for low-rise housing* Part 1 *Site investigation.*

15. As a result of the long history of geological mapping in the United Kingdom, there is considerable variation in the quality and coverage across the country. Geological maps are normally available at 1:63 360 (1-inch-to-the-mile) or 1:50 000 scales, but some parts of the country are covered at scales of 1:10 000, 1:10 560 (6-inches-to-the-mile), and 1:25 000. The availability of geological maps has been described by Thomas (1991).

16. Landslipping is shown on some geological maps, but not all landslips have been recorded and it is necessary to study aerial photographs. The preferred scale for aerial photography interpretation of this kind is 1:2000, although material up to a scale of 1:10,000 may be useful. Black and white photographs are the most readily available, and should be viewed in stereo for the best effect. Photographs taken at different times should be examined. Geological and geotechnical literature can be useful in identifying landslips.

17. *Head* is a geological term for an unstratified deposit which, it is conjectured, has been produced on a sloping site by the creep of water saturated ground when subject to alternate periods of freezing and thawing. Geotechnical properties of these types of subsoil will be very variable. Drift maps, especially at large scale, may identify significant thicknesses (greater than 1 to 2 m) of Head, but will not necessarily give any warning of the presence of thin layers of such materials; this must be checked by trial pitting.

18. Sources of information on old mining activities include records in local archives (eg County Records Offices) and records held by central government (eg Mining Records Office of Health and Safety Executive).

19. The Coal Authority was established by Parliament in 1994 to undertake certain activities formerly carried out by British Coal including:
❐ handling subsidence damage claims which are not the responsibility of licensed coalmine operators;
❐ providing public access to information on past and present coal mining operations.
The Coal Authority maintains the national coal mining database and provides a coal mining report service for properties in Great Britain. Information about the Coal Authority can be found at *www.coal.gov.uk*

20. BRE Digest 348 *Site investigation for low-rise building: the walk-over survey.*

21. BRE Digest 411 *Site investigation for low-rise building: direct investigations.*

22. BS 1377 (1990) and BS 5930: 1999. European test standards are in preparation.

23. Details of the various laboratory test techniques can be found in text books devoted to the subject, such as the three volume work of Head (1992, 1994, 1998).

24. Atkinson (2003) recommends that, where possible, trial pits should be at least 3 m from proposed dwellings.

25. BRE Digest 381 *Site investigation for low-rise building: trial pits.*

26. Safety is discussed in BS 5930: 1999, BS 6031: 1981 and BS 10175: 2001. Where necessary, and in accordance with current safety regulations (including the Construction (Design and Management) Regulations 1994 to the extent that they may be relevant to the works), the sides should be supported by timber shoring or trench sheeting to prevent collapse. When the pits are to be entered for logging or testing purposes, fully supported sides are essential. For a deep trench the possibility of gas should be considered.

27. The soil description should follow BS 5930. See also BRE Digest 383.

28. The recovery of completely undisturbed samples is virtually impossible because the sampling process will always cause some disturbance. The important consideration is that the sampling process should deliver samples in which the degree of disturbance does not significantly alter the soil property that is to be measured.

29. The Association of Geotechnical and Geoenvironmental Specialists (1998c) has provided guidance on the selection of laboratory testing.

30. A test specification for the laboratory vane is given in BS 1377: Part 7: 1990 clause 3.

31. *A simple guide to in-situ ground testing* published by BRE in 2003
Part 1: *What is it and why do it?*
Part 2: *Cone penetration testing*
Part 3: *Flat dilatometer testing*
Part 4: *Dynamic probing*
Part 5: *Pressuremeter testing*
Part 6: *Large-diameter plate loading tests*
Part 7: *Geophysical testing*

32. A test specification for the SPT is given in BS 1377: Part 9: 1990 clause 3.3. BS 5930: 1999 also contains information on the test. A European test standard has been prepared.

33. More information about these types of correlation and important limitations to their use can be found in Tables 5.4 and 5.5 of Coduto (1998). It should be recognised that different authorities have put forward somewhat different correlations and the correlations given in Coduto are not identical with those given in Section 5.7.

34. Corrections can be applied to the SPT blow count N for such factors as overburden pressure and energy transfer. Many SPT-based design correlations were developed using hammers with an efficiency of about 60% and Skempton (1986) has described a method for converting the blow count N recorded in the field into a normalised N_{60}. A further normalisation of N to an overburden pressure of 100 kPa is represented as $(N_1)_{60}$. For $I_D > 0.35$, an approximate correlation between N and density index is:
$$I_D = [(N_1)_{60}/60]^{0.5}.$$

35. BS 1377: Part 9: 1990 clause 3.2 gives two alternative DP specifications using 90° cones of base area 15cm² and 20 cm² respectively. BS 5930: 1999 and *A simple guide to in-situ ground testing*: Part 4: *Dynamic probing* list five test specifications; in the lightest (DPL) a hammer of 10 kg mass falls 0.5 m and in the heaviest (DPSH) a hammer of 63.5 kg mass falls 0.75 m. A European test standard has been prepared.

36. The dynamic point resistance q_d is calculated:
$$q_d = 10\, N_{10}\, (Mgh/A)\, (M/M+M')$$
where:
N_{10} is the number of blows per 0.1 m advance into the ground corrected for rod friction
M is the mass of the hammer in kg
M' is the mass of the extension rods and anvil in kg
g is the acceleration due to gravity in m/s²

GROUND ASSESSMENT

h is the drop height of the hammer in m
A is the projected area of the cone tip in m^2

37. A test specification for the CPT is given in BS 1377: Part 9: 1990 clause 3.1. See also BS 5930: 1999. A European test standard is being prepared.

38. Humpheson *et al* (1992).

39. More information on the advantages of the CPT is given in Part 2 of the BRE *A simple guide to in-situ ground testing*.

40. These correlations are summarised by Lunne *et al* (1997).

41. A specification for the field vane test is given in BS 1377: Part 9: 1990 clause 4.4. See also BS 5930:1999.

42. *A simple guide to in-situ ground testing*: Part 7: *Geophysical testing*, published by BRE in 2003, provides a brief but useful introduction to this large subject. For more information, see McDowell *et al* (2002) *Geophysics in engineering investigations*.

43. There has been substantial growth in the use of subsurface radar since the 1980s and the method has been introduced to site investigation in the UK (McCann and Green, 1996). Radar waves are electromagnetic waves, similar to radio and light waves, which within a certain frequency band can propagate appreciable distances through soil and rock. The method is more effective in searching for a suspected target than as a general reconnaissance tool and can be used to search for cavities and buried objects, such as pipes, at relatively shallow depths.

44. Resistivity is the electrical resistance of an element of unit cross-sectional area and unit length, and its value is a measure of the capability of the soil to carry electrical currents. Electrical resistivity is an indicator of corrosive potential and BS 1377: Part 9 clause 5.1 presents a test procedure for determining the apparent resistivity of soil in order to assess the corrosivity of the soil towards various metals. Resistivity has also been related to soil properties such as water content, degree of saturation and permeability (Kalinski and Kelly, 1993). Resistivity measurements have been used to detect deposits of sand and gravel and identify areas of clay; sands and gravels have a higher resistivity than clay.

45. The spectral analysis of surface waves (SASW) is a non-intrusive seismic test for determining wave velocity profiles (Matthews *et al*, 1996). The method uses a hammer or other type of impact as an energy source. SASW is based on the principle that the depth of the soil profile sampled by surface waves varies with frequency and hence wavelength. In most soils the Rayleigh waves travel at a depth below ground surface of between one-half and one-third of the wavelength. Rayleigh waves of different wavelengths propagate at different depths and if the stiffness of the soil varies with depth, surface waves of different wavelengths will propagate at different velocities. The method uses the spectral analysis of the propagating Rayleigh wave to determine the frequency wavelength dispersion.

46. In the continuous surface wave system (CSW) the wave source is a frequency controlled vibrator, which makes it possible to derive the relationship between Rayleigh wave frequency and wavelength and hence calculate velocity and depth. While this method requires a relatively costly energy source, it has the advantage of good frequency resolution. Matthews *et al* (1996) have discussed the relative merits of the two surface wave methods.

CHAPTER 6: FOUNDATION DESIGN

The foundations of a building transmit the loads from the structure into the ground, at the same time, transmitting the effects of ground movements back into the structure. Total collapse of a building due to foundation failure is rare but ground movements sufficiently large to cause some damage to a building are not uncommon. This chapter discusses foundation movement due to deformation of the ground attributable to building loads. Ground movement due to ground processes which are not directly connected with the weight of the building are discussed in Chapter 7. The foundation design aims to limit ground movements, keeping the cost of the foundations as low as possible while meeting the required deformation criteria. Rarely will all measurable movements be prevented, unless the ground consists of rock or very stiff soil. Chapter 8 deals with problems specifically associated with brownfield land, and Chapter 9 with situations where ground treatment is required prior to construction to reduce the potential for ground movement.

6.1 Introduction

The structural elements that connect buildings to the ground are called *foundations*[1]. The need for a solid foundation had long been appreciated but until the development of soil mechanics there was no adequate theoretical basis for design: reliance had to be placed on experience gained from trial and error. The choice of foundation type was little influenced by soil conditions and sites with very poor ground conditions were generally not built on. The increasing use of sites with difficult soils has serious implications for foundation design and requires a better appreciation of the way soil reacts to applied loads and natural forces. Although it is now possible to design foundations that are both safe and economic with much greater confidence, and total collapse of a building due to foundation failure is very unusual, ground movements sufficiently large to cause some damage or distress to a building are not uncommon.

Foundation movement can occur due to deformation of the ground attributable to:
❐ building loads which are applied to the ground through the structural foundations;
❐ processes not directly connected with the applied loads, such as changes in water content within the surface layer of a clay soil due to evaporation and transpiration.

The foundations of a building transmit the loads from the structure into the ground and, at the same time, transmit the effects of ground movements back into the structure.

Foundation design, and where necessary ground treatment (Chapter 9), aim to limit and control building movements but rarely will all measurable movements be prevented, unless the ground consists of rock or very stiff soil, or unless expense is no object! The inevitability of some ground movement occurring in response to foundation loading may seem obvious, but the widely used concept of an allowable bearing pressure sometimes has obscured this. It has been suggested that the scientific stage of foundation engineering

FOUNDATION DESIGN

came into existence when foundation engineers at last recognised that every load produces some settlement, regardless of the allowable bearing pressure of the ground[2].

The foundation design should ensure that the movements which are transmitted to the superstructure are acceptable and that consequent building distortion and tilt never exceed tolerable levels. It is necessary, therefore, not only to predict the nature and magnitude of the foundation movements but also to assess how much movement is tolerable. While it might seem advisable to set very low limits for acceptable movements, grossly underestimating the tolerability to ground movement of a building can have a severe financial penalty when unnecessary and excessive foundation costs are incurred. A case has been reported where a concrete raft nearly 1.5 m thick was required for a log cabin built adjacent to the Thames in Staines[3]!

This chapter describes the geotechnical aspects of different types of foundation that are commonly used and the influence of the ground conditions on the selection of the foundation type. It does not deal with the structural aspects of foundation design[4]. Table 6.1 summarises some ground-related hazards that commonly affect low-rise buildings. The most acute foundation problems on poor ground are often associated with small buildings for which deep foundations are not an economically viable solution, but stiff foundations can be provided for small buildings relatively cheaply. Acceptability of ground movements is discussed. The principles of geotechnical design are examined focusing principally on bearing capacity and settlement. Foundations for extensions present particular problems. Finally, the implications of ground chemistry are considered.

Table 6.1 Some ground-related hazards for foundations of low-rise buildings

Effect	*Possible causes*
Differential settlement or heave of foundation or floor slab	● Soft spots under spread footings on clay ● Growth or removal of vegetation on shrinkable clays ● Mining subsidence ● Collapse settlement on pre-existing made ground ● Self-weight settlement of poorly compacted fill ● Floor slab heave on unsuitable fill material
Soil failure	● Failure of foundations on very soft subsoil ● Instability of temporary or permanent slopes
Chemical processes	● Groundwater attack on foundation concrete ● Reactions due to chemical waste or domestic refuse

6.2 Types of foundation

Foundations differ in the area of ground over which they distribute the load, and the depth at which they apply load to the ground. It is useful to draw a distinction between shallow and deep foundations. A shallow foundation transmits structural loads to the near-surface soil or rock and can be defined as being less than 3 m below ground level[5]. However, where the depth D to width B ratio of a shallow foundation is large, it may need to be designed as a deep foundation; a deep foundation with a small depth to width ratio may need to be designed as a shallow foundation[6].

Strip foundation
When foundations are excavated for low-rise construction, much reliance is placed on the experience, judgement and local knowledge of the builder and the building control service[7]. For many low-rise buildings, spread footings at the base of the walls provide an adequate foundation. The wall rests on a width of ground greater than the thickness of the wall above the footing; this compensates for the difference between the bearing capacity of the ground and the greater strength of the wall. There are two types of spread footing: the traditional strip footing and the trench-fill foundation.

The strip foundation applies wall loadings to the ground over a limited width at shallow depth and is likely to be the most economic foundation where the ground has adequate bearing capacity and stability. Strip foundations are normally provided under each loadbearing wall and should be located centrally under the wall. Table 6.2 gives recommended minimum foundation widths.

Table 6.2 Minimum width of strip foundations[8]

Type of ground	Condition of ground	*Minimum width (mm) for wall loading of (kN/ linear metre)*				
		20	*30*	*40*	*50*	*60*	*70*
Rock		------------------ width of wall ------------------
Gravel, sand	Compact	250	300	400	500	600	650
Clay, sandy clay	Stiff	250	300	400	500	600	650
Clay, sandy clay	Firm	300	350	450	600	750	850
Sand, silty sand, clayey sand	Loose	400	600	*Specialist advice is needed*
Silt, clay, sandy clay, silty clay	Soft	450	650	*Specialist advice is needed*
Silt, clay, sandy clay, silty clay	Very soft	*Specialist advice is needed*

For low-rise buildings with a typical wall loading of 50 kN/m, strip foundations are likely to be the obvious choice where the surface ground is either a weathered rock or a well-compacted sand or gravel. It is necessary to excavate only to a depth of about 0.45 m to ensure that the foundations are unaffected by frost[9]. In other ground conditions, alternative types of foundation, such as pad, raft or piles, may be technically superior and sometimes more economical. Concreting should take account of weather conditions. In some cases reinforcement may be required.

Where a shallow foundation will be near an existing service trench, it is normally recommended[10] that the foundation depth D should be such that a line from the bottom edge of the foundation to the bottom edge of the service trench is at an angle β which is not greater than 45° to the horizontal (Figure 6.1).

FOUNDATION DESIGN

Figure 6.1 Shallow foundations close to existing drain

Strip footing

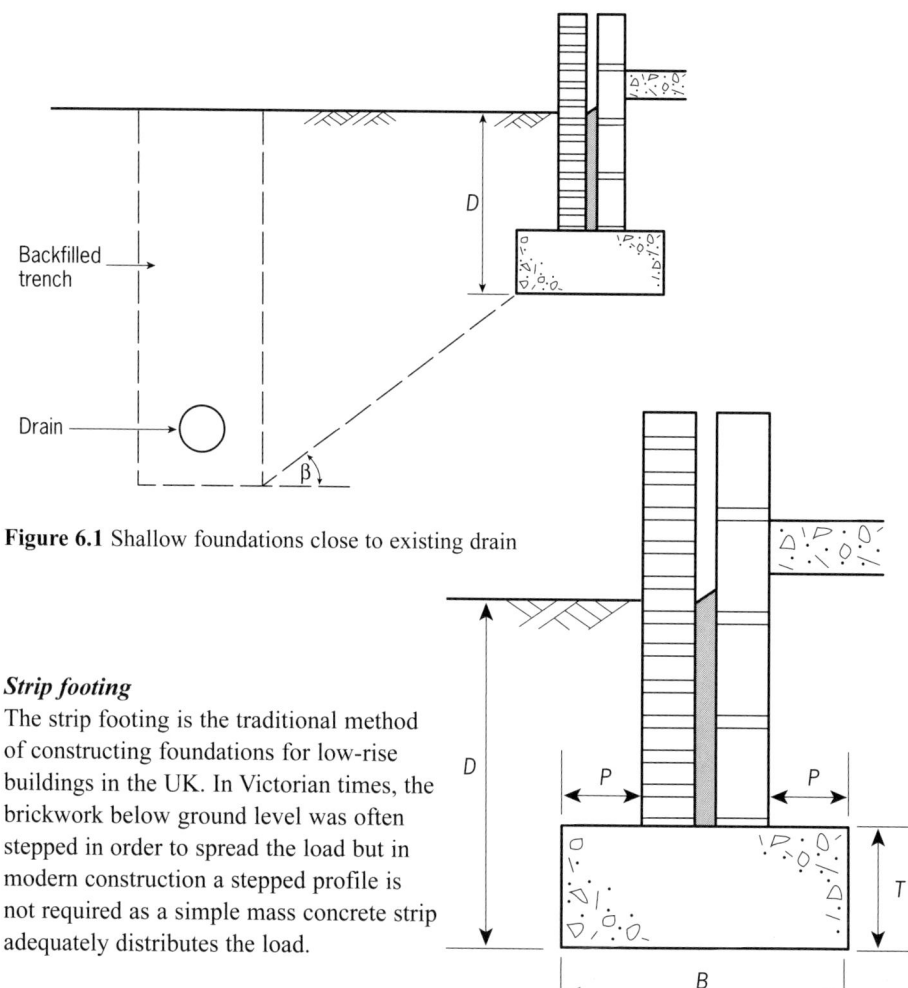

The strip footing is the traditional method of constructing foundations for low-rise buildings in the UK. In Victorian times, the brickwork below ground level was often stepped in order to spread the load but in modern construction a stepped profile is not required as a simple mass concrete strip adequately distributes the load.

The footing typically consists of a 0.3 m-thick layer of concrete which is cast into the bottom of a narrow trench to provide support for the brickwork or masonry walls (Figure 6.2). As a minimum, the thickness of the footing T should be equal to the larger of P or 150 mm[11]. Because of the difficulties of laying bricks or concrete blocks in a very narrow trench, the minimum practical width for a strip footing B is about 0.6 m. Where it is necessary to use a wider footing to spread the load over a greater area, the footing may need to be reinforced to prevent cracking.

Figure 6.2 Strip footing

Traditionally, the depth of the trench was determined by the exposed soil, and excavation would continue until an apparently competent stratum was encountered. As a minimum, it was considered necessary to remove topsoil or other obviously compressible material and, in speculative housing, the underside of the footing might have been as little as 0.35 m below ground level, although a depth of about 0.45 m was more typical. For building on clay soils, it is now recognised that deeper foundations are required to isolate the building from the swelling and shrinking that occurs in the ground as a result of moisture variations. A minimum depth D for building on clay soils of 0.9 m is usually recommended[12] and, where there are trees or other large vegetation, greater depths are likely to be required. This important subject is covered in Section 7.5.

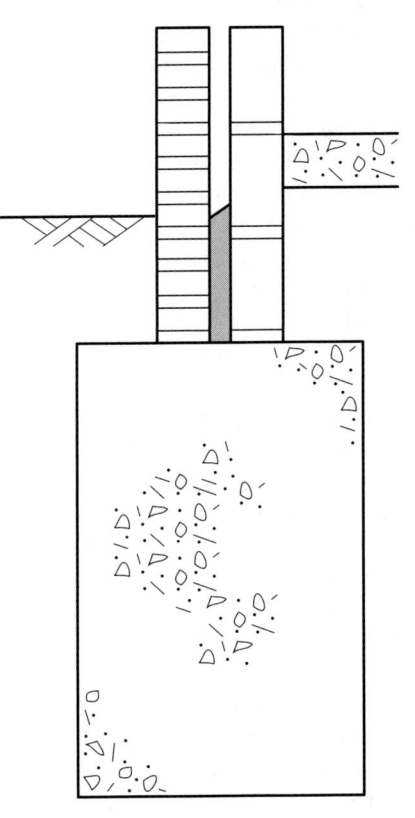

Trench-fill foundation

This type of strip foundation is formed by filling a trench with concrete immediately the trench has been dug by a machine, such as a backhoe. Concrete should be placed to within a few tens of centimetres of ground level (Figure 6.3). This form of construction has become increasingly attractive as labour costs have increased and ready-mix concrete has become widely available. An important safety consideration where a foundation depth of more than 1.2 m is required is that the need for working below ground level is minimised. Since human access is not necessary, the width of the trench-fill foundation is often dictated by the practicalities of setting out and the width of bucket of the type of mechanical excavator normally available on a building site. In good ground conditions, a typical width for a trench-fill foundation is in the range of 0.4 to 0.5 m and accurate setting-out is required.

Trench-fill foundations are commonly used to depths of 3.5 m below ground level[13], although they become increasingly uneconomic, even technically undesirable, as the required foundation depth is increased. Their main advantages over piled foundations are that the ground conditions can be inspected visually and specialised drilling machines are not needed.

Figure 6.3 Trench-fill foundation

FOUNDATION DESIGN

Pad foundation
Where the building loads are carried by isolated columns rather than by loadbearing walls, it may be preferable to excavate and cast individual pads rather than a continuous footing. This form of construction becomes more economic when the distance between adjacent pads is much greater than the dimensions of each pad. Panels of brick infill between the columns can be supported on ground beams spanning between pads or on separate footings, which may be at the same depth as the pads or shallower.

Pads may be circular, square or rectangular in plan and are usually formed using mass concrete of uniform thickness. Where it is necessary to distribute the load from a heavily loaded column, the pad may be haunched or stepped, and may require reinforcement.

Where a deeper foundation is required to transfer the foundation loads to a stiffer or more stable stratum, the pad, together with the buried column that it supports, is often referred to as a pier foundation. The distinction is not a clear one, but the term pier is commonly used to describe the larger and more massive foundations that are required to support a bridge or other large structure.

Raft foundation
A raft foundation consists of a reinforced concrete slab, whose thickness and stiffness are designed to spread the applied wall and column loads over a large area. This type of foundation can transmit the weight of an entire building over a much larger area than can strip or pad footings. Rafts are used where it is necessary to limit the load applied to the underlying soil or to reduce the effects of differential foundation movements as a result of variable soil conditions or variations in loading. Two types of raft are shown in Figure 6.4: a plane raft, sometimes called a flat-bottomed raft (a), and an edge-beam raft, sometimes called a semi-raft or a pseudo-raft (b) . Two particular applications should be noted:
❏ In coal mining areas where subsidence strains may be caused by longwall mining (Section 7.7), the typical foundation for low-rise housing is a plane raft[14]. The raft is usually placed on a polyethylene DPM laid on a 150 mm-thick bed of compacted coarse sand to reduce the friction between the base of the raft and the ground. Reinforcement is normally placed in the centre of the raft, rather than top and bottom.
❏ For low-rise housing on fills (Section 7.4), raft foundations usually consist of a stiffened edge beam with an integral ground supported floor slab (Figure 6.4b). The depth of the beam is usually at least 0.45 m to provide protection against frost heave.

Where a raft is founded well below ground level to form a basement, it is sometimes termed a buoyancy raft or box foundation. For some high-rise applications, this form of foundation offers the additional benefit of reducing the ground loading, because the weight of the foundation is considerably less than the soil it replaces.

Typical
thickness
150–300mm

Reinforcement (top and bottom)

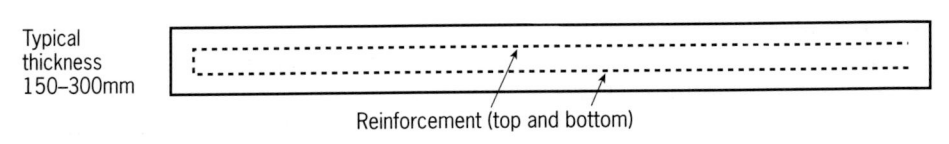

(a) Section through a typical plane raft

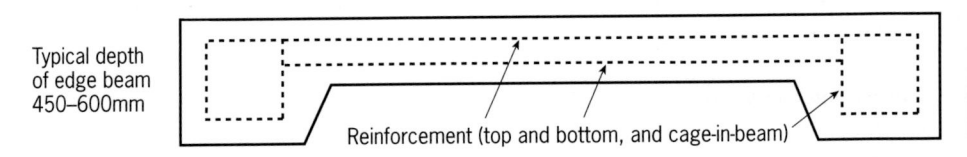

Typical depth
of edge beam
450–600mm

Reinforcement (top and bottom, and cage-in-beam)

(b) Section through a typical edge-beam raft

Figure 6.4 Raft foundations

Piled foundation

Piles are long and slender foundation elements, which can transmit the loads applied by the weight of the building through a surface layer of soft or unstable soil to an underlying competent stratum at significant depth below the base of the building. They can transmit loads to much greater depths than do other types of foundation and are often the most economical way of supporting very heavy loads and resisting uplift and lateral forces. Piles can limit the settlements of large structures on stiff clay. Small piles can be used to isolate houses and light industrial buildings from shrinkage and swelling in the surface layer of clay soils.

Many different types of pile and methods of installation have been developed for a wide range of uses and soil conditions[15]. There are two basic types: displacement piles[16] and replacement piles[17].

❐ Displacement piles are installed in the ground without excavation or removal of ground material. Piles made from steel or concrete can be driven or jacked into the ground. Steel piles are available in various shapes and sizes; driven piles can also be made of timber[18]. Driven cast-in-place piles are formed by driving a steel tube into the ground and filling with concrete.

❐ Replacement piles are formed by boring a hole in the ground and filling with plain or reinforced concrete. Bored cast-in-place piles consist of concrete which is placed into a pre-formed hole before the insertion of reinforcement. With continuous flight auger piles concrete is continuously pumped into the void through the hollow stem of the flight auger as the auger is retracted.

FOUNDATION DESIGN

Ground supported floor slab

Traditionally, most houses used timber floors consisting of boards supported by joists spanning between footings or sleeper walls. Modern construction techniques favour the use of concrete as a building material and, where a foundation other than a raft is used to support a building, floor loading and non-loadbearing walls can be supported on a separate ground bearing slab. Where it is not possible to use a ground supported floor slab (for example on swelling soil), a beam-and-block suspended floor spanning between the foundations can be constructed using precast concrete beams to support lightweight concrete blocks.

Precast concrete foundation

The use of precast concrete foundations for low-rise housing is increasing. A number of systems are available and the specialist contractor can provide a complete foundation package to finished floor level. The ground beams can be supported either on piles or at shallow depth; some systems involve post-tensioning the ground beams. These types of precast foundation can minimise site preparation and the amount of excavated material that has to be removed from the site; this can be particularly important on brownfield sites.

6.3 Acceptable ground movements

The following factors affect how much foundation movement is acceptable for a building:
- visual appearance;
- serviceability;
- structural integrity and stability.

In many cases, allowable foundation movements for low-rise buildings are limited by their effect on visual appearance and serviceability, rather than the integrity or stability of the structure. For domestic buildings, visual appearance may be critical; cracking of walls and cladding materials and distortion, with consequential sticking windows and doors, should be avoided. Serviceability is often the prime consideration for industrial buildings; it depends on the function of the building, the reaction of the user and owner, and economic factors such as value and insurance cover. In other cases, the level of acceptable movement will be dictated by a particular function of the building or one its services (for example the proper operation of overhead cranes, lifts, precision machinery, or drains). The significance of foundation damage is examined in Chapter 10.

Differential settlement is the primary concern because structural distortion results in damage. Excessive tilt can also cause problems related to visual perception and impairment of functionality and, ultimately, can imperil stability. Total settlement is also important for the performance of underground utility connections.

Cracking of walls is dependent on a wide range of factors and, in particular:
- the length to height ratio of the wall (L/H);
- the mode of deformation, whether the wall is hogging or sagging.

Distortion can be defined in various ways. The most meaningful parameter in the context of cracking is deflection ratio, Δ/L; this can be defined as the maximum vertical deviation from original construction divided by the length of the affected portion of the wall. However, foundation movement is normally quantified in terms of differential settlement δs, which is the maximum vertical displacement of one part of the building with respect to another. Figure 6.5 shows the relation between settlement and deflection ratio for a building undergoing sagging over the length L_{AD} and hogging over the length L_{DF}. Since differential settlement can produce either tilt or distortion, it cannot be directly related to the onset of cracking. The importance of horizontal extension in initiating damage must also be recognised.

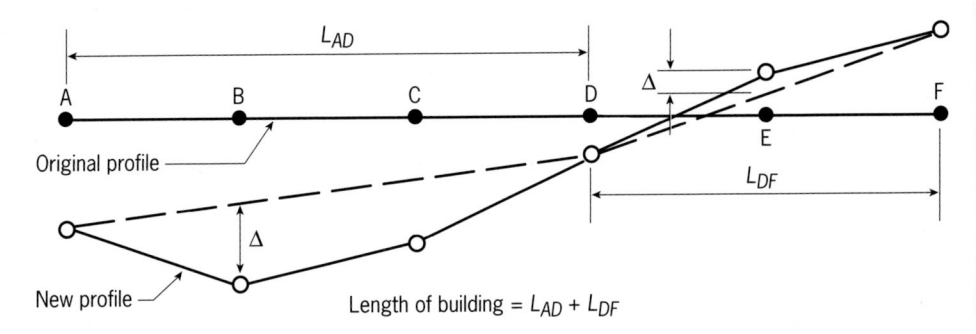

Figure 6.5 Deflection ratio in sagging and hogging – *after Burland and Wroth, 1975*

There have been a number of investigations of building damage caused by ground movement[19];examination of the available evidence has led to the following conclusions:
- load-bearing masonry structures are less tolerant of movement than frame and panel-infill structures;
- long, low buildings are more vulnerable to damaging movements than short, tall buildings;
- buildings distorted in a hogging mode suffer more damage than buildings distorted to the same degree but in a sagging mode.

Some tentative values for the distortion at which the onset of damage may occur have been put forward; Figure 6.6 shows criteria for the onset of visible cracking in rectangular beams in terms of deflection ratio expressed as a function of L/H. In practice, however, it

FOUNDATION DESIGN

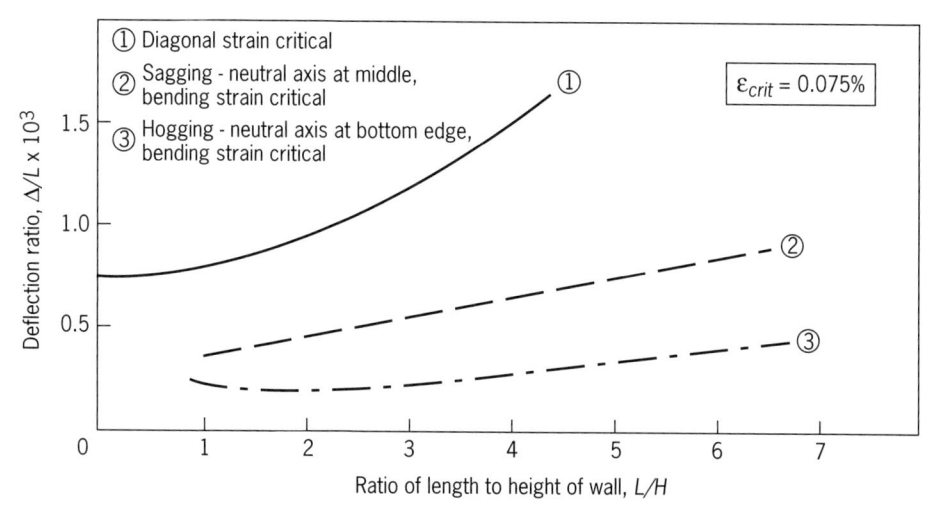

Figure 6.6 Criteria for onset of cracking in rectangular beams – *after Burland and Wroth, 1975*

is difficult to predict differential settlement, let alone deflection ratio. In specifying acceptable deformation criteria, there is a trade-off between simplicity and relevance. Total settlement is the simplest form of movement to predict and correlations between measurements of maximum total settlement and manifest damage have been attempted, but their reliability for design purposes has been questioned[20].

It is difficult to determine the differential movement that a building will tolerate before cracking or some other unacceptable event occurs. No two buildings are identical and the evidence of case records is extremely limited. The wide diversity of structural form and materials makes the likelihood of reliable predictions of building response to settlement rather small; for small buildings simple empirical guidelines continue to be used. It is widely held that a typical masonry, two-storey dwelling will not tolerate differential settlement exceeding about 25 mm. In small buildings with stiff raft foundations, large differential settlements may occur without any distortion of the building; here, tilt becomes the critical factor. Tilt of two-storey housing is likely be noticed at about 1/250 so an appropriate design limit is 1/400[21].

6.4 Basic features of design

With adequate knowledge of the site conditions and the nature of the building development, the most suitable type of foundation can be established. The different types of foundation have been described in Section 6.2 and the required performance for the foundations of low-rise buildings has been discussed in Section 6.3.

A basic requirement of the foundation design is to ensure that the weight of the building, as transmitted by the foundations through to the ground, will not exceed the bearing resistance, or bearing capacity, of the ground and that the settlement of the ground induced by the weight of the building is acceptably small. For simple low-rise construction, there is often insufficient soils data to predict how much settlement will occur; the bearing pressure applied to the ground by the building should be substantially smaller than the bearing capacity of the ground in order that the ground deformations are small. The evaluation of bearing resistance is described in Sections 6.5 and 6.6, and settlement in Sections 6.8 and 6.9.

The calculation of the loads acting on the foundations should include both dead load (the weight of the permanent construction including floors, walls, roof, partitions and services) and imposed load (load assumed to be produced by the intended use of the building). In practice, the loading applied to the ground by the foundations of a one to three-storey dwelling is relatively low; the combined dead and live loads are typically no more than 50 kN per metre run of wall. Hence, the bearing pressure under a 0.5 m wide foundation will usually not exceed 100 kPa.

In designing the foundations of a low-rise building, it must be determined whether the pressure acting below a minimum width foundation of, say, 0.5 m is likely to exceed the stress that the ground can safely sustain without excessive deformation. If this allowable bearing pressure is exceeded, a wider or deeper foundation will be needed. Most British soils are capable of sustaining stresses of this magnitude without excessive compression or consolidation. The most economical foundation is generally a strip footing or trench-fill foundation at the minimum depth to provide protection against seasonal moisture and temperature variations. Where buildings are to be sited on weak, highly compressible soils, the most economic foundation design will depend on the variation of allowable bearing pressure with depth and across the area of the building.

The minimum depth for a foundation depends not only on the requirements associated with bearing pressure and settlement but also on the need to found at a depth where the soil is unaffected by changes in moisture (Section 7.5) or frost heave. There also may be structural requirements, such as a basement or underfloor heating.

In certain ground conditions there is a possibility of substantial movements that are unrelated to foundation loading. Some of the common processes that can cause excessive ground movements such as compression of fill, clay shrinkage, collapse of underground cavities and landslip, are discussed in Chapter 7. Unless the foundation design takes proper account of the potential ground movements associated with these processes, there is a risk of unacceptable damage. Where ground movements is confined to the surface soil, deeper foundations can take advantage of the underlying stable stratum, or a rigid foundation can be used to even out differential movements.

For most ground conditions, the presence of a high water-table is unlikely to have a significant effect on the foundations, with the possible exception of loose coarse soil. However, it can complicate the construction process and may constrain foundation design. For example, in water-logged ground, driven piles may be chosen because of the difficulties of excavating below the water-table.

The following sections of this chapter describe some of the fundamental principles of foundation design. Although foundations for low-rise housing are often carried out using simple empirical guidelines[22], there are many situations, including those considered in Chapter 7, when advice from an appropriately qualified and experienced geotechnical engineer will be required.

6.5 Bearing resistance of shallow foundations

Bearing resistance, or ultimate bearing capacity as it is often termed, is the loading intensity at which the ground fails in shear. The bearing resistance q of a strip foundation of width B and at a depth D, can be expressed as follows:

$$q = cN_c + 0.5\gamma BN_\gamma + \gamma DN_q \qquad (6.1)$$

where:
  c is the cohesion intercept
  γ the unit weight of the soil
  the bearing capacity factors N_c, N_γ and N_q are functions of the angle of shear resistance for the soil ϕ.

Equation 6.1 permits the calculation of the gross bearing resistance. The net bearing resistance is obtained by subtracting the overburden pressure γD from q.

For undrained loading of relatively impermeable fine soils, such as clays, the bearing resistance of a strip foundation can be expressed in terms of total stress with the soil strength expressed in terms of the total stress parameter c_u and with $\phi_u = 0$; equation 6.1 then becomes:

$$q = c_u N_c + \gamma D \qquad (6.2)$$

where:
  $N_c = (2 + \pi) = 5.14$, $N_\gamma = 0$, $N_q = 1$.

For a footing of length L and width B, equation (6.2) can be expressed as:

$$q = c_u N_c (1 + 0.2[B/L]) + \gamma D \qquad (6.3)$$

For low-rise buildings on fine soils, inadequate bearing resistance is likely to be a significant problem only with soft clays and silts and peat and organic soils[23] (Table 6.2 and Section 7.2

A drained solution in terms of effective stress parameters should be used for coarse soils, which are sufficiently permeable to allow the excess pore pressures generated by the application of load to dissipate during the construction period. For drained loading of a strip foundation, with a water-table well below foundation level and $c' = 0$, equation 6.1 becomes:

$$q = 0.5\gamma BN_\gamma + \gamma DN_q \qquad (6.4)$$

The variations of N_γ and N_q with ϕ' are shown in Figure 6.7[24]. The gross bearing resistance has slightly different values for circular and rectangular footings and shape factors should be applied. For a footing of length L and width B, equation 6.4 can be expressed as:

$$q = 0.5\gamma BN_\gamma (1 - 0.3[B/L]) + \gamma DN_q (1+ \sin \phi' \, [B/L]) \qquad (6.5)$$

Where the water-table is close to the ground surface, it is necessary to allow for the much lower submerged unit weight of the soil below the water-table. More complex expressions are required for q than those given in equations 6.1, 6.4 and 6.5. A rise in groundwater level reduces the bearing resistance.

For coarse soil, the calculated bearing resistance is normally sufficiently high not to be a critical consideration except for very narrow footings on loose sand with a high water-table. Settlement is normally the critical consideration for shallow foundations on coarse soils. The effects of eccentric or inclined loading can be dealt with by increasing the width of the foundation.

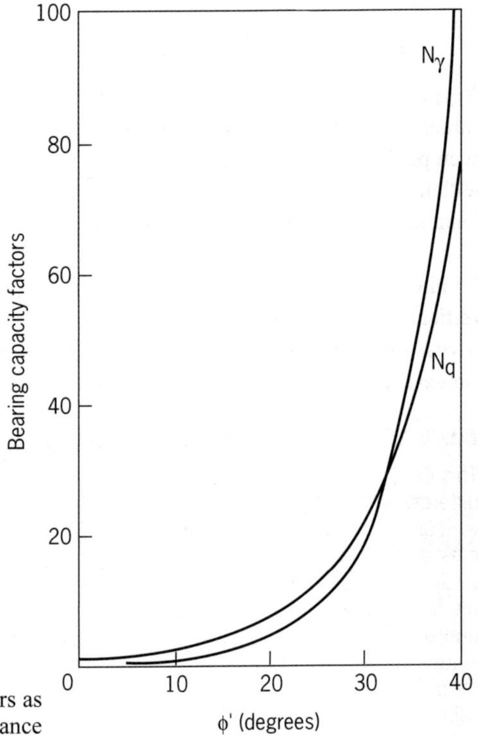

Figure 6.7 Bearing capacity factors as a function of angle of shear resistance

FOUNDATION DESIGN

Equations 6.1 to 6.5 provide ways of calculating bearing resistance but for design purposes a safety margin must be introduced. There are two ways of doing this:
- Applying a factor of safety to the ultimate bearing capacity in order to calculate an allowable bearing pressure;
- Using partial factors in load and resistance factor design.

In the past, the normal way of introducing a safety margin has been by the use of an allowable bearing pressure q_a, which is defined as the maximum intensity of foundation loading that can be safely applied to the ground without causing settlements that will cause distress to the building[25]. The allowable bearing pressure is, therefore, a function both of the material properties of the ground and the amount of distortion that can be tolerated by the structure. Both the load required to cause failure in the soil and the deformation characteristics of the soil must be considered. For rock strata, the allowable bearing pressure will usually be in excess of 1000 kPa. Dense granular soils and very stiff clays will typically have allowable bearing pressures greater than 300 kPa.

The usual procedure for deriving a design value for the allowable net bearing pressure q_{an} is to apply a factor of safety F to the calculated net bearing resistance q_n of the soil:

$$q_{an} = \frac{q_n}{F} = \frac{(q - \gamma D)}{F} \tag{6.6}$$

A typical factor of safety of 2.5 to 3 should ensure that no local yielding occurs; the lower values tend to be used where there are good site investigation data or a loading test has been performed. If the anticipated settlement is greater than the estimated acceptable settlement, the allowable bearing pressure must be reduced accordingly. Where the foundation loading exceeds the allowable bearing pressure, some re-design is required.

An alternative method to applying a safety margin is to apply partial factors to the applied vertical loads and to the bearing resistance. The new European standard, prEN 1997-1 *Geotechnical design – general rules*[26], uses partial factors; this approach is likely to be increasingly used. The approach is outlined in Box 6.1.

Box 6.1 Load and resistance factor design

This design method has the advantage that different partial factors can be applied to the various soil properties and loads, and it presents a more rational approach to the design problem. It should be recognised that the factor values are chosen principally to prevent failure and a separate assessment of settlement will be required in many cases (eg for soft ground) (Section 6.8). Characteristic values of soil strength parameters are evaluated as cautious estimates of the operational strength in the field at failure. Design values can then calculated by applying a factor to the characteristic value. From these design strengths, the bearing capacity factors can be determined and the bearing resistance calculated. Finally it has to be confirmed that the design value of the vertical loading is smaller than the design value of the bearing resistance.

6.6 Bearing resistance of deep foundations

Where shallow foundations are inadequate to support the structural loads, deep foundations must be provided. The expression *deep foundations* usually implies *piles*. The design of piled foundations is specialised and much of the expert knowledge lies with specialist piling firms. A brief account of some of the principal features is given here[27]. For deep foundations there are two components of bearing capacity (Figure 6.8):
- shaft resistance q_s resulting from frictional forces acting on the vertical face of the shaft of the pile;
- base resistance q_b resulting from bearing forces acting on the underside of the base of the pile.

In calculating the load carrying capacity of a pile, it should be remembered that the shaft resistance and the base resistance are mobilised at very different rates. For example, with a large diameter bored pile it can take more than 0.1 m of settlement to mobilise the full base resistance, whereas peak shaft resistance is typically mobilised at a small fraction of this movement. In these circumstances, it may be necessary to calculate the safe working load on the basis of the full shaft resistance plus a small proportion of the base resistance.

Base resistance

Base resistance is calculated in the same way as for a shallow foundation, although the values of the coefficients N_c, N_γ, and N_q need to be modified to take account of their variation with depth[28]. Where B is small, the influence of the N_γ term is also small and is normally neglected for deep foundations.

For clayey soil it is common practice to assume that the surcharge is equal to the weight of the foundation, so that the base resistance $q_b = c_u N_c$ for undrained loading in terms of total stress. For piles it is usual to assume $N_c = 9$.

For coarse soil, the base resistance simplifies to $\sigma'_v N_q$ for drained loading in terms of effective stress where:

σ'_v is the vertical effective stress at foundation level.

In principle, allowance should be made for both the weight of the pile and for the

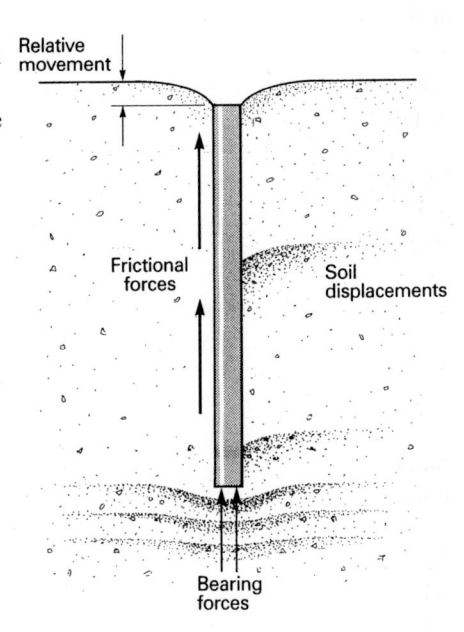

Figure 6.8 Bearing capacity of pile

overburden pressure at the base of the pile but, in many cases, these two effects will balance each other. The dependence of N_q on ϕ' results in the prediction of very high base resistances for dense sand and gravel. In practice the base resistance is found to reach a limiting value at penetrations of 10 to 20 pile diameters that is not exceeded by further penetration. The design of piles can be based directly on the results of cone penetration test data (Section 5.7).

Shaft resistance

Shear stress, τ, on the pile shaft is dependent on the horizontal effective stress acting at the pile-soil interface. In clay it is difficult to estimate the changes that occur during installation and loading of the foundations. Therefore, many designs are based on a total stress approach where the maximum shaft stress, τ_f, is related to the undrained shear strength using an empirical factor α:

$$\tau_f = \alpha . c_u \tag{6.7}$$

Values of α which have been derived from loading tests of various types of pile in differing ground conditions show a wide range of values[29]. A number of factors affect α: they include the surface finish of the pile, the pile type, the method of installation, disturbance of the soil during pile installation and subsequent consolidation, and the method of measurement of c_u. For bored piles in London Clay, $\alpha = 0.45$ has often been adopted and, typically, in over-consolidated marine clays α is between 0.4 and 0.6. Wherever possible, the final design should be based on values of α derived from load tests on prototype piles at the site.

In coarse soil, τ_f can be expressed in terms of the horizontal effective stress and an interface friction angle δ; for convenience, the horizontal stress is normally described as the product of the vertical effective stress, σ'_v, and a coefficient of earth pressure K so that:

$$\tau_f = K \sigma'_v \tan \delta = \beta \sigma'_v \tag{6.8}$$

A value for $\beta = K \tan \delta$ has to be assigned to the soil[30]. This method can also be used for clay soils.

Downdrag

Where a deep foundation passes through a layer of fill or soft soil, the surface soil may settle relative to the foundation. This is due to processes such as consolidation under the weight of fill used to raise ground levels, groundwater lowering, and disturbance of the soil caused by pile driving. Any frictional forces associated with this settlement will induce a downward loading on the piled foundation (Figure 6.9). This process is known as negative skin friction or downdrag.

Downdrag can cause large settlements and can even over-stress the piles. A peak value for downdrag at a given point on the foundation can be estimated in a manner similar to the calculation of the supporting shaft resistance, which has been described in the previous paragraphs. Estimating the total load induced by downdrag is complicated because the variation in the magnitude of the relative movement along the length of the pile embedded in the fill or soft clay means that the peak friction will not be mobilised simultaneously along the pile. Near the surface, where the soil movement is greatest, there will be a tendency for the skin friction to pass through a peak and to reduce to a residual value. Near the bottom of the soft layer, the movement will be insufficient to mobilise the peak friction and, if there is any settlement of the pile, the friction may act in support. A degree of engineering judgement is required to decide what proportion of the maximum skin friction is used in design.

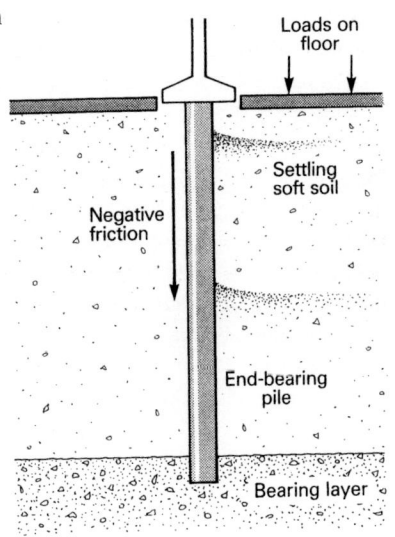

Figure 6.9 Downdrag on pile

Where short bored piles are used to support a low-rise building sited near trees on shrinkable clay, the piles will be subjected to cycles of uplift and downdrag as the clay shrinks and swells seasonally. This is examined in Section 7.5.

Downdrag can be reduced by a bitumen coating or by sleeving the upper part of the pile with a low-friction material. This can be expensive and it is normally preferable to increase the penetration depth to increase the load-carrying capacity of the pile.

6.7 Use of elastic theory in settlement calculations
In routine foundation design for low-rise buildings, settlement calculations are based on a number of simplifying assumptions[31]. Elastic theory is commonly used and some consideration of this is helpful prior to examining the estimation of the settlement of shallow foundations on fine and coarse soils.

When calculating the settlement likely to be caused by the weight of a building, some assessment is necessary of the stress distribution in the ground beneath the load. Figure 6.10 shows the elastic stress distribution, often termed a pressure bulb, beneath a flexible square foundation[32]. The approximation of real soil behaviour to that of a homogeneous, isotropic, elastic medium may appear crude but the vertical stress distribution is not sensitive to the assumption of linear elasticity and a more sophisticated analysis is not usually justified. However, where a relatively stiff soil overlies soft soil, the stress distribution will be significantly modified with the stiff soil acting to spread the load.

The contours in Figure 6.10 represent the increase in vertical stress caused by the foundation loading. If the soil actually behaved as a linear elastic material, the deformations of the soil would depend solely on the magnitude of these stress increments. However, soils are not linear elastic materials and the deformations of the soil depend not only on the magnitude of the stress increments but also on the overburden pressure, that is the vertical stress in the soil prior to construction.

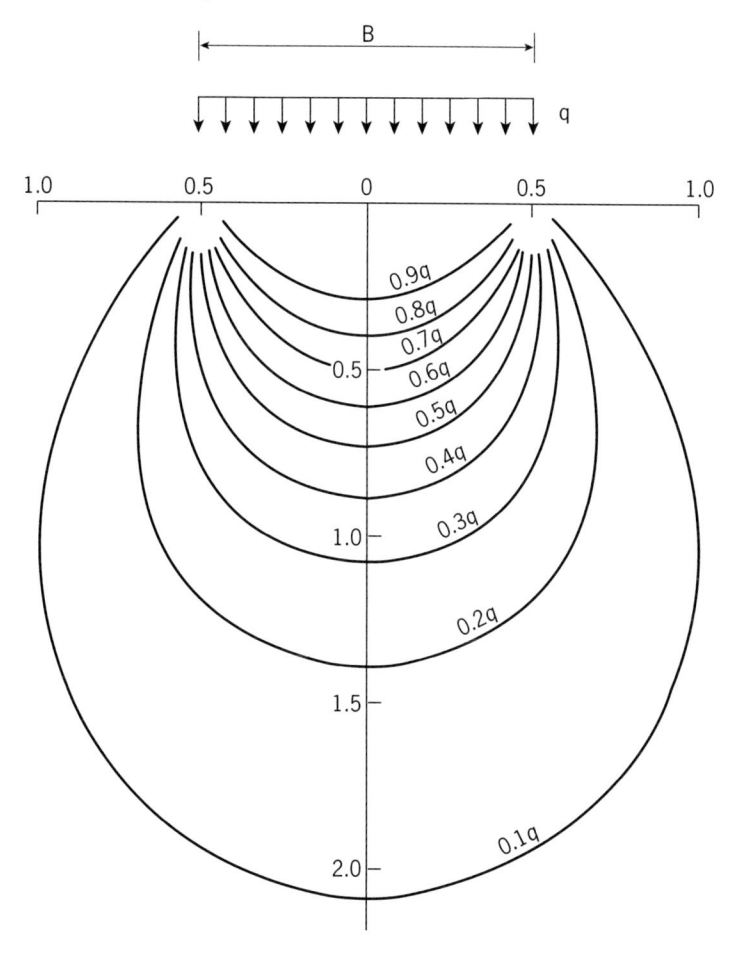

Figure 6.10 Elasctic vertical stress distribution for a flexible square foundation
– *after Coduto, 1998*

Figure 6.11 shows the distribution of vertical stress $\Delta\sigma_v$ with depth, on the centre-line of three strip foundations that apply a uniform stress to homogeneous ground with linear elastic properties. The footing widths B vary from 0.5 m to 1.5 m for a net loading q of 50 kN/m run; it can be seen that 1 m below foundation level there is little difference in vertical stress despite large differences in footing widths. The overburden pressure is also plotted in Figure 6.11 and is designated by:

$$\gamma(z+D)$$

where:
> γ is the bulk unit weight of the fill
> z is the depth below foundation level
> D is the depth of the foundation.

The foundation load is likely to cause appreciable compression of the fill only at depths where the increase in vertical stress $\Delta\sigma_v$ due to the applied load is a significant proportion of the existing in-situ stress, $\gamma(z+D)$, say $\{\Delta\sigma_v/\gamma(z+D)\} > 0.20$. This assumes that

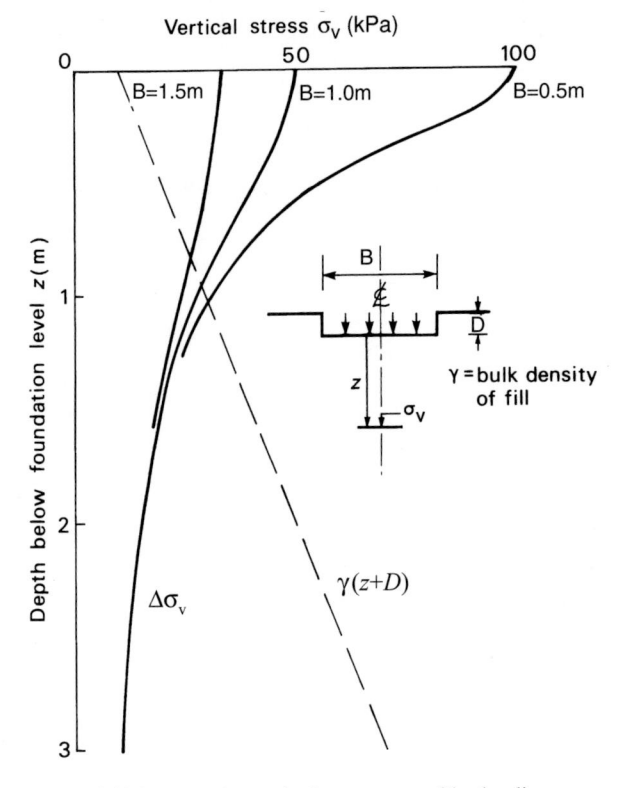

Figure 6.11 Increase in vertical stress caused by loading compared with overburden presssure

FOUNDATION DESIGN

groundwater level is below the depth of influence of the foundation loading. It can be seen that, for a foundation load of 50 kN/m run, the ground is significantly stressed by the foundation load to a depth below foundation level of only about 2.5 m.

An initial estimate of the significant depth z_e (m) affected by various loading conditions can be obtained from the following relationships:

- applied pressure q kPa on a very large raft foundation: $\quad z_e = 4(q/\gamma)$ \hfill (6.9)
- line load p kN/m acting on a strip foundation: $\quad z_e = 1.5\,(p/\gamma)^{0.5}$ \hfill (6.10)
- point load P kN acting on a small pad foundation: $\quad z_e = 1.2\,(P/\gamma)^{0.33}$ \hfill (6.11)

These relationships assume that the foundation is on uniform soil of bulk unit weight γ(kN/m³). Where there is a high groundwater level, the submerged unit weight γ' should be used instead of γ.

Where the stiffness of the soil is reasonably uniform over the depth of soil that will be influenced by the foundation loading, the settlement s resulting from an applied uniform stress q can be calculated using elastic theory:

$$s = f_s f_d\, q\, B\, (1 - \upsilon^2)/E \qquad (6.12)$$

where:
  E and υ are the Young's modulus and Poisson's ratio appropriate to the loading conditions
  B is the width of the loaded area
  f_s is a shape and rigidity factor
  f_d is a foundation depth factor.

The modulus should be determined at a relevant depth over an appropriate stress range. Even with a constant modulus, 80% of the settlement of a circular footing of diameter B occurs above a depth $z = 2B$; in the more realistic situation where the modulus increases with depth, 80% of the settlement may occur above a depth $z = B$.

Table 6.3 gives the shape and rigidity factor f_s for some common shapes of footing. Values are quoted for flexible and rigid footings; in general the settlement of a rigid foundation is 0.8 times the settlement at the centre of a flexible foundation.

Table 6.3 Shape and rigidity factor f_s for centre of footing[33]

Footing shape	Footing dimension B	Footing rigidity	f_s
Circular	Diameter	Flexible	1.0
Circular	Diameter	Rigid	0.79
Square	Side	Flexible	1.12
Square	Side	Rigid	0.89
Rectangular ($L/B = 3$)	Width	Flexible	1.78
Rectangular ($L/B = 3$)	Width	Rigid	1.41
Rectangular ($L/B = 5$)	Width	Flexible	2.10
Rectangular ($L/B = 5$)	Width	Rigid	1.70
Rectangular	Width	Flexible	$1.12(L/B)^{0.39}$
Rectangular	Width	Rigid	$0.90(L/B)^{0.38}$

The foundation depth factor f_d is defined as the ratio of the settlement of a foundation at depth D to the settlement of the same foundation at ground surface. An approximate expression for f_d is as follows[34]:

$$f_d = 1 - 0.08(D/B)[1 + (4/3)(B/L)] \qquad (6.13)$$

The expression is said to be valid for aspect ratios (L/B) up to 6 but, while it gives some indication that at greater foundation depths the effect of embedment is more pronounced, in reality the effects of embedment are more complex than this simple factor based on footing geometry suggests.

6.8 Settlement of shallow foundations in response to loading

The extent to which the ground deforms under an increase in loading depends on the magnitude of the increase in bearing pressure, the soil stiffness and the dimensions and stiffness of the foundations. Not even uniformly loaded homogeneous ground settles evenly and the complex properties of soil make it difficult to predict foundation settlement. Furthermore, the response of the ground to foundation loading will depend on the permeability of the soil. It is helpful, therefore, to examine separately the settlement of shallow foundations on fine soil and the settlement of shallow foundations on coarse soil.

Shallow foundations on fine soil

The settlement of shallow foundations on low permeability fine soils, such as clays, consists of three components:
- immediate undrained settlement s_o;
- primary consolidation settlement s_1 (movement which occurs due to compression associated with the dissipation of the excess pore water pressure produced by the loading);
- secondary settlement s_2 (movement which occurs due to compression at constant effective stress subsequent to the dissipation of excess pore water pressure).

Settlement predictions are based on estimating each of these components and summing them to arrive at the total settlement. Immediate settlement can be calculated using elastic theory and, since the deformations occur under undrained conditions with zero volume change, the undrained Poisson's ratio $\upsilon = 0.5$. Where the undrained Young's modulus E_u can be assumed to be reasonably constant throughout the loaded stratum, the immediate undrained settlement of a foundation s_o with a typical dimension B under an applied load q can be approximated to:

$$s_o = 0.75 f_s f_d q B / E_u \qquad (6.14)$$

It is usual to derive an appropriate value of E_u from undrained triaxial compression tests. Where no measurements of E_u are available, a crude estimate can be derived from correlations with undrained shear strength. The stiffness of many soils increases markedly with depth and for large foundations, such as rafts, it is inappropriate to assume a uniform modulus. However, it is possible to adapt the method described above to calculate the settlement on a soil divided into layers, each of which is assigned a modulus.

The settlement of a layer of saturated soil of low permeability is controlled by the flow of water out of the affected area, which is a function of the hydraulic gradient generated by the excess pore pressure and the drainage conditions at the soil boundaries. As the water flows away from the loaded area, the pore pressure dissipates and the soil compresses. This process is known as primary consolidation; it slowly continues until all the excess pore pressure has dissipated. The rate of settlement can be assessed from measured values of the coefficient of consolidation c_v using equation 4.16 (Section 4.5).

Estimates of the magnitude of the consolidation settlement s_1 are based on the coefficient of volume compressibility m_v measured in the oedometer test (Section 4.5). For a soil divided into a series of layers of equal compressibility with thickness Δz, the consolidation settlement s_1 is given by:

$$s_1 = \mu \, \Sigma \, (m_v \, \Delta\sigma'_v \, \Delta z) \qquad (6.15)$$

where:
    $\Delta\sigma'_v$ is the additional vertical effective stress applied to a layer as a result of the foundation load
    μ is an empirical coefficient to account for differences between boundary conditions in the oedometer test and those that exist in the field.

Values of $\Delta\sigma'_v$ can be estimated from an elastic stress distribution for the foundation. Values of μ range typically from 0.7 to 1.0 for normally consolidated clays and from 0.3 to 0.7 for overconsolidated clays[35].

Primary consolidation is followed by secondary compression, which continues under conditions of constant effective stress. Estimates of secondary compression are normally based on empirical data, assuming a linear relation between creep settlement and the logarithm of time that has elapsed since the load was applied (Section 4.5). The continuing rate of compression of clay soils in one-dimensional compression can be described by the coefficient of secondary compression C_α, which can be defined in terms of vertical strain[36] as shown in equation 4.17. The creep settlement s_2 of a layer of thickness h that occurs in a time period from t_1 to t_2 after application of the load is given by:

$$s_2 = C_\alpha h \log [t_2/t_1] \qquad (6.16)$$

Shallow foundations on coarse soil

Drainage occurs quickly in coarse soil, such as sand and gravel, and settlement is normally assumed to occur during construction. Where a deformation modulus can be measured, the settlement can be calculated using a similar equation to that described for fine soil. It is more appropriate to use the constrained modulus $D (= 1/m_v)$, rather than Young's modulus, to calculate the settlement:

$$s = f_s f_d \frac{qB}{D} \frac{(1-\upsilon)^2}{(1-2\upsilon)} \qquad (6.17)$$

Whereas Poisson's ratio υ can be assumed to be 0.5 for fine soil under undrained conditions, under drained conditions it is often assumed to be 0.15 for coarse soil and 0.25 for fine soil.

The difficulty of sampling coarse soil means that there are no routine procedures for measuring the compression characteristics in the laboratory and design tends to depend on semi-empirical methods based on the results of in-situ tests. The tests that are most commonly used are the standard penetration test (SPT)[37] and the cone penetration test (CPT)[38]; they are described in Section 5.7.

It has been generally found that settlement is related linearly to applied stress. Although this is often the case over a wide range of

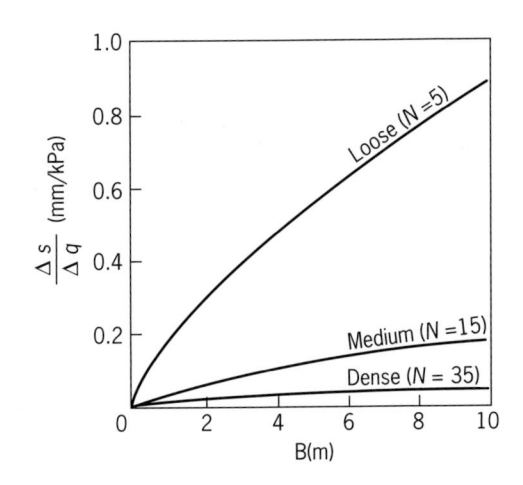

Figure 6.12 Settlement of foundation on sand as a function of foundation width and SPT N value

FOUNDATION DESIGN

working stresses, there will be some limiting stress where the linearity ceases as bearing failure is approached. In the situation where there is a large factor of safety against bearing capacity failure, an empirical relationship has been proposed for the calculation of the settlement of normally consolidated sand using the following equation[39]:

$$\Delta s/\Delta q = 1.71\ B^{0.7}/N^{1.4} \qquad (6.18)$$

where:
Δs is the settlement in mm produced by an increase in bearing pressure Δq kPa over a foundation of width B m.
N is the average SPT blow count over the depth of influence of the foundation[40].

N is not corrected for effective overburden pressure but corrections are made for fine and silty sand below the water-table and for gravels. In Figure 6.12, $\Delta s/\Delta q$ is plotted against B for three values of N typically representing a loose, medium and dense condition. The expected settlement is quite small except for loose sand with a high bearing pressure or a wide foundation.

6.9 Settlement of deep foundations in response to loading

Where piled foundations are installed, it is necessary to estimate the settlement that can be expected. In Section 6.6 it has been shown how piles transmit their load into the ground in two ways, through shaft resistance and through base resistance. Shaft resistance is likely to be mobilised at much smaller settlements than the base resistance and any form of settlement calculation will be difficult.

A pile load test can be used to determine the settlement of a single pile at working load, but it is necessary to assess the settlement of a group of piles as a whole, rather than that of individual piles. There are two approaches to the problem:

❏ The action of the pile group is to transfer the foundation load to some lower stratum; for the purposes of calculating settlement, the pile group can be represented as a raft at a depth equal to 2/3 of the pile length[41]. The settlement can then be calculated using the techniques already described.

❏ An alternative approach is to relate the settlement of a pile group of n piles to the settlement of an individual pile at the average load through an amplification factor R which is a function of the number of piles[42].

Owing to the relatively high bearing capacity of coarse soil, allowable bearing pressures for foundations on sand and gravel are nearly always governed by settlement, unless the soil is in a loose condition and the water-table is near the surface. Allowable bearing pressures for foundations on fine soil may be governed either by the bearing resistance or the amount of settlement that the proposed building can tolerate.

Where piles are designed to transmit the loads applied by the weight of the building to a rock or very stiff soil stratum, the pile will be end-bearing and settlement very small.

6.10 Foundations for low-rise building extensions
When providing foundations for extensions, the three principal aims should be:
- to support the new structural loads;
- to provide long-term stability and serviceability of the new structure;
- to ensure compatibility between old and new structures.

Extensions to low-rise buildings vary considerably in size and complexity. The extent and sophistication of the ground investigation that is required will show corresponding variations. While a simple one-storey, single-room extension may require only a visual inspection of the ground before a simple strip foundation is placed, a two-storey extension will require more detailed investigation of ground conditions. It will be particularly important to establish whether or not the ground conditions for the extension are similar to those of the existing building.

In addition to the ground investigation, some investigation of the existing structure and its foundations may be needed to answer the following questions:
- What is the type and age of the building? Buildings have different tolerances to ground movement. Old lime mortars can accommodate movement more readily than modern masonry.
- What is the current condition of the building? Is there evidence of damage or repair?
- What are the type and dimensions of the foundations?

While providing a suitably stable foundation for an extension will not usually cause undue problems, it may be difficult to predict how its performance will compare to that of the foundations of the existing building. Compatibility between old and new is seldom given the consideration that it deserves[43].

There can be serious difficulties with the need to minimise the differential movement between the building and the new extension. For an extension to move in harmony with a main building on shallow foundations, the foundations of the extension should be at a similar depth. However, many extension foundations are far deeper than those of the original building, creating the potential for differential movement. If the existing building has relatively shallow foundations on shrinkable clay, damage may occur at the interface between a stable extension and the main building which undergoes some seasonal movements. Figure 6.13 shows one approach to mitigating damage at the interface between an extension and the original building where the foundation depth of the old building is inadequate according to current standards.

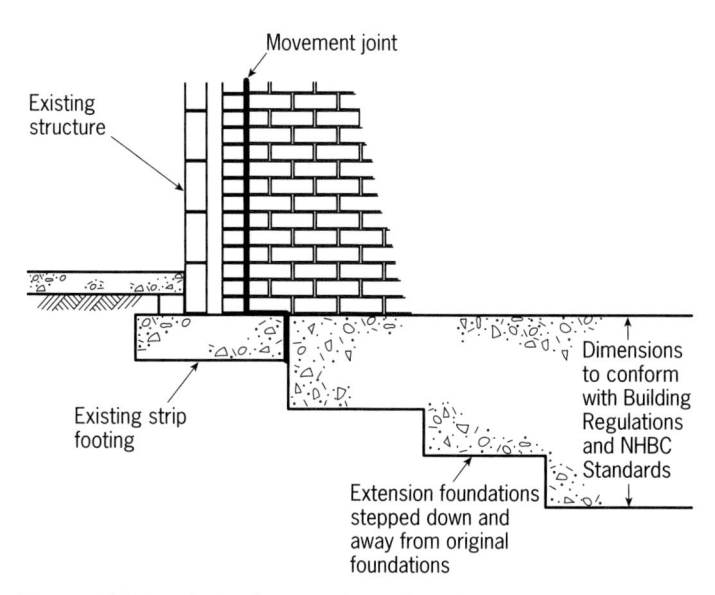

Figure 6.13 Foundation for extension to low-rise building where extension foundation is much deeper than foundation of original building

6.11 Ground chemistry

Although the detailed chemical composition of the ground is often of limited interest for foundation design, the presence of certain chemical constituents in soil can be significant. The interaction between foundation building materials and chemically aggressive ground is a major subject[44] and can only be briefly touched on here. Aggressive ground conditions are particularly common on former industrial sites: these hazards are described in Chapter 8. The effects on load-carrying characteristics of volume change owing to chemical reactions in fills are described in Section 7.4.

The effects of sulfates, acids and chlorides give rise to concern for both unreinforced and reinforced concrete. Magnesium, ammonia and phenol are also known to cause deterioration of concrete. High quality, dense concrete is the primary prerequisite for durability and there is no substitute for quality of the materials used. Ultimately, chemical attack may cause the collapse of a structure but such extreme examples are not often found. More common is localised deterioration and cracking resulting in loss of strength.

Poor accessibility makes it difficult to assess chemical attack below ground. With concrete piles there is the possibility that sulfate attack, and other forms of attack, could adversely affect the soil/pile interface so as to reduce shaft resistance in service. There could be similar effects if coating materials used to protect piles from a particular contaminant were themselves attacked by another contaminant. Damage has generally been found on the tops of exposed piles or foundation walls. The form of concrete is important in determining the risks of attack. Slender sections are more at risk than massive foundations. Modern housing, which tends to be of lighter construction, should be carefully designed and specified. Slabs and floors on the ground are at risk especially where they can dry from the top which encourages the movement of contaminants into the concrete from the ground.

The attack on a material buried in the ground depends not only upon the absolute aggressiveness of the substances or organisms present, and on the amount and concentration, but also on its availability. Where the aggressive agent is consumed by reaction with the material under attack, the rate of attack will be limited if the immediate area is depleted of the aggressive agent and the supply is not replenished. It may be expected that the potential for attack of a water-soluble aggressive agent would be greater in a highly permeable sandy soil than in a relatively impermeable clay soil. This is the reverse of the potential for corrosion of buried steel, for which sandy soils provide the preferable environment. The movement of ions through soils is not just related to permeability but also to the ability of the soil to bind the ions by various physico-chemical processes, the existence of concentration gradients, and other factors. The permeability of soils to organic solvents and alkaline solutions will not be the same as their permeability to water, and may be considerably higher.

Routine laboratory chemical testing is usually limited to organic content[45], carbonate content[46], sulfate content[47], pH value[48] and chloride content[49]. Corrosiveness to steel can be evaluated by means of resistivity tests[50] and determining redox potential[51].

Notes

1. *Foundation* can also refer to the ground underneath a building. Substructure refers to the part of a building below ground level and includes both foundations and basements.

2. At the Building Research Congress, held in London in 1951, Karl Terzaghi commented that; *"The scientific stage of foundation engineering came into existence as soon as it was realised by a foundation engineer that every load produces settlement, regardless of what the allowable bearing value of the subsoil might be"*. The need for such a comment is a reflection on the misunderstandings that have arisen between structural engineers and geotechnical engineers.

3. The property owner was reported as saying *"It seems a bit like overkill that we had to pour 350 tonnes of concrete for a building that weighs just 16 tonnes"* (Property section of Daily Telegraph 24th June 2000, p P12).

4. Atkinson (2003) and Curtin *et al* (1994) give details of the structural design of foundations.

5. Clause 3.1 of BS 8004: 1986 *Code of practice for foundations* defines shallow foundations as being foundations where the depth below finished ground level is less than 3 m.

6. Shallow foundations have been defined in terms of the ratio of depth to width. For example, Budhu (2000) p 319 defined a shallow foundation as a foundation where $(D/B)<2.5$. See also Atkinson (2003).

FOUNDATION DESIGN

7. Practical guidance on the design of strip foundations is in Approved Document A of the Building Regulations and NHBC Standards Chapter 4.4. A simple site guide to foundation construction has been provided by Martin (1996).

8. Table 6.2 is based on tables in Approved Document A of the Building Regulations and NHBC Standards Chapter 4.4.

9. Certain soils expand when frozen due to the formation of ice and the frost susceptibility of the ground can be a significant consideration in the design of shallow foundations in some situations. Although Clause 3.2.9.1 of BS 8004: 1986 states that in the British Isles a depth of 450 mm below ground level is generally sufficient for protection, this depth may need to be increased in upland areas and other areas subject to long periods of frost. See also Section 2E of Approved Document A of the Building Regulations.

10. Good Building Guide 39 *Simple foundations for low-rise housing* Part 2.

11. Section 2E of Approved Document A of the Building Regulations.

12. Clause 3.2.8.1 of BS 8004:1986 recommends that on clay soils strip foundations of traditional brick and masonry dwellings should be founded at a minimum depth of 0.9 m below finished ground level to reduce the relative movement to acceptable limits. NHBC Standards Chapter 4.4 specifies a minimum foundation depth in shrinkable soil varying from 0.75 m in a soil with low shrinkage potential to 1.0 m in a soil with high shrinkage potential. In Approved Document A, it is stated that in shrinkable clay soils the depth of strip foundations should not be less than 0.75 m.

13. NHBC Standards Chapter 4.4 requires that trench-fill foundations more than 2.5 m deep are designed by an engineer.

14. Details of the design of rafts in areas of longwall mining can be found in Atkinson (2003).

15. Some details of pile types can be found in BRE Digest 315 *Choosing piles for new construction*. BRE Report BR 470 *Working platforms for tracked plant* describes the design of ground-supported granular platforms which are normally required to maintain the stability of the heavy plant used to install piles.

16. BS EN 12699:2001 *Execution of special geotechnical work – Displacement piles* establishes general principles for the construction of piles that are installed without excavation or removal of ground material.

17. BS EN 1536:2000 *Execution of special geotechnical work – Bored piles* establishes general principles for the construction of piles that are formed by boring a hole in the ground and filling with concrete.

18. Timber piles have been used throughout human history. They are vulnerable to rotting within the zone exposed to seasonal changes.

19. Among the more important investigations of building damage caused by ground movement are; Skempton and Macdonald (1956), Burland and Wroth (1975) and Boscardin and Cording (1989).

20. Design limits for total settlement were proposed by Skempton and MacDonald (1956) as shown in the Table below. The lower maximum settlement limits for foundations on sand were due to the perception that variability was more likely and hence differential settlement would be a greater proportion of total settlement. The attempt to relate building distortion to maximum settlement has been strongly criticised.

	Clay	*Sand*
Isolated foundations	65 mm	40 mm
Rafts	65 – 100 mm	40 – 65 mm

21. BRE Digest 475 *Tilt of low-rise buildings with particular reference to progressive foundation movement* gives guidance on acceptable tilt.

22. BS 8103 *Structural design of low-rise buildings* Part 1: 1995 *Code of practice for stability, site investigation, foundations and ground floor slabs for housing* lists site hazards which require investigation by a suitably qualified engineer. BRE Good Building Guide 39 *Simple foundations for low-rise housing* also discusses the circumstances where foundation design using empirical guidelines is not appropriate and a suitably qualified engineer should be consulted. It provides guidance on minimum foundation widths in relation to soil type and loading.

23. For preliminary design purposes, some typical values of allowable bearing pressure for a range of ground types are given are given in Table 1 of BS 8004: 1986.

24. A number of different methods of calculating the bearing capacity factors have been proposed and at large values of ϕ' the differences can be significant. For a discussion of this, see, for example, Lancellotta (1995), Budhu (2000).

25. BS 8004: 1986 *Code of Practice for foundations* uses the concept of allowable bearing pressure.

26. The use of geotechnical load and resistance factor design has been reviewed by Scott *et al* (1993). The new European standard, prEN 1997-1 *Geotechnical design – general rules* will promote the use of this design method.

27. Design practice in the UK has been described by Findlay *et al* (1997).

28. See, for example, Lancellotta (1995), Budhu (2000).

29. Skempton (1959), Fleming *et al* (1992).

30. Burland (1973), Meyerhof (1976), Findlay *et al* (1997).

31. In routine settlement calculations no account is taken of the complex interaction between the soil and the structure. Only expensive major constructions merit those calculations of settlement and soil-structure interaction which require advanced computer analysis.

32. Poulos and Davis (1974) have provided elastic stress distributions for many loading configurations.

33. The circular shape and rigidity factors listed in Table 6.1 can be derived quite simply from elastic theory. For the square and rectangular foundations approximate methods have to be used; the values in the Table have been derived from the following sources: particular values for square and rectangular – Davis and Selvadurai (1996); general relationship for rectangular flexible - Giroud (1970); general relationship for rectangular rigid - Gazetas *et al* (1985).

34. Gazetas *et al* (1985).

35. This method of calculating primary consolidation settlement, based on Skempton and Bjerrum (1957), is described in Budhu (2000) pp 347-350.

36. C_α is sometimes quoted in terms of void ratio rather than vertical strain, in which case equation 6.17 requires modification.

37. Meyerhof (1965); Burland and Burbidge (1985).

38. Schmertmann (1970) proposed a method for estimating the settlement of sands from CPT data. See Lunne *et al* (1997) for a review of such methods.

39. Burland and Burbidge (1985) based this equation on an analysis of a large number of records of settlement on sand and gravel. They also give a method of relating settlement to CPT results.

FOUNDATION DESIGN

40. It should be noted that the empirical relationship described in equation 6.18 indicates that settlement is proportional to $B^{0.7}$ whereas elastic theory predicts that settlement and B are linearly related.

41. Tomlinson (2001).

42. It has been suggested that for a group of n piles the amplification factor $R = n^{\omega}$ where the exponent ω is between 0.3 and 0.6. For a discussion of the deformation of pile groups, see Fleming *et al* (1992).

43. More detailed guidance is in the BRE Good Building Guide 53 *Foundations for low-rise building extensions*.

44. Hawkins (1998) has examined the engineering significance of sulfates in dark pyritic mudrocks. See also Nixon (1978). Materials at risk include concrete, mortars, metals, plastics and masonry. General guidance on the performance of building materials used in contaminated land has been given in BRE Report BR 255, and by Garvin *et al* (1999) and Environment Agency and NHBC (2000). Particular hazards are associated with sulfate attack on concrete are described in BRE Special Digest 1. See also Hawkins (1997).

45. Clause 3 of BS 1377: Part 3: 1990 (see also clause 4 for loss on ignition).

46. Clause 6 of BS 1377: Part 3: 1990.

47. Clause 5 of BS 1377: Part 3: 1990 (but see also BRE Special Digest 1).

48. Clause 9 of BS 1377: Part 3: 1990.

49. Clause 7 of BS 1377: Part 3: 1990.

50. Clause 10 of BS 1377: Part 3: 1990 for laboratory test; clause 5.1 of BS 1377: Part 9: 1990 for in-situ test.

51. Clause 11 of BS 1377: Part 3: 1990 for laboratory test; clause 5.2 of BS 1377: Part 9: 1990 for in-situ test.

CHAPTER 7: FOUNDATIONS ON DIFFICULT GROUND

The requirement in the design of foundations that building loads are safely transmitted through the foundations and in such a manner that the resulting ground movements do not impair the stability of the building, or cause significant damage to it, was discussed in Chapter 6. This chapter examines some particular problems for building on ground with the potential for movements of a magnitude that could damage a building. Large ground movements can occur due to the high compressibility of soft clay and organic soil under the weight of the building and due to processes that are unconnected with the weight of the building, such as volume changes in clay soils, collapse compression of fills, down-slope movements and the collapse of underground mines and cavities. Where the potential for large ground movements is identified, there are a number of possible approaches for building development. Deep foundations may prevent ground movements affecting the building, but services may be adversely affected. Raft foundations can prevent building distortion, although excessive tilt is still a hazard. It may be possible to design the building so that the effects of large ground movements are mitigated. The particular hazards presented by brownfield sites are described in Chapter 8. Ground treatment, which can reduce the potential for ground movement, is described in Chapter 9.

7.1 Introduction

It is important to identify those processes that could give rise to large ground movements. Such movements can occur due to the high compressibility of soft clay and organic soil under the weight of the building and also due to processes that are unconnected with the weight of the building, such as volume changes in clay soils, collapse compression of fills on wetting, down-slope movements and the collapse of underground mines and cavities. In the UK, large ground movements are most commonly associated with:

❒ volume changes in shrinkable clay soils;
❒ collapse compression on wetting of fill;
❒ collapse of underground mines and cavities.

Where the potential for large movement is recognised, there are several approaches for building development:

❒ avoid siting a building at this location, but this may have severe financial implications;
❒ use deep foundations which can prevent ground movement affecting the building, although services still may be adversely affected by the movements;
❒ implement measures to prevent the occurrence of large ground movements, such as the use of an appropriate form of ground treatment;
❒ mitigate the effect on the building of large ground movement by using stiff foundations;
❒ design the building so that the effects of large ground movements are mitigated.

FOUNDATIONS ON DIFFICULT GROUND

7.2 Soft natural ground
Soft natural ground includes alluvial clay and silt, and deposits of highly compressible organic matter such as peat. It is typically found in lowland areas, such as river estuaries and floodplains, and near streams, lakes and the sea. Groundwater levels are likely to be close to the ground surface and to fluctuate seasonally. Tidal effects can influence groundwater levels. Deposits of soft ground may vary considerably in both lateral extent and depth. Small localised soft spots may result from flooding, leaking drains and other human activities, and these can occur almost anywhere.

Soft ground is characterised by low undrained shear strength, high compressibility and low permeability. Ground with an undrained shear strength between 20 and 40 kPa is considered to be soft, and very soft where the shear strength is less than 20 kPa. Ground can be extremely compressible if organic material is present. It is important to determine the type, extent, variability and depth of soft ground, and the nature of underlying soils.

Most soft clay has a desiccated crust, typically 0.5 to 2 m deep, which is much stiffer than the underlying soil. A crust is formed by seasonal desiccation caused by evaporation and removal of moisture by plant roots. If this strong crust is left largely undisturbed by foundation construction, it can have a beneficial effect on shallow foundations. However, while exploiting a stiff crust to spread the load is potentially a most advantageous foundation design method, it is often difficult to achieve in practice. Regulatory guidelines are likely to preclude shallow depth foundations because of the requirements for frost protection and the hazard of clay shrinkage. In the desiccated zone, the ground is commonly not adequately sampled and tested.

Where foundation loads are high, soft ground may experience excessive compression, which may occur over months or years. Bearing failure, differential settlement and tilting, and disruption between a building and its services are significant risks that should be minimised. Foundation loads must either be spread across a larger surface area or transferred down to more competent ground. The degree of difficulty involved when building on soft ground depends on the type and size of structure, and the level of experience and expertise of those involved[1].

To limit compression and avoid bearing failure, bearing pressures can be reduced by the use of wide strip or raft foundations. The depth to which a strip foundation produces a significant increase in stress is shown in Figure 6.11. Stiffened edge beam rafts, which are robust and able to withstand small differential ground movements, are widely used on a range of variable ground conditions. While a raft will greatly reduce the bearing pressure, load will be applied to softer soil at greater depth as illustrated in Figure 7.1. Where the soil has a desiccated crust, the strip footing may be founded largely on ground which is stiffer than the ground supporting the raft. Where a site is underlain by a thin layer of peat,

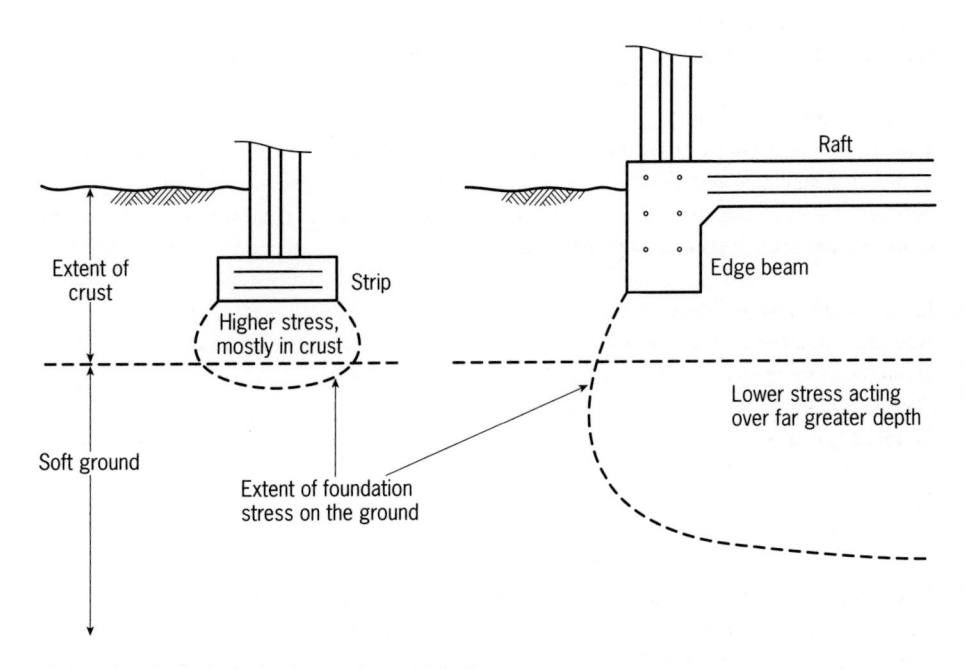

Figure 7.1 Ground stressed by strip and raft foundations

there may be no advantage in using a raft if the layer is deep enough to be outside the pressure bulb generated by strip footings. It has been common practice to apply an upper limit of 25 kPa for the allowable bearing pressure to ensure adequate performance by minimising settlements. However, this can only be regarded as a crude guide.

Where large amounts of differential settlement may occur due to a variation in thickness of a soft layer across the site, rafts can be provided with jacking points. Once most of the settlement is complete, the raft can be levelled by jacking against piers or piles; the resultant gap is then grouted. Another approach is to found on competent material beneath the soft ground using piles.

Ground treatment can be used in combination with shallow foundations. There are various methods that can be used to stiffen and strengthen the ground prior to building:
❐ consolidate the ground by installing vertical band drains and surcharging with fill;
❐ install vibrated stone columns (often termed *vibro*);
❐ install vibrated concrete columns;
❐ in-situ soil mixing;
❐ ex-situ soil mixing followed by replacement as engineered fill;
❐ reinforce with geotextiles.

FOUNDATIONS ON DIFFICULT GROUND

These treatment methods are described in Chapter 9. Replacing the soft ground with engineered fill is viable only with shallow depths and small areas of soft ground.

7.3 Loose natural ground

The bearing resistance of shallow foundations on coarse soil depends strongly on the width of the foundation. Where buildings are founded on loose sand and gravel, it may be necessary to use wide strip or raft foundations to obtain an adequate bearing resistance and to reduce the settlement to acceptable levels.

Loose, sandy soil may be subject to loss of strength and associated reduction in volume when subjected to dynamic loading. This can cause significant ground settlement. Dynamic loading can be associated with:
- earthquakes;
- traffic vibrations;
- machine foundation vibrations;
- pile driving;
- blasting.

When saturated sandy soil loses shear strength owing to an increase in pore pressure with a corresponding reduction in effective stress, the phenomenon is termed liquefaction. Density index I_D (Section 3.5) is an important parameter in determining resistance to liquefaction. The greater the density index, the greater the energy required for liquefaction and it is not likely to occur at all in dense soils owing to their tendency to dilate.

In very loose sand[2] (soils with an SPT N value less than 5) there is a particularly high risk of liquefaction under saturated conditions. Consequently, very loose sand should be treated prior to construction[3] (Chapter 9); alternatively, piles should be used to transfer the foundation loads to a deeper stratum.

Practical methods of evaluating liquefaction potential of sand deposits are of major concern in seismically active regions of the world. Since it is difficult to determine the in-situ density of sand in the field, density index and liquefaction potential are often inferred from penetration testing. In-situ test methods are described in Section 5.7. The UK is an area of low seismicity so the probability of earthquakes causing significant damage is relatively low[4].

7.4 Fill

An increasing proportion of building development in the United Kingdom takes place on made ground and some general features of fill materials have been described in Section 3.9. There are particular problems for foundations on fill[5] and it is helpful to compare the properties of fills with the properties of natural soils:

❑ *Nature of the material:* fill may be composed of the same type of material as a natural soil, such as clay, sand or rock, but it could be composed of something quite different, such as wastes from chemical and industrial processes.
❑ *Method of deposition:* fill may have been deposited in a manner that is quite similar to the deposition of some natural soils, such as slow sedimentation under water, but it may have been deposited in a quite different way, such as compaction with a heavy vibrating roller.
❑ *Age:* many fills have been placed recently so are much younger than most natural soils.

Fills exhibit a range of engineering properties as wide as that of natural soils. For example, there is a vast difference between the behaviour of an engineered heavily compacted sand-and-gravel fill and recently-placed domestic refuse. The engineering properties of hydraulic fill can be expected to be very different to fill placed at lower water contents in dry conditions.

Hazards associated with fill

Physical characteristics are generally the most significant in determining the load-carrying characteristics of a fill but, with some waste fills, chemical and biological properties and processes can be important. Hazards can include toxicity, carcinogenicity, corrosivity, combustibility, inflammability and explosiveness, and asphyxiation. The effects can be grouped under a number of headings:
❑ hazards for buildings from volume changes in fill caused by physical, chemical or biological reactions;
❑ hazards for construction materials, because of their vulnerability to aggressive ground conditions;
❑ hazards for buildings and occupants, arising from combustion and gas generation;
❑ health and safety hazards for people on site;
❑ environmental hazards.

Ground movements within fills which may require particular consideration are:
❑ volume reduction caused by collapse compression on wetting (Section 4.9);
❑ volume reduction caused by biodegradation (Section 8.4);
❑ volume expansion caused by chemical reactions (Section 8.4).

Foundation solutions

The difference between engineered and non-engineered fills is described in Section 3.9. Where an engineered fill has been placed under carefully controlled conditions, no special considerations may be needed because the loadbearing characteristics of well-compacted fill are likely to be superior to many naturally occurring soils. However, where a non-engineered fill has received little compaction, the allowable bearing pressure of the surface soil is likely to be limited by consideration of long term settlement of the fill. Furthermore,

a non-engineered fill may exhibit extreme heterogeneity. In such circumstances, there are essentially three options available to the foundation designer[6]:
- to use piers or piles to transfer the foundation loads to a deeper, more competent stratum;
- to use wide strip footings or a raft foundation to reduce the bearing pressure of the foundations, and to stiffen the structure against differential compression of the fill;
- to use a ground treatment technique to increase the strength and decrease the compressibility of the fill (Chapter 9).

Which option proves the most effective and economical will depend on the thickness of the fill and its compressibility and variability. Where the fill is variable in thickness or composition, footings may need reinforcement to help bridge over soft spots. Wide strip footings and rafts may be ineffective at reducing foundation movement on fill with a significant organic content, which is likely to settle appreciably as a result of biodegradation and self-weight compression. Pile and pier design in such soils must take into account negative skin friction (Section 6.6). The practicalities of ground improvement techniques depend on the properties and composition of the fill (Chapter 9).

Fill materials containing organic matter are likely to produce methane gas as they degrade so precautions in the form of natural or forced ventilation below ground floor slabs may be needed (Section 8.7)

Where sites have been previously built on, it may be necessary to remove old foundations and to infill carefully such features as basements and wells. For smaller features, it will be difficult to backfill under controlled conditions. The ground conditions across the site are likely to be variable and it may be necessary to reinforce strip footings to allow them to bridge soft spots. In some cases, it may be more practicable to use a suitably reinforced raft foundation. Alternatively, piles may be used to ensure that the new building is founded well below the influence of the previous foundations.

The problems associated with shallow foundations for low-rise buildings on made ground can be minimised by:
- avoiding building across the edges of filled areas where the structure would be partly founded on fill and partly on undisturbed natural ground[7];
- restricting construction to small units and not building long terraces of houses.

7.5 Shrinkable clay and the effect of trees
Trees can damage buildings and services by indirect or direct action. Indirect action means, in particular, problems associated with shrinking and swelling clay, which is the subject of this section. Direct action includes the growth of roots or the trunk lifting or distorting buildings and services, but this is a relatively uncommon cause of damage.

There has been a high incidence of subsidence damage of buildings on clay soils. The subject is of particular importance for the UK because of the concentration of population in the south-east of England where shrinkable clay is widespread[8]. A brief account of the shrinkage and swelling of clay is in Section 4.8.

The surface layer of a clay soil is inherently unstable because of its tendency to change in volume as a result of variations in water content. Evaporation and transpiration can cause significant ground movement that can damage buildings on shallow foundations. Most of the overconsolidated clays prevalent throughout south-east England can be classed as shrinkable; the distribution of these clays is shown in Figure 7.2[9]. Most glacial deposits are found north of the indicated limit. Table 7.1 describes the typical volume change potential for the strata shown in Figure 7.2 (see Table 4.4 for a categorisation of volume change potential).

Table 7.1 Typical volume change potential for some common clays

Clay	Volume change potential
Glacial till	Low
Kimmeridge clay	High
London clay	High to very high
Lower Lias clay	Medium
Gault clay	High to very high
Mercia mudstone	Low to medium
Oxford clay	High
Weald clay	High

Much heavily overconsolidated shrinkable clay near the surface is classified as firm, with typical allowable bearing pressures in the range 75 to 150 kPa. Bearing resistance is, therefore, unlikely to be an important consideration in designing a foundation for a low-rise building. The depth of the foundations is critical; foundations must be deep enough to be protected from the volume changes in the soil associated with the desiccation produced by evaporation and moisture extracted by vegetation[10]. Desiccation reduces with depth, but is enhanced by the presence of vegetation.

Ground shrinkage

Shrinkage and swelling can be estimated from compressibility data for saturated clay soil using the coefficient of volume compressibility m_v over an appropriate stress range, which can be measured in a standard oedometer test (Section 4.5). The bulk modulus K is related to m_v and Poisson's ratio υ as follows:

$$m_v = \frac{(1+\upsilon)}{[3K(1-\upsilon)]} \tag{7.1}$$

Using a value for υ of 0.2, this simplifies to $m_v = 1/2K$. The volume change ΔV resulting from desiccation can be assessed from the magnitude of the suction p_k and the bulk modulus or coefficient of compressibility of the soil:

$$\Delta V = p_k/K = 2\, p_k\, m_v \tag{7.2}$$

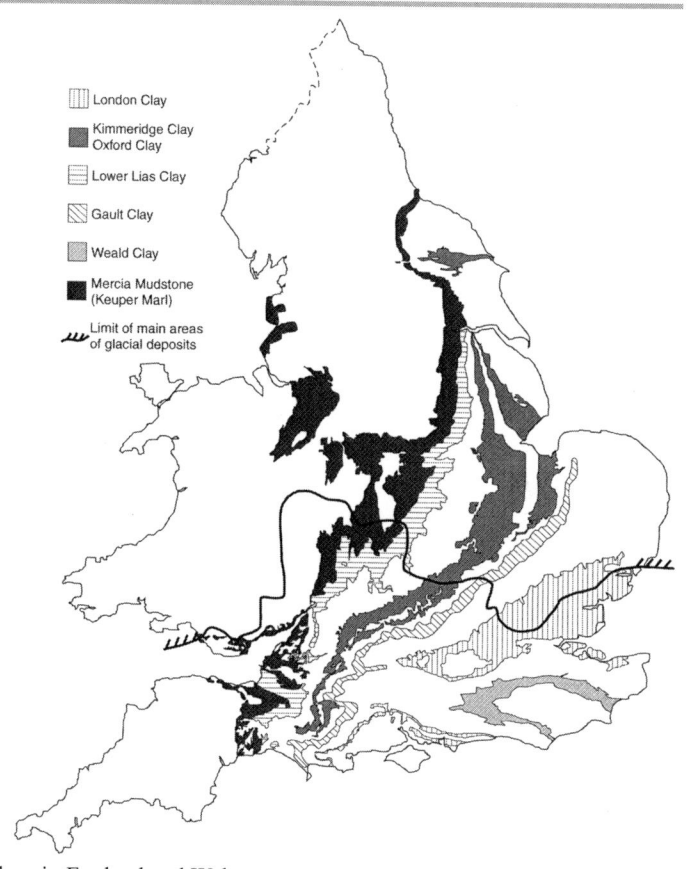

Figure 7.2 Shrinkable clays in England and Wales

Assuming that the soil remains saturated, which is valid only for relatively low levels of desiccation, this volume change can be related to the change in water content:

$$\Delta V = \frac{\rho_s(w_o - w_d)}{(1 + \rho_s w_d)} \qquad (7.3)$$

where:
    w_d is the gravimetric water content[11] of the desiccated soil
    w_o is the gravimetric water content of the undesiccated soil
    ρ_s is the particle density.

In equations (7.2) and (7.3), ΔV, w_o and w_d are expressed as decimal quantities and not as percentages.

Unlike the oedometer compression test, not all the volume change in the ground manifests itself as vertical strain. Moreover, high levels of desiccation cause clays to crack and desaturate, resulting in a significantly smaller volume change than might be anticipated from the measured change in water content or suction. An empirical factor, known as the water shrinkage factor S_f is, therefore, needed to relate vertical movements to volume change. An average value of $S_f = 4$ can be used for estimating the movement that occurs close to the ground surface[12].

Effect of trees
Desiccation is enhanced by the presence of vegetation, particularly large trees[13]. In grass-covered areas, the effects of evaporation and transpiration by light vegetation are largely confined to the upper 1.0 to 1.5 m in firm shrinkable clay soils. Where there are trees and, to a lesser extent hedges and large shrubs, moisture can be extracted from greater depths; a large oak tree can desiccate the soil to 6 m or more. The shrinkage typically produced by a large tree is shown in Box 7.1.

Box 7.1 Example of clay shrinkage
For a suction of 300 kPa associated with a large tree in London Clay with a coefficient of compressibility m_v of 0.1 m²/MN, the tree would be expected to cause an average reduction in volume of 6 %. For London Clay, ρ_s is typically 2.75 and $w_o = 30\%$; therefore a 6 % reduction in volume reduces the water content by about 4%.
A vertical movement of 90 mm at the ground surface will be associated with the large tree, which produces an average volume shrinkage of 6% to a depth of, say, 6 m.

High plasticity clay tends to have a very low permeability, so rainfall during winter months cannot fully replenish the moisture removed by large trees during the summer. A zone of permanently desiccated soil develops under the tree. As the tree grows, the desiccated zone increases in depth and width, producing more settlement, which is likely to affect nearby structures.

The extent of the desiccated zone depends on the moisture demand of the tree. In general, broad-leaf trees have a greater moisture demand than evergreens; oak, elm, willow and poplar are notorious for causing damage. However, trees with lower moisture demands, especially plane, lime and ash, are now more likely to be planted close to buildings and, therefore, will more frequently be the cause of damage. The degree of desiccation that a tree generates depends also on the availability of water in the ground[14]. This depends in turn on a number of factors, some of which, such as the permeability of the surface layer and drainage, can be controlled to some extent. It is impractical to control such factors as the slope of the ground and the shelter provided by the house and other nearby buildings.

The seasonal ground movements which have been monitored at a high plasticity clay site are described in Box 7.2. Figure 7.3 shows the vertical ground movements measured at various depths in open ground over a period of three years.

Box 7.2 Seasonal movements at a high plasticity site[15]

Measurements of ground movement at various depths at a clay site with very high shrinkage potential ($I_p > 50\%$) in a grass-covered area confirmed that ground movements were largely confined to the surface metre of soil, although unusually dry weather in 1989 and 1990 produced movements of 6 mm and 13 mm respectively at a depth of 1 m. Movements in an area 5 m away from large poplar trees were larger and, at a depth of 1 m, exceeded 35 mm even in an average year, such as 1988. Although water content and suction profiles indicate that the soil was desiccated to 4.5 m, very small movements were recorded at a depth of 2 m during 1988. In an average year, the desiccation between 2 m and 4.5 m was relatively insensitive to seasonal variations. To assess the potential effect of the ground movements on buildings, measurements were made on 1.5 m-deep concrete pads positioned 5 m from large poplars. Cumulative moisture losses produced a ratcheting effect, with settlements of more than 50 mm over a period of three years. These movements would be in addition to any long-term subsidence associated with the growth of the trees and would be capable of causing serious damage in most buildings, particularly if only part of the building were affected, maximising differential movements and, hence, distortion and damage.

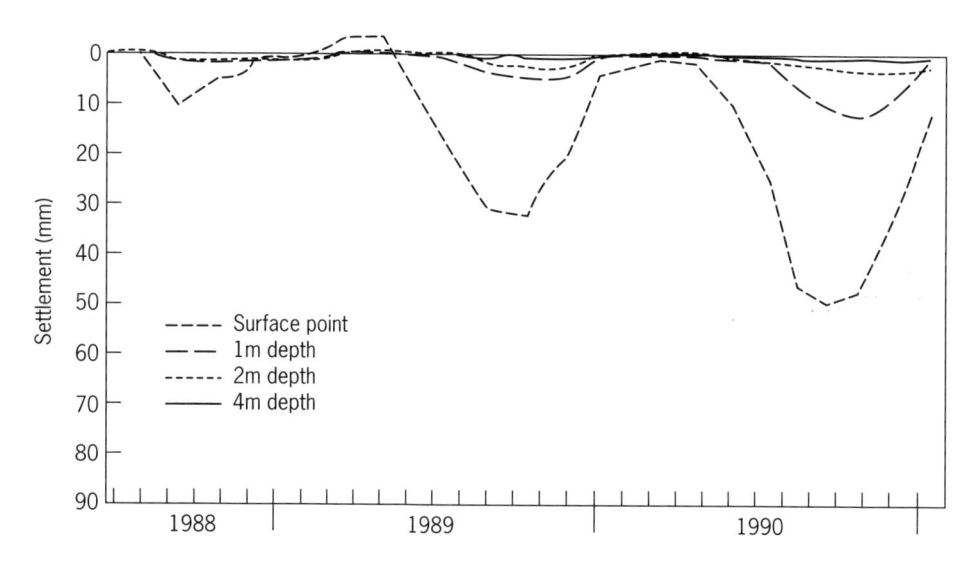

Figure 7.3 Ground movements in high plasticity clay – *after Freeman et al, 1992*

GEOTECHNICS FOR BUILDING PROFESSIONALS

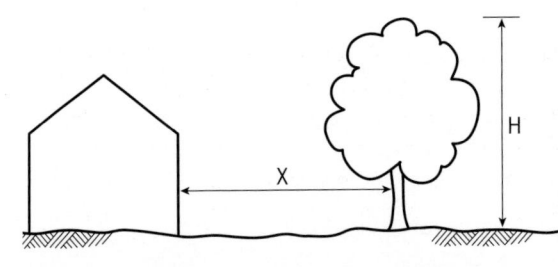

Figure 7.4 Safe distance between tree and house

One way to avoid problems is to maintain an adequate distance between the tree and the building (X) so that the tree cannot influence the soil beneath the building (Figure 7.4). It is generally considered that the safe distance is the expected maximum height (H) of the tree, but in many cases this is excessive. The trees most likely to cause damage are listed in Table 7.2; this gives the distance between each type of tree and the building within which 75% and 90% of the reported cases of damage occurred. For trees with a lower water demand, it has been suggested that the distance can be reduced to half the mature height. These recommendations take no account of the shrinkage potential of the soil or the fact that the leaf area of the tree, rather than its height, determines its water demand; they should, therefore, be treated with caution.

Table 7.2 Risk of damage by different varieties of tree in shrinkable clay[16]
– in descending order of threat

Species	Maximum height (m) of tree	Distance between tree and building (m) 75 % of cases of damage	90 % of cases of damage
Oak	16 – 23	13	18
Poplar	24	15	20
Ash	23	10	13
Elm	20 – 25	12	19
Horse chestnut	16 – 25	10	15
Hawthorn	10	7	9
Lime	16 – 24	8	11
Willow	15	11	18
Beech	20	9	11
Plane	25 – 30	8	10
Apple, Pear	8 – 12	6	8
Maple, Sycamore	17 – 24	9	12
Cherry, Plum	6 – 12	6	8
Birch	12 – 14	7	8
Cypress	18 – 25	4	5
Rowan	8 – 12	7	9

Figure 4.11 illustrates the deep zone of desiccated soil that a tree can produce as it reaches maturity. When the tree is felled or dies, or excavation for new foundations severs a substantial part of the root system, the removal of water ceases and the suction in the desiccated soil begins to draw moisture from undesiccated clay at the extremity of the zone and from any free water entering from cracks above. The low permeability of very plastic clay makes the re-hydration extremely slow[17].

FOUNDATIONS ON DIFFICULT GROUND

Foundation solutions

Where there are no trees or large shrubs, a foundation depth of 0.9 m is usually adequate to prevent damage to a conventional masonry structure on a shrinkable clay[18]. However, foundations designed to this minimum make no provision for future tree planting. Many home owners will want to plant small trees near the house so it is prudent to provide foundations that are somewhat deeper than the recommended minimum. Trench-fill foundations are normally used if a foundation depth of between 0.9 and 1.5 m is required.

It may be necessary to use deeper foundations where there are large, broad-leaf trees, such as oak, willow, elm or poplar, or where it is known that such large trees will grow[19]. The depths of desiccation under a large tree can be considerably greater than the recommended foundation depth; in London clay, for example, desiccation to depths of 6 m is not unusual. Nevertheless, the desiccation in the deeper soil is largely unaffected by seasonal variations and is fairly constant unless the tree grows significantly or is removed. Consequently, the depth of the foundations could be substantially less than the depth of desiccation and still provide adequate stability for the building. Large trees should be left in place but this is not possible where the tree is on, or very close to, the proposed site for the building. In such circumstances, a piled foundation should be considered; it can be designed to offer a far higher margin of safety against movement and is often no more expensive than deep trench-fill.

BRE has long advocated the use of pile and beam foundations on clay sites near major vegetation[20]. However, some builders are unwilling to install this type of foundation, preferring to use a deep trench-fill foundation. This can present problems during construction and, more importantly, subsequently.
❒ Construction difficulties include the large quantities of soil to be removed from the foundation excavation and the risk of trench collapse.
❒ The pressure exerted by a soil that is trying to swell can be very large if the clay is initially very dry, but this high pressure is exerted only where there is sufficient resistance to prevent swelling. Even with the deep trench-fill foundation in Figure 7.5, there is unlikely to be much horizontal movement of the foundation. Where substantial swelling is likely, it is advisable to limit the swelling pressure exerted on the foundation by using a sheet of compressible material, such as low-density expanded polystyrene, between the foundation and the clay[21].
❒ In heavily desiccated soil, a deep trench-fill foundation inevitably upsets the equilibrium in the soil by cutting through tree roots, even if no trees are removed. Lateral movement could occur in the situation in Figure 7.6 where deep trench-fill foundations have severed the roots of trees that are outside the footprint of the building. A sheet of compressible material should protect against lateral displacement of the foundation.
❒ In swelling soils, deep trench-fill foundations are vulnerable to vertical uplift stresses generated on the large surface areas of the foundation which are in contact with the

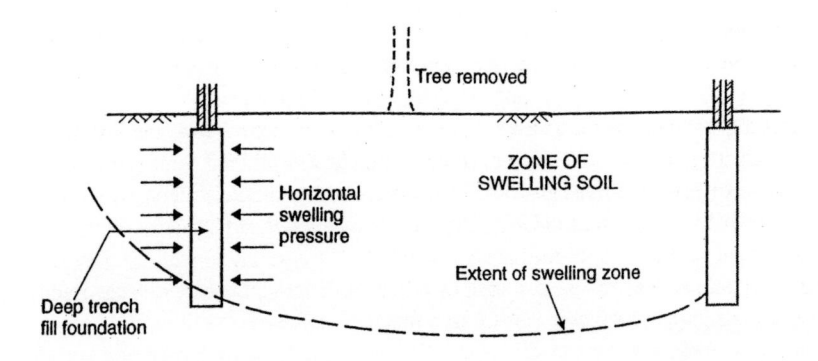

Figure 7.5 Horizontal pressure on trench-fill foundation in swelling soil

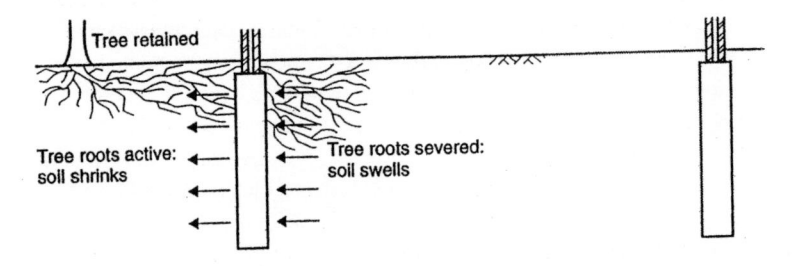

Figure 7.6 Horizontal swelling caused by severance of tree roots

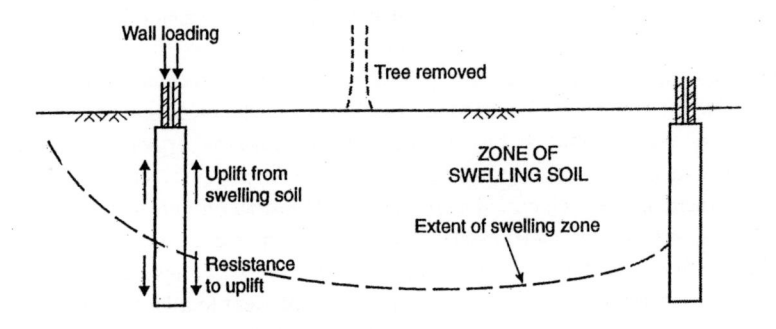

Figure 7.7 Uplift stresses on trench-fill foundation in swelling soil

ground. This is shown in Figure 7.7. Sometimes a slip surface (such as a plastics sheet) is installed down the trench sides to try to reduce these uplift stresses.

Piled foundations are less vulnerable to lateral movement than trench-fill foundations and their installation has a far smaller effect on the equilibrium in the soil. Cast-in-place bored piles with diameters of 250 to 500 mm are used for many domestic applications. The piles are positioned at corners and at other critical locations, such as each side of a door opening. Load-carrying capacities seldom exceed 150 kN; this is small compared to most large building applications. The required length of pile is correspondingly less; piles used for houses are commonly referred to as short bored piles.

In heavily desiccated soil, total load-carrying capacity is seldom the critical factor determining the required length of pile. The pile should be long enough to provide an adequate anchorage in underlying soil that is either undesiccated or where the desiccation remains practically constant. The resistance on the lower half of the pile should be capable of resisting the forces acting on the upper half; the length of the pile is, therefore, likely to be twice the depth recommended for trench-fill foundations. Detailed design may have to take account of the uplift generated by swelling soil and/or the downdrag generated by shrinking soil. Sleeving materials can be used to reduce these forces and consequently reduce the required pile length. It may be preferable to increase the anchorage length of the pile rather than specify sleeving that may be difficult to install and whose performance may be uncertain.

Raft foundations are sometimes used for buildings founded on shrinkable clay. The raft is normally cast onto a bed of compacted hardcore or other suitable granular material that replaces a layer of excavated shrinkable material[22].

Since floor slabs are susceptible to damage by clay shrinkage and swelling, it is prudent to use a suspended floor with an adequate void[23] under it, rather than a ground-bearing slab.

Because evaporation and transpiration depend primarily on temperature, wind and sunlight, they tend to be greatest on the south side of buildings. Therefore clay shrinkage is likely to cause the building to tilt slightly towards the south during summer months. The differential movements are likely to be greatest where the south side of the building is exposed and the remaining walls receive shelter from neighbouring buildings or from garden walls. The effects of evaporation are largely limited to the surface metre of soil and it is unusual for this tilt to cause any noticeable damage, even for buildings on very shallow foundations. Occasionally the movements may be sufficient to affect a particularly vulnerable structure, such as the bay windows to the front of Victorian houses, which were often built on shallower foundations than the rest of the house

No increase in foundation depth is required where buildings are to be sited near trees on non-shrinkable soils, such as sand, gravel and chalk. However, special care may be needed to prevent the excavation for the foundations severing too many roots and thereby threatening the health and stability of the tree. It may, for example, be necessary to hand-excavate trenches and retain any roots more than 50 mm diameter. These roots should then be 'boxed out' before casting the foundations. Damaged roots should be cut back to a smooth surface and treated with a fungicidal sealant[24].

Drainage and run-off
Ground that slopes steeply downwards away from foundations tends to increase desiccation, by increasing run-off and lowering the ground water-table locally. Conversely, ground that slopes downwards towards foundations tends to reduce desiccation by increasing the supply of water to the soil under the foundations.

Relatively impermeable coverings, such as asphalt or concrete, may help to reduce evaporation but will also encourage run-off; this may, in turn, force trees and other large vegetation to extend their root systems to maintain a supply of moisture. Consequently, where there is shrinkable clay, consideration should be given to the long-term changes in desiccation and associated ground movements that might occur as a result of surrounding large trees with extensive areas of concrete or asphalt. Where there is a risk of causing damage to nearby buildings, consideration should be given to using more permeable alternatives, or to grading the coverings so that the water is allowed to enter the ground where it has the most benefit.

Improving surface water drainage can reduce the supply of water to established trees and thereby trigger long-term movements as the equilibrium in the ground is adjusted. Where a grassed area is being replaced by a patio or similar feature, a soakaway may be provided in a position where existing trees can exploit the supply of water rather than channelling the run-off into a drain. It may be beneficial to improve the supply of water to the soil under the foundations by grading the patio towards a flowerbed or gravel surround to the house, but this may encourage the growth of roots in this area, creating problems if there is a prolonged period of dry weather.

7.6 Unstable slopes
Slope instability can cause ground movements that are a threat to buildings. Unstable slopes are found in many different types of ground and there are many ways that a slope can fail. Where the unstable slope is very long compared to its height, the failure surface is often parallel to the ground surface at a depth where the strength of the surface soil or rock has been weakened by weathering. However, where the slope is of limited length, the failure surface tends to be circular (Figure 7.8a) unless there are existing planes of weakness or less competent strata in the ground; these will influence both the location and

FOUNDATIONS ON DIFFICULT GROUND

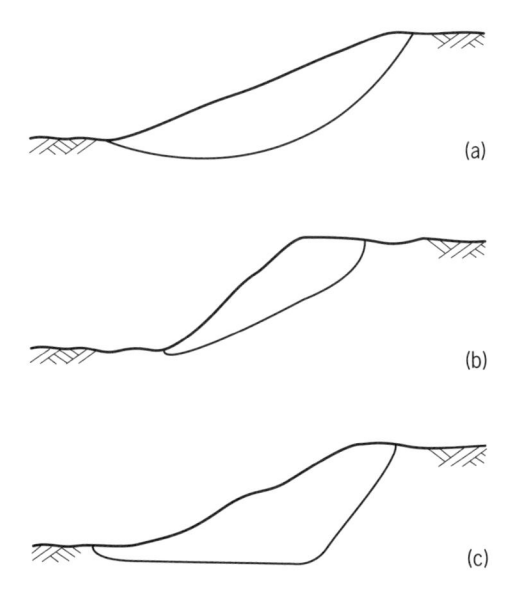

Figure 7.8 Slope instability
    (a) circular slip surface in homogenous ground
    (b) non-circular slip surface with dipping strata
    (c) non-circular slip surface with horizontal strata

the shape of the failure surface (Figure 7.8b and 7.8c). In rock, the failure surface tends to follow pre-existing defects, such as bedding planes, joints and faults and, where soil overlies a sloping rock surface, there is a tendency for the interface between the two materials to form the slip surface.

Hazards posed by unstable slopes
Many natural and man-made slopes exist in a state of marginal stability: down-slope movements of the ground can be triggered by a relatively minor change of circumstances. Instability of this kind, normally referred to as landslide or landslip, may result either in a sudden movement of a large mass of soil or rock, or in large ground strains. Any building in the path of the movement is likely to act as a retaining structure, albeit inadequately, unless it has been specifically designed for this purpose.

Coastal landslips can pose a threat to houses, particularly in parts of the country where the coastline is heavily populated[25]. Coastal protection is a major and, in some locations, a contentious subject.

It is rare in the UK for low-rise properties to be affected by large, irresistible landslides. A far more common cause of damage is the insidious process of slope creep: a phenomenon that affects the surface metre or two of certain natural clay slopes, which migrate slowly down the hill over long time scales. It is possible that, rather than being a continuous slow activity, the movements may be triggered by certain climatic conditions and may, therefore, occur intermittently. One possibility is that slope creep is related to exceptional desiccation during prolonged periods of dry weather; in particular, the down-slope movements may be produced by the incomplete closure of desiccation cracks, which would explain why only the surface layer of soil is affected. In London clay, slope creep can occur on slopes of 7°.

In addition to causing long-term damage to houses and other structures on shallow foundations, slope creep produces noticeable ripples in roads and other asphalt surfaces. However, as the soils that experience slope creep tend to be shrinkable clays, the effects are often difficult to distinguish from clay shrinkage.

The cost of stabilising a creeping slope is likely to be prohibitive and, if such areas cannot be avoided, an appropriate foundation design for the houses or other buildings is required which will take account of the movements. This can be done in one of two ways:
❒ using deep foundations to anchor the structure in the underlying stable soil;
❒ using a raft to allow the whole structure to move as a single unit with provision for correcting its level if necessary (Chapter 11).

Causes of slope instability
Movement occurs in a slope when the total disturbing force acting on a block of soil or rock exceeds the total resistance in the ground. The disturbing forces can include the weight of the block, seepage forces, seismic forces and superimposed loads, while the resisting force is the shear strength of the ground acting over the total area of the potential failure surface. Slope instability commonly occurs as a result of a reduction in the resisting force, which can be brought about in a variety of ways (Box 7.3).

The weight imposed on a slope by buildings is normally insignificant when compared to the weight of the soil mass and is therefore unlikely to cause more than a localised failure. However, excavation of foundations and basements near the toe or crest of a slope may reduce the area of the potential sliding surface (Box 7.4).

Where the modifications to stability induced by a triggering event are insufficient to cause an instantaneous failure, substantial movements may occur several years later. The most common cause of failure is an influx of water, which tends to increase the total weight of the soil mass, reduce effective stresses and increase the seepage force. Special care is needed in the design and maintenance of drainage systems for buildings sited on

Box 7.3 Mechanisms which reduce slope stability
Slope instability can be inadvertently reduced by:
- an increase in pore water pressure which reduces the effective stress on the slip surface;
- excavation at the toe of the slope which reduces the area of the potential failure surface;
- subsurface erosion by groundwater flow;
- weathering of the surface layer of soil or rock, leading to a reduction in strength;
- progressive failure in which recurrent small strains lead to a long-term reduction in strength; this is particularly relevant to overconsolidated clay where the reduction in strength from peak to residual can be large;
- leaching of natural cementing material into the ground by seepage water; this is particularly relevant to some soft clays;
- removing vegetation and killing the roots which have been binding the surface soil;
- high frequency vibration causing a loss of strength in loose granular soil owing to a build-up of pore water pressure.

Box 7.4 Building activities that can affect the stability of the slope
The stability of a slope may be inadvertently reduced during building operations:
- cut-and-fill construction techniques, which are commonly used for low-rise construction on sloping ground, can change the stresses in the soil and result in a local steepening of the slope;
- adding fill material to the top of a slope can lead to instability of the slope, as well as the fill itself;
- adding fill material to the toe of a slope and/or removing it from the top of a slope usually reduces the likelihood of instability;
- construction of retaining walls may interrupt existing drainage.

potentially unstable slopes, particularly with regard to stormwater drainage. Soakaways should not be used on sloping ground if there is even a small likelihood of instability. Special care is also needed where sewers, drains and water supply pipes cross areas where there is evidence of past movement or a likelihood of even small movements in the future. In such circumstances, flexible joints are essential to eliminate the possibility of a cracked pipe leaking water directly into the ground near the potential failure surface.

Identifying slope instability
The general procedures for identifying unstable ground during the site investigation are described in Chapter 5. Signs of instability include cracked or hummocky surfaces, crescent-shaped depressions, crooked fences and hedgerows, bogs or wet ground in elevated positions, plants normally associated with wet ground (for example, reeds and bullrushes) growing on a slope, and water seeping from the ground. Trees disturbed by shallow movements early in their life often end up with a characteristic bend in their trunks close to the ground surface, as the tree continues to grow vertically. In urban areas, the effect of unstable slopes is reflected in uneven road surfaces, leaning lamp-posts and damage to buildings. Local knowledge can help identify when the movements occurred. Slope angles should be measured and compared to existing slopes in the same soil which are known to be stable, or to calculated or documented values for marginal stability.

Calculating the stability of slopes is normally carried out using well-proven methods embodied in commercially available computer programs[26]. The profile of strength with depth has to be determined at representative locations. Ideally, effective stress parameters should be determined using the techniques described in Section 4.6 but, in some cases, the cost of the necessary sampling and testing programme may be prohibitive. Reliance then has to be placed on qualitative assessment and precedent to determine whether a given slope requires stabilisation. Boreholes drilled to obtain soil samples should be carefully backfilled to avoid creating damaging seepage paths.

Slope stabilisation
It is usually more practicable to take appropriate measures to improve the stability of a marginal slope rather than design foundations to withstand or accommodate the down-slope movements. Where slope stabilisation is impracticable, two options might be explored:
- the building could be founded on a rigid raft so that the entire structure moves as a single unit;
- where the slip surface is relatively shallow, the building could be founded on piles designed to withstand the lateral forces generated by the sliding soil.

Slope stabilisation is a specialised subject that requires geotechnical engineers with relevant knowledge and experience[27]. As well as taking steps to improve the stability of marginal slopes, care should be taken in the design of the building, including its foundations, services, drives, pathways and any landscaping, to ensure that the effect on slope stability is minimised.

Where building on unstable slopes cannot be avoided, there are a number of techniques that can be used to improve the stability of the slope; they include:
- regrading;
- use of anchors;
- soil nailing;
- construction of a retaining wall;
- grout injection;
- planting vegetation;
- installing drainage.

Often the most effective way of improving the stability of a slope is to regrade it using cut-and-fill techniques but the practicality of this approach is likely to be limited in areas with existing buildings. There are three options:
- grade to a uniform, flatter angle;
- concentrate on filling at the toe of the slope, creating a step or berm in the cross-section;
- reduce the overall slope height while keeping the profile unchanged.

Reducing the overall slope angle tends to be the most effective solution where the potential failure plane is near the surface. Adding material to the toe or removing material from the crest tend to be more effective for resisting deep-seated failure. The choice between a cut or a fill solution is normally constrained by the need to maintain either the crest or the toe at a specified position; where this is not the case, a combination of cut-and-fill is normally adopted to obviate the need to import or remove material. Cut slopes can normally be left a little steeper than filled slopes, reflecting the greater strength of the material in its undisturbed condition. Fill slopes can be steepened by importing a suitable material. For natural slopes with irregular and complex slip surfaces, cut-and-fill solutions are complicated by the risk that placing fill in one area may reduce stability elsewhere.

The stability of rock slopes can be improved by anchoring the unstable surface to the ground below the slip surface. The design of an anchoring system is a complex process, requiring the estimation of the total force that is needed to stabilise the slope. There are two types of anchors:
- unstressed anchors, which rely on a dowelling action to increase the resistance to sliding;
- stressed anchors[28], which increase the effective stress acting on the slip surface and can also offer a component to counterbalance the disturbing force.

In the last few years in the UK there has been an increased use of soil nails to stabilise embankments, cuttings and natural slopes[29]. Steel rods are inserted into the slope to reinforce it.

Retaining walls can be used to stabilise slopes by transferring load from the unstable soil or rock to the underlying undisturbed ground. The forces and moments which act on the wall can be very large and can be calculated only if a full stability analysis is performed incorporating the properties and stress conditions in the ground. Unless the wall is supported by prestressed anchors, a certain amount of deformation will be needed to mobilise its resistance. Sheet pile walls may be used as a temporary expedient, but in the majority of cases their flexural resistance and stiffness are too small to provide permanent support. They have the advantage that, unlike concrete walls, their resistance is mobilised immediately upon insertion.

Injecting cement and chemical grouts into the ground has been successful in arresting movement in highway and railway slips where there are significant voids or zones of very weak ground into which the grout can penetrate. The benefit of grouting clay soil is less obvious[30]. There is a risk in permeable soils that grout may block natural drainage pathways and, in urban areas, may intrude into buried pipes; these possibilities should be carefully considered before grout is used.

Vegetation can contribute to both stability and instability. For example, grass grown on suitably prepared slopes encourages run-off during storms, helps prevent scour, and physically binds the soil together. However, a layer of topsoil bound together by grass can itself be unstable and can slide like a carpet on underlying gravel or weathered rock. Although the physical strength of roots is not generally a key factor, the desiccation caused by shrubs and trees tends to improve stability by reducing the water content and increasing the strength of the surface soil. However, trees may be a liability when they become very large as their weight and the action of wind on them contribute to localised instability. Moreover, in jointed rock, the roots may open up the joints, allowing water to flow in and eventually detach blocks.

Monitoring slope stability
Slope stabilisation is expensive and, in cases where there is no risk of a sudden irresistible movement, it is often preferable to monitor the rate of down-slope movement to determine the likelihood of the movements causing unacceptable damage to buildings before deciding on the scope of remedial work[31]. A slip indicator, described in Box 7.5, provides a cheap and simple means of detecting slope instability; a more satisfactory, though expensive method of monitoring down-slope movements, is to use an inclinometer[32].

Information is sometimes required on the pore water pressure in the ground, for example to enable a full stability analysis to be performed or to determine the rate of consolidation under a surcharge. Measurements of this kind can be made with piezometers. The simplest piezometer is the Casagrande standpipe type, which consists of a plastic or metal tube with a porous tip[33].

While the standpipe piezometer is adequate for measuring slow changes in groundwater level, it cannot be used to measure more rapid changes in pore water pressure and, in general, is not suited to remote or automated reading. There are, therefore, a number of alternative types of piezometer, which have more rapid response times and which can be read remotely. These include electrical, hydraulic, pneumatic and vibrating wire (acoustic) piezometers[34].

7.7 Underground mines and cavities.
The collapse of natural cavities and old mine workings can cause sudden ground movements without warning; this can have a severe affect on nearby buildings[35].

> **Box 7.5 Slip indicator**
> The simplest way of detecting down-slope movement is to use slip indicator; it consists of a small-diameter tube grouted into a pre-drilled hole. Any movement causes the tube to kink; the depth of the slip surface can then be determined by lowering a weight down the tube until it meets the obstruction. A second weight attached to a thin cord can be left down the hole below the slip surface to measure the thickness of the shear zone. The slip indicator may fail to detect small movements and gives no measure of the amount of movement that has occurred.

FOUNDATIONS ON DIFFICULT GROUND

Hazards posed by underground mines

There are a very large number of old mine workings in the UK, many of which are uncharted. The majority of these are the legacy of coal extraction during the 17th, 18th and 19th centuries; other resources, such as silver, lead, tin, iron ore, flints, limestone, shale, salt and refractory clays have also been mined at various times. Many of the coal seams that were worked during the industrial revolution now underlie urban areas.

Many old mines were left with pillars supporting the roof of the overlying strata, a technique referred to by such terms as pillar-and-stall and room-and-pillar mining. At depths greater than about 60 m, the pillars tended to collapse soon after abandonment; at shallower depths, however, the overlying strata can be supported for several hundred years, until gradual degradation, and the erosion of infill material by groundwater, leads to failure of the pillars, producing a general subsidence of the ground surface. At the same time, roof strata between pillars can collapse producing a larger dome; this in turn collapses, allowing the void to migrate upwards. This process continues until the void reaches the ground surface creating a crown hole, unless the collapsed material, which bulks as it breaks up, completely fills the void or the dome reaches a stable and competent stratum. The likelihood of the void reaching the surface reduces with increasing amounts of cover, but is also dependent on the nature of the overlying soil. Loose, water-bearing coarse deposits, such as gravel and sand, may accentuate the ground disturbance, while stiff cohesive deposits, such as glacial till, are likely to provide an arching action.

Room-and-pillar workings, particularly the deeper ones, are prone to collapse at any time, producing large and sudden deformations at the surface, which can have severe effects on local buildings. The collapse may occur as a single event or sporadically over a period of time and is not necessarily associated with the application of foundation loads, although construction activity, particularly piling, can promote the conditions that lead to instability. Rules based on experience and practice are sometimes used to assess the likelihood of a mine collapse causing surface subsidence, but they are of limited applicability[36].

A large number of shafts are associated with old mine workings. Many of these shafts were loosely backfilled and covered, making their location and identification difficult. The modern practice is to seal redundant shafts at the surface of the bedrock using a reinforced concrete slab with a thickness at least twice the diameter of the shaft. In older shafts, the seal usually consisted of a timber platform covered with uncompacted fill materials to restore the general ground level. Sometimes, the shaft was sealed using whatever materials were to hand. Even where the platform remains intact, there is still a possibility that failure may occur in the supporting shaft lining or surrounding rock. The collapse of the shaft or the gradual decay of a timber platform can cause a sudden failure, resulting in the appearance of a crater at ground level. Where rockhead is near the surface and the collapse arises from the failure of the platform supporting the fill material, the resulting subsidence is likely to be severe but localised. Where the collapse has been caused by a breakdown of the shaft lining, or where the shaft is covered by a thick deposit of unconsolidated material, the subsidence may be less severe but will affect a larger area.

Bell pits, which date back to the 13th century, consist typically of an unsupported shaft 1.0 m to 1.2 m in diameter widened to a diameter of up to 20 m at its base. Pits were sunk as often as required to exploit a seam. The depth was generally limited to about 12 m, although some later pits were as deep as 30 m. With the advent of mechanical pumps and winding gear in the 18th century, deeper working became practicable, and mines tended to consist of one or two vertical shafts providing access and ventilation to a number of horizontal or near-horizontal galleries. The collapse of abandoned mine shafts or bell pits can cause problems similar to those created by the collapse of solution features.

Recent deep mining of coal in the UK consisted almost entirely of the longwall method, where the coal seam is removed completely over a wide front and the mine then collapsed deliberately. Longwall mining produces a wave of subsidence; its characteristics are well established and largely predictable. Damage is often associated with the lateral strains generated in the surface soil, rather than a loss of support; any building in the path of this wave experiences first a tensile force at foundation level followed, as the mine advances, by a compressive force. The surface movements associated with longwall mining are generally complete 12 to 18 months after the advancing face has passed; it is, therefore, more economic to carry out permanent repairs once the movements have stopped, rather than attempting to design foundations which provide complete protection[37].

Hazards posed by natural cavities

The principal cause of natural cavities in the UK is gradual dissolution by percolating groundwater, a process which affects such carbonates as chalk and limestone. The cavities produced by this process are referred to as solution features and are typically caves or vertical pipes; they may subsequently have been filled with loose soil. Collapse of the infill material leading to sudden movement of the ground surface can be triggered by the application of foundation loads or, more commonly, by a sudden influx of water, for example, following a period of heavy rainfall.

Badly sited soakaways, broken drains or fractured water supply pipes may aggravate existing solution features and are a common cause of subsidence damage for buildings founded on chalk or limestone. The risks are greatest where the chalk or limestone is overlain by a thin layer of relatively permeable sediment, such as sand. In these circumstances, the enlargement of the solution feature is likely to go unnoticed until it reaches a size where the material above it can no longer arch across the hole and it collapses creating a crater, or sinkhole, several metres in diameter.

Identifying the hazard

The general procedures for identifying unstable ground during the site investigation are described in Chapter 5. Geological maps can be used to identify outcrops of chalk and limestone, and the existence of solution features may be acknowledged in the relevant sheet memoirs. They can often be identified on black-and-white aerial photographs, where the surface depression appears as a light-toned feature.

Geological maps can be used to identify strata which contain coal and other exploitable resources, and the position of old mine shafts are sometimes marked on contemporary topographical maps (Section 5.3). Unfortunately, there are no records for many of the older mines and, even where records exist, they may not correctly represent the room-and-pillar geometry, which has a direct bearing on the stability of the workings[38].

Comparison of bench marks on sequential editions of Ordnance Survey maps may enable the degree and extent of any subsidence to be estimated. Aerial photographs can be useful in locating surface depressions and other irregularities which may indicate that some unusual form of surface movement has occurred.

Solution features are often marked by surface depressions and streams that disappear into the ground. Signs of mining activity include old mine buildings or winding gear, surface depressions and spoil heaps. In urban areas, the effects of mining subsidence are also likely to be apparent in the damage to houses and other structures.

There have been recent advances in the use of geophysical techniques such as ground radar for locating underground cavities.

Although relatively expensive, boreholes offer the most reliable way of locating and mapping man-made cavities; they are essential if it is planned to consolidate any voids. They also provide useful information on the nature and characteristics of the overlying ground. The extent of the investigation is likely to be determined by the nature of the ground conditions, the level of existing knowledge and the importance of the structures that are to be protected.

Where there is little previous knowledge, a number of key boreholes are needed to identify the geological sequence and to identify those strata which are likely to be of particular interest. Where the presence of old room-and-pillar mines is suspected, it is generally prudent to continue these boreholes to a depth of 60 m to ensure that all strata containing uncollapsed workings are intercepted. The extent of the investigation should be greater than the plan area of the buildings to be protected and it is usual to position boreholes to a distance beyond the location of any proposed buildings equal to one-third the depth of the deepest workings. Rotary or cable-percussion techniques permit undisturbed sampling of the strata penetrated. Typically, rotary core drilling using water flush is used to obtain cores about 75 mm in diameter, which can then be examined to identify the condition and extent of the workings as well as the presence of any voids in the overlying material.

Once the key boreholes have been completed, a series of proving boreholes are excavated on a regular grid to determine the depth and thickness of the workings; to minimise cost, these boreholes are usually drilled without core recovery and the driller depends instead on an examination of the returns, the rate of penetration and the feel of the drill bit.

Deeper boreholes are likely to be needed where it is planned to locate and fill an abandoned shaft, which could be 200 m deep. To search for shafts, boreholes are usually drilled in a spiral sequence starting at the estimated location; this technique can be more cost effective than simply using a grid pattern. In open ground, a series of deep trenches can be excavated, and this is often a more cost-effective option.

During current longwall mining operations, surface movements can be monitored by periodic surveys using optical levelling techniques; associated damage to buildings can be monitored using the techniques described in Chapter 10. Similar techniques can be applied to the movements caused by the collapse of natural cavities and old mine workings; these movements are seldom anticipated and the primary purpose of any monitoring is, therefore, restricted to establishing whether the movement has stabilised. Detailed studies of the deformations that occur in the ground as a result of mine collapse are rare.

Approach to building development
Areas thought to contain solution features or old mine workings should be properly surveyed to locate any cavities. The positions of the proposed buildings should avoid any cavities that have been identified.

A number of general principles can be adopted to help reduce the level of damage[39]:
❐ A shallow plane raft foundation laid on a slip membrane and sand cushion is the best method of protection against tension or compression strains in the ground surface caused by future longwall mining.
❐ Structures should be completely rigid or completely flexible. Flexible superstructures should be used whenever possible.
❐ Long blocks should be divided into units separated by movement joints.
❐ Small buildings should be kept separate from one another, avoiding linkage by connecting wing walls or outbuildings.

Foundation solutions
Where construction in areas known to contain underground cavities is unavoidable, appropriate foundation design should be adopted to minimise the susceptibility of the

FOUNDATIONS ON DIFFICULT GROUND

structure to movement. In certain circumstances, for example where a major development has to be protected, it may be cost effective to stabilise or consolidate the cavities in order to reduce to acceptable levels the potential for future movement. The cavities should be filled using the techniques described in Section 9.8. The most common solution is to fill the workings with a pfa:cement-based grout via boreholes.

Where filling the cavities is impracticable, buildings can be founded on piles taken down to a stable stratum below the level of the mining activity or on a heavy raft foundation which is capable of bridging over a local collapse. If the former solution is adopted, the piles must be protected from the potential downdrag forces that would be generated by a subsidence of the overburden supported by the mine workings. This can be achieved by providing a space between the shaft of the pile and the overburden filled with a suitable low friction material, such as bitumen. Piled foundations are not suitable for areas where mining subsidence causes lateral ground movements that can either shear the piles or cause tensile failure of the ground beams connecting the piles.

Raft foundations should be as shallow as possible to limit the horizontal stresses produced by lateral ground movements acting on the edge of the raft. The raft should also be cast onto a membrane to reduce the friction on the underside of the raft. For similar reasons, where strip footings are considered adequate, they should be cast onto a layer of sand with provision at the ends of the foundation trenches for longitudinal movement.

Notes

1. General information on the problems with soft ground is in BRE Digest 471 *Low-rise building foundations on soft ground*.

2. Sandy materials are most susceptible, while an increased fines content reduces the tendency to liquefaction. The particle size distribution for the most easily liquefiable soils is between 0.07 mm and 0.6 mm and the boundary for potentially liquefiable soils is sometimes quoted as between 0.02 mm and 2.0 mm.

3. When ground improvement is required prior to construction, it is sensible to select an improvement technique that is closely related to the perceived hazard. If liquefaction under dynamic loading is the major hazard, an improvement technique involving dynamic loading such as vibrocompaction or dynamic compaction is appropriate.

4. Prior to 1970, the only structures built in Great Britain which allowed for significant earthquake forces were a number of dams in an area of Scotland that had suffered lengthy and repetitive swarms of moderately sized earthquakes. Following on from the work of the nuclear industry, seismic risk is increasingly being addressed in the design of major structures. A review of seismic hazard and the vulnerability of the built environment was carried out by Ove Arup and Partners (1993). A seismic zone map of the UK has been provided by Musson and Winter (1996 and 1997) with contours of peak ground acceleration (PGA) for earthquakes with an annual exceedance probability of 1×10^{-4}.

5. BRE has produced a number of guidance documents dealing with made ground including BRE Report BR 424 *Building on fill: geotechnical aspects* and BRE Digest 427 *Low-rise buildings on fill*. See also Charles and Watts (1996).

6. In Section 2E of Approved Document A of the Building Regulations it is stated that there should not be non-engineered fill within the loaded area.

7. Charles and Skinner (2001).

8. Guidance on building on clay soils is in BRE Digests 240, 241 and 242 *Low-rise buildings on shrinkable clay soils*. See also Dreiscoll and Crilly (2000). NHBC Standards Chapter 4.2 *Building near trees* has guidance for building near trees, particularly in shrinkable soils. See also Cutler and Richardson (1989) and Biddle (1998).

9. Figure 7.2 gives only a general indication of the location of shrinkable clays; more detailed information can be obtained from the maps and accompanying memoirs published by the British Geological Survey. Some shrinkable clays occur further north than the areas indicated in the figure; in the north however, surface clays are generally sandy and their potential shrinkage is therefore smaller. Soft, alluvial clays, which are found in and around estuaries, lakes and river courses, have a crust of firm clay that has been strengthened by overconsolidation brought about by successive seasons of wetting and drying; clay shrinkage is not the only foundation problem in these areas since excessive settlement due to loading the underlying softer clay and peat can also occur (Section 7.2).

10. The significance of desiccation in causing building damage is described in BRE Digest 412 *Desiccation in clay soils* which gives much fuller guidance on estimating ground shrinkage and heave potential than is presented in Section 7.5. The filter paper method of soil suction determination to establish the state of desiccation in clay soil profiles is described in BRE Information Paper IP 4/93 *A method of determining the state of desiccation in clay soils*. See also Chandler *et al* (1992) and Chandler and Gutierrez (1986).

11. Although water contents are usually measured by engineers in gravimetric terms, they can be expressed in volumetric terms: the volume of water expressed as a proportion of the total volume occupied by the soil.

12. The water shrinkage factor S_f of a soil layer is defined as the ratio of the water deficiency to the vertical ground movement that occurs in that layer. Crilly *et al* (1992) have suggested that it is not a constant, but varies with depth, reducing from a high value near ground surface to about 1 at depth. However, it appears that an average value of S_f of 4 is reasonable for estimating heave at 1 m below ground level (BRE Digest 412). Calculations made using a variable S_f, as shown in the Table below (D is the depth of significant dessication), can be compared to those based on a constant value of 4; calculating the change in height of individual layers allows the ground movement at any given depth to be calculated and thus the movement of a given depth of foundation or underpinning.

Depth of layer	$0 - 0.25D$	$0.25D - 0.50D$	$0.50D - 0.75D$	$0.75D - 1.0D$
S_f	10	5	2	1

13. Design guidance on minimising the effect of trees on foundations in clay soils is in BRE Digest 298 *The influence of trees on house foundations in clay soils*.

14. Driscoll (1983) suggested that significant desiccation, corresponding to pore water suction values of 100 kPa or more, may be present if $w < 0.4\ w_L$. This test was presented as an approximate rule-of-thumb for use in highly shrinkable clay and became widely used even in soils of lower plasticity where it has been found to be unreliable.

15. For many years, BRE has studied the seasonal ground movements at a site at Chattenden in Kent where there is high plasticity London clay (Crilly *et al*, 1992; Driscoll and Chown, 2001).

16. The ranking of trees most likely to cause damage (see Table 7.2) is in BRE Digest 298. It is based on information collected by the Royal Botanic Gardens during the 1970s (Cutler and Richardson, 1989) The recommendations in the table should be treated with caution and more specific advice on safe planting distances can be found in BS 5837:1991. Further information on the mature height of

FOUNDATIONS ON DIFFICULT GROUND

trees is in NHBC Standards Chapter 4.2.

17. BRE has carried out long-term monitoring at several sites. Cheney (1988) has reported a case where Elm tree removal resulted in swelling that lasted for more than 25 years and caused building heave of 160 mm.

18. Section 2E of Approved Document A of the Building Regulations states that the depth to the underside of foundations on clay soils should be not less than 0.75 m. For a foundation on shrinkable clay soil outside the zone of influence of any tree, NHBC Standards Chapter 4.2 relates minimum foundation depth to volume change potential as follows:
- low volume change potential; 0.75 m;
- medium volume change potential; 0.9 m;
- high volume change potential; 1.0 m.

Volume change potential is defined in terms of a modified plasticity index and is similar to the classification given in Table 4.4 here. Tree planting must be restricted to ensure that foundations with these minimum depths do not subsequently come within the zone of influence of a tree.

19. Detailed recommendations for foundation depths when building near trees are contained in NHBC Standards Chapter 4.2 where it states that trench-fill foundations deeper than 2.5 m should be designed by an engineer.

20. BRE Digest 241 *Low-rise buildings on shrinkable clay soils* Part 2.

21. In NHBC Standards Chapter 4.2, heave precautions for trench-fill foundations in shrinkable soil, when building within the zone of influence of trees which are to remain or be removed, include the provision of compressible material against the inside faces of all external wall foundations greater than 1.5 m deep.

22. There can be difficulties in controlling the compaction of fill over a limited area. NHBC Standards Chapter 4.2 stipulates that the depth of fill placed under a raft on shrinkable soil should not exceed 1.25 m. The application of rafts is restricted to where the recommended depth for a trench-fill foundation would be 2.5 m or less.

23. NHBC Standards Chapter 4.2 provides guidance on the depth of void required.

24. Further details are in BS 5837.

25. The collapse of the Holbeck Hall Hotel at South Bay, Scarborough, provided a graphic example of the seriousness of coastal landslips. Guests and staff were evacuated from the hotel on 4 June 1993. Before the landslip, the hotel, built in 1883, was on a 60 m high sandy boulder clay cliff and was 60 m back from the cliff face. The slip involved an estimated 1×10^6 tonnes of earth. See Department of the Environment (1994).

26. The rudiments of slope stability calculations are give in most standard soil mechanics text books. More detailed information is provided by Bromhead (1992).

27. A number of instructive case histories are in Chandler (1991).

28. Stressed anchors are formed by grouting a steel bar or tendon into a pre-drilled hole. In soft rocks and soils, the pull-out resistance can be increased by enlarging the end of the drill-hole using a reaming tool. In more competent rocks, a mechanical device is often used to expand the anchor to grip the sides of the drill hole. The bar or tendon is tensioned at the surface using jacks or by turning a nut on a threaded bar.

29. BS 8006:1995 *Code of practice for strengthened/reinforced soils and other fills* describes the design of reinforced soils including soil nailed walls.

30. The grouting of clay may have several effects. Ion exchange in the pore water can cause a change in mineralogy, which can produce an improvement in shear strength. Grout may also be able to penetrate slip surfaces producing a cementing action. However, the benefit to the soil mass as a whole is likely to be limited by the relatively low permeability of the soil.

31. Details of instrumentation techniques are in Dunnicliff (1988).

32. An inclinometer makes possible the accurate measurement of horizontal movement. The device consists of a 1 m-long torpedo, instrumented with a sensitive strain-gauged pendulum. The torpedo is fitted with small wheels at each end, which locate in specially slotted aluminium or plastic tubes that are permanently grouted into pre-drilled vertical holes. Readings are taken at 1 m intervals as the torpedo is lowered down the tube; this enables the deviation of the tube from the vertical to be computed. The tubes have four equally spaced slots, which allow the inclinometer to profile the verticality of the tube in two orthogonal planes. The torpedo can also be reversed to average out instrument errors. The main disadvantages of the inclinometer are the cost of the boreholes and the demountable instrumentation, and the risk of the tubes being distorted by even small movements and rendered impassable.

33. Standpipe piezometers can be driven into soft soil. Alternatively, a standpipe can be installed into a borehole and the filter formed by pouring a layer of sand around the tip; the remainder of the borehole can then be filled with a cement and bentonite grout, placed via a tremie pipe. The water rises in the standpipe until it reaches equilibrium level. The piezometer is read simply by lowering an electrical dipper down the standpipe; when it reaches the level of the water, it buzzes.

34. While all these devices respond far more quickly to a change in pore water pressure than a standpipe piezometer, they are more susceptible to erroneous readings as a result of desaturation of the porous tip; to combat this, a ceramic filter with a high air entry value is normally used. A detailed account of the advantages and limitations of these piezometers is in Dunnicliff (1988).

35. Buildings affected by such collapses are reported in the technical and national press each year as the following two examples illustrate:
'Sixteen families were homeless this week after a crater opened up in a Reading street damaging the carriageway and nearby houses. It is thought that the 200 m^3 hole resulted from the collapse of an old chalk mine or historic fortifications.' (New Civil Engineer, 20 January 2000, p6).
'A Belgian town is in danger of collapse because of an extensive network of tunnels built by Allied troops during the First World War, scientists said yesterday. They believe that the problem, the result of 85-year-old shoring timbers starting to collapse, affects much of the former Western Front.'(Daily Telegraph, Wednesday 13 December 2000).
A description of the hazard posed by the abandoned limestone workings in the West Midlands is in Whittaker and Reddish (1989) pp 388-393.

36. Rules based on experience and practice to assess the likelihood of a mine collapse causing surface subsidence include the safe depth approach, where it is assumed that no damage will result if old workings are covered by at least 15 m of rock, and the 100-year-rule, where it is assumed that if no subsidence has occurred within 100 years, no subsidence will occur thereafter. Such rules are of limited applicability, their use is not recommended. See also Atkinson (2003).

37. *Subsidence of low-rise buildings* (Institution of Structural Engineers, 2000).

38. There are no records for many of the older mines, particularly before 1872, when legislation was introduced making it compulsory to keep plans of all mine workings.

39. Atkinson (2003), Tomlinson (2001).

CHAPTER 8: BROWNFIELD LAND

Building on brownfield sites is in the public interest and can present considerable opportunities for building developers, but there can be particular risks associated with the previous use of the land. Ground-related hazards for the built environment include poor load carrying properties, detrimental chemical interaction with building materials and gas generation from biodegradation of organic matter. Where such hazards are present but not identified, or are misdiagnosed, remedial measures may be required with significant implications for whole-life costs. The chapter provides an overview of the hazards which may be encountered on brownfield sites. While concentrating principally on the risks to the built environment, it also considers the hazards to human health and groundwater associated with contamination.

8.1 Introduction

It is widely accepted that as much new building development as possible should be located on brownfield sites[1]. Maximising the re-use of previously developed land[2] should help promote urban regeneration, and minimising the amount of greenfield land being taken for building development should provide environmental benefits. The term *brownfield* has been widely adopted to describe previously developed land. It represents a broader concept than either derelict land or contaminated land since a brownfield site is not necessarily either derelict or contaminated (Box 8.1).

The redevelopment of sites in urban and industrial areas is not, of course, new; it has been a feature of urban life throughout recorded history. However, the scale and speed of building developments on brownfield land has greatly increased in the last few years. It has been questioned whether current approaches to the redevelopment of such sites are always the most appropriate. In some cases, hazards may be overlooked or their significance may not be recognised; in other cases, solutions may be over-engineered with unnecessary expenditure[3]. The way that hazards and the consequent risks are perceived can be of crucial importance[4].

While building on previously developed sites is in the public interest and presents major opportunities for developers, there can be particular risks associated with the previous land use. Ground-related hazards for the built environment include:

Box 8.1 Brownfield land
Although the term *brownfield* is widely used, it has no universally recognised definition. It has sometimes been regarded as synonymous with contaminated land; more commonly it has been understood as signifying the opposite of *greenfield* in planning terms, where greenfield is taken to mean land that has not been previously developed. Brownfield land should be regarded as any land that has been previously developed, particularly where it has been occupied by buildings or other types of permanent structure and the associated infrastructure. Brownfield land may contain contamination but it represents a much broader concept than contaminated land.

- poor load-carrying properties of the ground;
- detrimental interaction between building materials and chemically aggressive ground conditions;
- gas generation from biodegradation of organic matter within the ground;
- underground fires associated with combustible materials.

Where such hazards are present but are not identified or are misdiagnosed, remedial measures may be required with significant implications for whole-life costs. This becomes more difficult and complex where brownfield sites are heavily contaminated and there are hazards for human health and detrimental effects on the natural environment and such resources as groundwater.

Each site should be treated on its merits and there should be consultation as early as possible between the planners, architects, valuers, building control officers, environmental health officers, geotechnical engineers and developers. Even a cursory study of previous use and a rudimentary preliminary survey may give an indication of the extent of any contamination. In addition to chemical contamination, there may be serious physical impediments to redevelopment, such as large concrete structures below ground and demolition rubble and domestic and commercial waste spread across the site. Brownfield hazards can be classified into two broad categories:
- Geoenvironmental hazards are principally concerned with contamination and are linked to the chemical and biological properties of the ground and the groundwater conditions. The range of problems associated with chemical contamination is vast and chemical contamination can present an immediate or long-term threat, not only to human health via ground or groundwater, but also to plants, amenity, construction operations and buildings and services. Biodegradation of organic matter leads to gas generation and settlement.
- Geotechnical hazards, which may include buried foundations and settlement of fill, are essentially linked to the physical properties of the ground and impact principally on the built environment.

Contaminated land requires careful assessment; building developments on such sites require the help of appropriate specialists. Many consulting engineering firms can provide specialist advice. However, the non-specialist will find some background knowledge valuable when facing a multiplicity of technical options presented by the professional in contaminated land matters[5]. It is important that the type and level of the contamination is evaluated in relation to the intended end-use of the development. The protection of groundwater from pollution is a major concern. A brief outline of the statutory requirements relating to buildings and contaminated ground is in Chapter 2.

8.2 Types of brownfield site

Not all brownfield land is contaminated but many sites are and such sites can present significant challenges for redevelopment. Heavy industries have contracted substantially under the pressure of international competition and there is much interest in the redevelopment of heavily polluted sites. Large-scale contamination is likely at chemical works, gasworks, oil refineries, petroleum storage sites, scrapyards, sewage works, steel works and many other sites where there have been industrial and manufacturing processes[6]. Pollution may be encountered at large vehicle depots or where there has been refuelling of cars, lorries, aircraft and other vehicles and where fuel has leaked into adjacent ground.

Ideally, decommissioning, dismantling and decontamination should be carried out carefully with a view to the future re-use of an industrial site. In practice, this degree of care in the demolition process is not always evident and sometimes the problems for redevelopment are exacerbated by contaminants spread around the site during demolition. Records of the previous development may be lacking and the demolition work may not have been reliably recorded.

Coal gas (town gas) manufacture was widespread in the UK for over a hundred years until, with the advent of natural gas in the early 1970s, it virtually ceased. The legacy of the industry is evident in hundreds of former gasworks sites which are among the most numerous and widely distributed examples of contaminated land (Box 8.2). Similar contaminants can occur on other sites where coal carbonisation processes were operated, for example coke ovens, tar distillation plants and by-products works. The presence of contaminants may limit the range of further uses for these sites.

Box 8.2 Gasworks sites[7]

Both the soil and groundwater are contaminated in many gasworks sites. The contamination has been derived from the products, by-products and waste materials of the coal carbonisation process. Spent oxide, which has high sulfate content, is the most frequently encountered solid waste and will attack building materials. Other contaminants include coal tar, oil and diesel and ammoniacal liquor. Chemical attack is a hazard for building materials. Physical hazards buried in the ground include gasholder bases, tanks and pipework. The location of pollutants may not be known. Further problems include potential combustibility and emissions of toxic or flammable gases.

Scrapyards, which have been used for breaking-up redundant or obsolete manufactured items such as motor vehicles, electrical equipment, and machinery, are among the most commonly encountered sites with contaminated land in and around urban areas. The contamination encountered at such sites is described in Box 8.3 and the various hazards may should be taken into account when housing, agricultural or amenity redevelopment schemes are planned.

Contaminated sites may present hazards additional to those directly associated with

> **Box 8.3 Scrapyards**
> Contamination of scrapyard sites is likely to be severe; contaminants can include metals, oils and solvents. The contamination can be in the form of finely divided solids or of liquids from spillages, leakages and waste disposal. Although a scrapyard may be small in area, extreme variability can make it difficult to detect all the contaminants. General untidiness and lack of control over the activities carried out on the site may have encouraged illicit tipping and the disposal of a large variety of hazardous wastes.

the contamination. At gasworks sites (Box 8.2) old foundations and buried tanks are a hazard. In addition to the presence of various contaminants, the generation of landfill gas presents a serious hazard on or adjacent to landfill sites (Box 8.4) and similar problems can occur at the sites of old sewage works (Box 8.5).

Landfilling has been the most widely used refuse disposal technique in the UK and the variation in composition of refuse during the twentieth century is shown in Figure 8.1. The composition of household refuse has changed markedly over the years; prior to 1960, the inert ash content was over 50% but between 1960 and 1970 this reduced to about 20% with subsequent further reduction. Landfill placed before 1960 not only has had a long period for biodegradation to take place, but it was an inherently better material than landfill placed since 1970. Natural ground conditions cannot always be relied upon to do the required containment and increasing use is being made of deep cut-off walls, which are taken down into impermeable strata to prevent lateral migration of leachate and gas. The migration of landfill gas into buildings can be a hazard, not only for building development on the landfill site but also for buildings on adjacent sites

Landfill sites can pose severe problems (Box 8.4) and are generally unsuitable for many forms of building development unless appropriate remedial treatment is carried out beforehand; such treatment is likely to be technically complex and expensive. Building development on sites where biodegradation is continuing is generally avoided.

The provision of sewage treatment facilities in large centralised works of considerable capacity has meant that many older sites have become available for redevelopment. The sites range from those formerly containing works serving a few hundred homes, to large city treatment works which, with their

> **Box 8.4 Landfill sites**
> Landfill sites contain biodegradable materials; during the long process of decomposition, the large reduction in the volume and consequent settlement of the landfill is accompanied by the generation of gas and the formation of leachate. Large settlements continue over many years and present a serious problem for the design of foundations and services; gas emission is usually the most significant hazard. The rate at which emission occurs will be related to the age of the site, but gas emission at an old site can be reactivated by disturbance from construction processes. Leachate from the landfill can contaminate the groundwater and pollute aquifers as well as causing other environmental problems. Other hazards include substances harmful to health and plant growth, chemical attack on services, combustibility of material, and odour and drainage problems.

BROWNFIELD LAND

associated effluent disposal areas and farmland, may cover many hectares. Possible hazards to human health and toxicity to plants may be encountered on agricultural or horticultural land that has been treated by the repeated application of sewage sludge or effluent, often over many years. Such land may be adjacent to the sewage works and form an integral part of its operations, or may be on farms that have received the sludge for its value as a fertiliser.

Brownfield sites which are not chemically contaminated may present serious physical hazards, particularly where the site is covered by a substantial depth of fill. On many industrial sites, large quantities of solid wastes have accumulated; in many cases they have been tipped in high lifts without any compaction and the materials in the deposit may be very variable. Such non-engineered fills have poor load-carrying properties with the potential for significant settlement (Sections 3.9 and 7.4).

> **Box 8.5 Sewage works**
> The range of possible contaminants is wide at the site of a sewage works; solid, semi-solid and liquid wastes may be present. Deposits of materials on the land surface should be easily identifiable, but the presence of wastes in lagoons, drying beds and humus tanks may be less obvious when their surfaces have dried out and become vegetated giving the appearance of solid ground. Waste sludges are the greatest hazard and can contain a wide range of metals and other elements in varying and high concentrations. Disease-producing organisms may be present including bacteria, viruses and the eggs or cysts of parasites. Although these hazards are likely to be small on old sites, caution is required. Sewage sludge contains a significant proportion of organic matter, which may decay to produce methane and other gases.

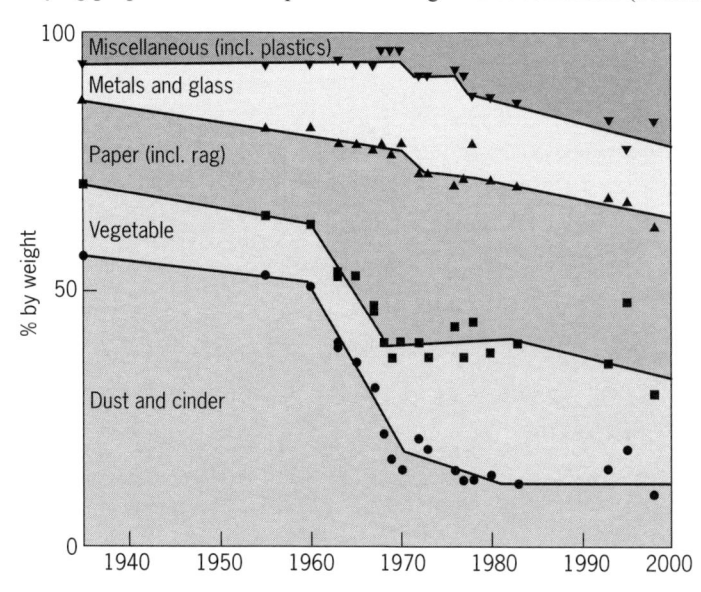

Figure 8.1 Composition of domestic refuse in the UK 1935 - 2000 – after *Watts and Charles*, 1999

In the coal mining regions of the UK there are large quantities of colliery spoil and the regeneration of brownfield sites in these areas, following the demise of coal mining, frequently involves building on sites covered by a substantial depth of colliery spoil (Box 8.6). The problems are likely to relate principally to the geotechnical properties of the spoil. The suitability of colliery for use as hardcore is described in Table 12.1.

In the past, many colliery sites have been reclaimed for soft uses (for example to improve the landscape); in such cases the colliery spoil is unlikely to have been placed in thin layers with the heavy compaction necessary for an engineered fill on which buildings will be erected. Figure 8.2 shows colliery spoil being placed in 2 m-high lifts with only minimal compaction from the trucks bringing the spoil onto the site. Colliery spoil so placed has potential for substantial settlement, particularly when water infiltrates into it (Sections 4.9 and 7.4). Figure 8.3 shows colliery spoil being end-tipped into a disused dock. This method of placement will result in the fill being in a very loose condition, but the high water-table should ensure that collapse compression is not a problem.

Box 8.6 Colliery spoil
In 1974, there were some 3000 million tonnes of colliery spoil in the UK and, ten years later, 50 million tonnes were still being placed in spoil heaps every year. The rapid decline of the coal industry in the 1990s led to a dramatic reduction in the production of these wastes. For many years there have been reclamation schemes to improve the landscape and the environment. The major programme for the regeneration of colliery sites, which began in the late 1990s, involves substantial building development on the sites of old collieries.

Opencast coal mining has left extensive sites where the natural consolidated strata have been replaced by large depths of fill material. Although generally free of contamination, at many sites the opencast backfill was not systematically compacted during placement. Where such sites have been restored to agriculture, they may not be identified as brownfield land but there can be potential for large settlements to occur (Section 4.9).

Figure 8.2 Colliery spoil being placed in 2m-deep lifts

BROWNFIELD LAND

Figure 8.3 Colliery spoil end-tipped into disused dock

The manufacture of iron and steel has given rise to large quantities of ferrous slag (Box 8.7). Where buildings are placed on such material, or where the material is used as hardcore under a ground supported floor slab, the main hazard is associated with expansion of the slag, which can be large[8]. Some slag contains sulfates which attack concrete[9]. While much of this material is located close to the iron and steel works in which it was produced, ferrous slag has been widely transported for use as fill material. Slags have been crushed for use as aggregates[10]. The suitability of slag for use as hardcore[11] is described in Table 12.1.

Box 8.7 Ferrous slag
Two main types of ferrous slag are produced during the manufacture of iron and steel: blastfurnace slag when iron ore is smelted to produce pig iron, and steel slag when pig iron is converted into steel. Steel slag poses the greater hazard with respect to expansion but old blastfurnace slag, which differs in composition from modern slag, may also be expansive. When building development commences on an apparently stable slag deposit, expansion may be triggered by site activities which lead to the mixing of reactive constituents or the access of air and water. Volume expansion in steel slag can be large and there have been failures of structures built on such slags. Upward movement of floor slabs placed on blastfurnace slag has occurred due to sulfate attack in the concrete, but this should not be confused with heave of the slag.

Where a major building development is carried out on an extensive brownfield site[12], different parts of the site may have had very different forms of previous development and included within the site there may be some greenfield land. It may be possible to locate the more sensitive parts of the development, such as low-rise housing, on the areas where ground conditions are most favourable.

8.3 Systems at risk
There are three different types of receptor[13] which may be vulnerable to brownfield hazards:
- people - the human population (Box 8.8);
- nature - the natural environment, including surface water and groundwater (Box 8.9);
- buildings - the built environment (Box 8.10).

These three systems are, of course, interdependent. The threats from the various hazards to the different types of receptor in the human population, the natural environment and the built environment should not be considered in isolation from each other; an integrated approach is required.

In an increasingly risk-averse society, matters concerning health and safety receive much attention and publicity. As a consequence, and since brownfield sites are frequently contaminated, hazards associated with contamination and the risks posed to health often have been the dominant issues in the redevelopment of derelict land and brownfield sites.

Box 8.8 Risk to human population
The hazard on a site may be associated with the presence of a contaminant but, in order for the contaminant to pose a risk, there must be a reasonable pathway by which the contaminant could be transferred to a person present on the site. From the early stages of investigation through to the final use of the site, many people may be at risk. With low-rise housing, the occupiers could be the people most exposed to many of the hazards. While there is no evidence to suggest that chemical contamination on brownfield sites has posed a major threat to human life, at least in the short term, long-term health and quality of life could be affected and this is difficult to evaluate.

Box 8.9 Risk to natural environment
Threats to the natural environment are often narrowly believed to concern soil and groundwater contamination. However, a wider range of issues may need to be examined in the light of growing concern over degradation of the natural environment highlighted by policies promoting such concepts as biodiversity. Proposals to reclaim derelict land may meet opposition from those who consider that ecosystems established amidst the dereliction, and present as a direct result of it, should be preserved. The construction process is likely to cause disturbance, dust and noise. It is difficult to establish an appropriate balance between economic well-being and the natural environment because, unlike matters of human health and building damage, there are no widely agreed objectives or ground rules.

BROWNFIELD LAND

> **Box 8.10 Risk to built environment**
> The site and the building development form an interactive system and it is important to evaluate the risk of adverse interactions during the lifetime of the development. Hazards to the built environment on a brownfield site can be physical, chemical or biological in character; concerns can include poor load-carrying properties of the ground and the interaction of building materials and services with aggressive ground conditions.

A proper balance is needed between the need to regenerate communities blighted by the demise of once dominant industries and the environmental impact of such regeneration. An indication of some of the issues that can arise is in Box 8.9.

Such issues as the degradation of the natural environment, sustainable construction involving the rate of use of renewable and non-renewable resources, recycling waste materials, energy consumption in buildings and the emission of pollutants, are important but the primary objective is to build safe, durable and economic structures. While human health and environmental issues are of great concern on some brownfield sites, the ground-related risks to the built environment, which includes buildings, transport infrastructure, services, gardens and ancillary buildings, should not be overlooked (Box 8.10).

8.4 Ground-related hazards

A wide range of hazards may be encountered on brownfield sites. Many of them are related to the previous use of the site but there may also be problems associated with the original state of the ground; these can include various types of ground movement and aggressive ground conditions, described in earlier chapters. Problems associated with rising groundwater levels due to reduced abstraction by declining manufacturing industries (Section 10.4) and with flooding hazards on floodplain developments affect both greenfield and brownfield sites.

While the geotechnical or physical hazards can present major problems for the built environment, the geoenvironmental or chemical hazards can have a major impact not only on the built environment but also on the human population and the natural environment. Some of the ground-related hazards encountered on brownfield sites are shown in Figure 8.4. Hazards for the built environment on a brownfield site can include inadequate properties of the ground for supporting foundations and services, chemical attack on foundations and services, explosion of gas generated from biodegradation of organic matter and from other deleterious substances in the ground, and combustion. In providing value for money at a suitably low level of risk, the designer of a building development also has to address the hazards posed by contamination for the human population and the natural environment.

The ground-related hazards can be physical, chemical or biological in character and associated with four aspects of the ground conditions:
- solid phase of the ground, for example ground movements can damage buildings;

❏ liquid phase of the ground, for example groundwater and leachate can be carriers of aggressive contaminants and can affect geotechnical behaviour;
❏ gaseous phase of the ground, for example gas migration through and from landfill sites and into buildings has caused explosions;
❏ fire; subterranean fires most commonly occur in colliery spoil and domestic refuse.
Although the hazards that are described here are often found on brownfield sites, they are not necessarily confined to such sites. For example, aggressive ground conditions can occur on greenfield sites.

Human health
The presence of substances that may be toxic to human, animal or plant life is of particular concern. The principal hazards to human health can be categorised by whether their effect is immediate or longer term. This distinction can be illustrated, firstly by acid or alkaline contaminants which may cause burns or irritation if they come into direct contact or are inhaled and, secondly by asbestos fibres which, when inhaled, can cause longer term problems. A significant number of chemicals have been identified which can have a direct impact on health or which may pollute groundwater supplies[14]. The hazard on a site may be associated with the presence of a contaminant but, in order for the contaminant to pose a risk, there must be the potential for the presence of a receptor (human being) and a reasonable pathway by which the contaminant can be transferred to the receptor.

Groundwater pollution
Contamination of groundwater and the pollution of aquifers and watercourses by release of mobile contaminants can affect both the human population and natural environment.

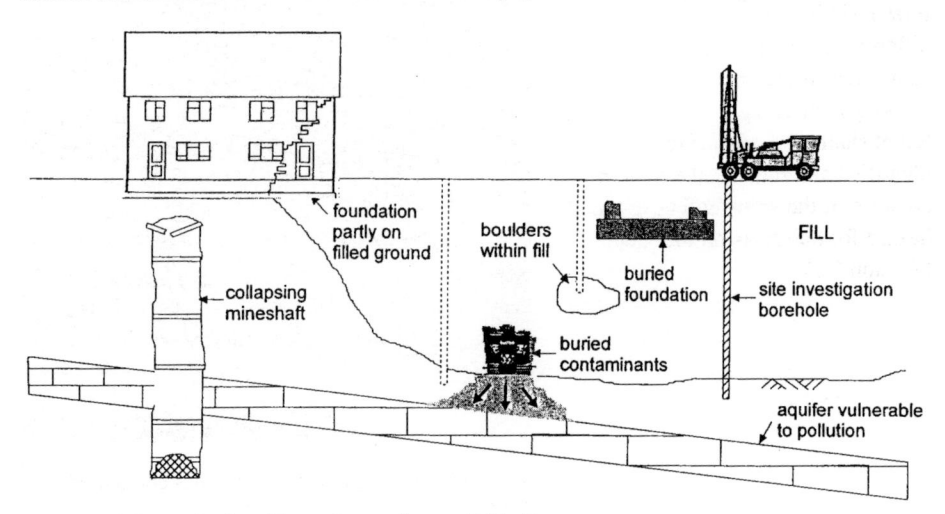

Figure 8.4 Ground-related hazards on a brownfield site

Contamination may spread into an aquifer and the risks posed by groundwater contamination may cover a wide area. The site owner could be required to clean up contaminated groundwater even if it has migrated off the site being used for building development. Contaminated groundwater may require remediation to standards set by the Environment Agency. The possibility of the contamination of groundwater by remediation or construction works is a critical issue in the determination of an appropriate strategy for building development.

Ground movement
Previous human activity on the site may have created ground-related problems by the uncontrolled placing of fill material (Section 3.9). Problems for building development frequently occur where fill is deeper and more variable than anticipated and where the limits of backfilled areas, such as infilled quarries and pits, have not been correctly located. In addition to these non-engineered compressible fills, expansive fills (Box 8.7) and biodegradable fills (Box 8.4) can also cause significant ground movements.

Other physical hazards in the ground include buried foundations, old pipework and tanks remaining from previous site usage. These may form obstructions when excavating for foundations and, if left in place, can form local hard spots which cause differential settlement (Figure 8.5). In some areas there may be old mine workings where large voids have been left at shallow depths. If these are not identified and infilled, the sudden collapse of the voids will be a major hazard for buildings on the site (Section 7.7).

Figure 8.5 Underground obstructions and buried foundations at brownfield site

Chemical attack on foundations and services
At brownfield sites, building materials are often subjected to aggressive environments which cause physical or chemical changes and result in loss of strength and other deleterious changes. This results in increased maintenance requirements, a reduction in the service life of materials and buildings and, ultimately, risks to health and safety or the environment.

Predicting the durability of building materials in aggressive ground conditions is difficult because of the wide variety of chemicals and their concentration that may be encountered and the variability of the ground in transmitting the chemicals[15]. Some of the hazards associated with chemical attack on building materials and services are listed in Box 8.11.

Building sub-structures which are in contact with the ground include foundations, floor slabs, basements, and water and gas-proofing membranes. Chemical attack on the sub-structure by aggressive ground conditions can cause movement and structural instability. Materials include reinforced and unreinforced concrete, steel, masonry and geosynthetics. The effects of sulfates, acids and chlorides give concern for both reinforced and unreinforced concrete foundations. Sulfate attack on concrete is characterised by expansion, leading to loss of strength and stiffness, cracking, spalling and eventual disintegration. Heave of floor slabs can occur due to sulfate attack[16]. Magnesium, ammonia and phenol are also known to cause deterioration of concrete. High quality dense concrete is the primary prerequisite for durability. The consequences of chemical attack on concrete foundations are difficult to assess. Ultimately, chemical attack may cause the collapse of a structure but such extreme examples are rare. More commonly, localised deterioration and cracking results in loss of strength.

Concrete, metal, plastics and ceramic pipes may be vulnerable to damage and degradation. Both physical and chemical damage need to be considered. A damaged water supply pipe could permit the entry of contaminants into the water supply. The migration of organic compounds through plastics, particularly MDPE, water pipes is also a key issue since it can cause tainting of the water. Chemicals may pass through

Box 8.11 Hazards for building materials
The principal concerns arising from inadequate long-term durability and performance of building materials are:
❑ Chemical attack of foundation concrete affecting structural integrity.
❑ Movement of concrete floor slabs and oversite concrete compromising serviceability of buildings.
❑ Damage to impermeable membranes installed in floors of buildings allowing ingress of contaminants as liquids, volatiles and gases.
❑ Entry of contaminants into water supplies through damaged and degraded plastics, or inadequately protected pipes leading to risks to health.
❑ Permeation of organic compounds (for example phenols and diesel) into MDPE pipes carrying potable water causing tainting of the water.
❑ Damage to gas supply pipes leading to release of gas with potential for explosions.

plastics pipes without damaging them. A damaged gas pipe could lead to the release of gas into the surrounding ground. Backfilled service trenches, old pipework and granular surrounds can form migration pathways for contaminants and gas.

Materials which are used in buildings or for services and which are at risk from contaminated land are summarised in Table 8.1.

Table 8.1. Building materials at risk from contaminated land

Elements	*Materials*
Foundations: strip, trench-fill, pad, raft, pile	Concrete, reinforced concrete, steel, masonry (mortar, bricks, concrete blocks)
Floor slabs, oversite concrete	Concrete
Basements	Concrete, masonry
Moisture and gas protection measures within building	Geomembranes, tanking materials, concrete
Water supply, drainage, gas, electricity and telephone services, inspection chambers	Concrete, plastics, ceramic and metal pipes, coatings for pipes, sealing rings for pipe joints, sheathing and insulation of electric cables

Gas migration

Gases emanating from the ground can present hazards, including asphyxiation, poisoning and explosion. Where fills or natural soils contain biodegradable material, gas generation can be a significant hazard for building developments on the site and for buildings adjacent to the landfill, which can be affected by gas migration[17]. Stringent precautions will be required for many types of occupied building (Section 8.7). Problems can also arise as a result of the migration of the vapour phase following petroleum and solvent spillages or from leaking supply pipes.

Combustion

Wherever combustible material is present below ground, there is a possibility that fires may start and propagate beneath the surface. The majority of such fires have occurred in sites containing domestic refuse and colliery waste[18]. Because the supply of oxygen at depth is limited, such fires proceed slowly by smouldering or charring; only occasionally does a fire break through to the surface and flames appear. Slow combustion may proceed for long periods with little evidence at the surface. There may be occasional emissions of steam or smoke, or vegetation may die or become blackened, but such signs are not always recognised. The vegetation may be particularly lush above the fire owing to the increased soil temperatures.

The seat of the fire may not be directly under the building. A fire that has started at a location distant from the building can still pose a threat due to lateral migration through the combustible material of the site if there are no barriers to halt the passage of the flame front. An underground fire can have several serious effects on a building:

❐ Production and release of toxic, asphyxiant or noxious gases which can travel considerable distances through the ground; depending on the nature of the combustible materials, a smouldering combustion can produce such gases as carbon dioxide, carbon monoxide, sulfur dioxide, hydrogen sulfide and hydrogen chloride.

❐ Creation of voids with consequent subsidence of the burnt ground, causing physical damage and impairing the stability of buildings or other structures on or near the site and the deformation and fracture of service pipes; hidden cavities may make the extinguishing process and land reclamation more hazardous.

❐ Heat damage to buried structures and such services as power supply cables; the passage of a fire beneath a building can expose the construction materials and services to temperatures as high as 1000°C, causing concrete to spall plastics service pipes to melt; some subterranean fires are known to have been burning for more than twenty years and attacks on the integrity of building materials could occur over long periods.

Underground fires may be very difficult to extinguish. In the early stages when the area affected is limited, excavation, accompanied by the application of water, may extinguish the fire. Once the fire is well established, this process may increase the air supply and cause the fire to advance faster than the digging. The construction of barriers to control the spread is expensive and frequently difficult, especially if the fire is large or the combustible material is deep. In some cases grouting techniques (Section 9) can be used to limit the spread of combustion.

8.5 Identification of contamination
General aspects of site investigation are described in Chapter 5. In this section, the particular problems involved in locating and identifying contamination are described.

The processes that were undertaken on the site before it became available for re-development should be identified at an early stage. As described in Chapter 5, there are many possible sources of historical information about a site and recourse to these is essential. However, before serious planning of the redevelopment takes place, a specialist investigation is required to determine more precisely the nature, extent and concentration of any contaminants present in the ground. Investigation of contaminated land is a skilled task which should be undertaken by suitably experienced site investigation contractors. There may be health hazards for the operatives during the site investigation.

A risk-based approach can be used to identify and quantify any hazards and the the risk they may pose[19]. The principle of defining and revising a conceptual model[20] for the site is a key theme. The design and execution of field investigations, including suitable sample distribution strategies, sampling and testing, require specialist knowledge and experience.

BROWNFIELD LAND

The objectives of the investigation should be clearly set out and they depend on the intended use of the land and the development stage. The overall objectives are first to characterise the type and concentration of the contaminants present, and then to identify pathways and receptors for the risk assessment.

Geotechnical and geoenvironmental investigations involve broadly similar activities; a desk study and site reconnaissance, exploratory, main and any supplementary investigations, and reporting[21]. Combined investigations benefit from an early assessment of the objectives of the investigation and from careful planning.

When investigating a potentially hazardous site, a preliminary investigation, consisting of a desk study and walkover survey, should first be carried out to assess potential hazards. If the presence of hazards is suspected, a more detailed investigation is required to characterise the hazards and design appropriate remediation and foundation solutions[22]. If no hazards are suspected, this should be confirmed by a basic investigation.

The preliminary investigation, which includes a site reconnaissance, should provide information on the past and present uses of the site, the surrounding area and the nature of the hazards and physical constraints. An initial conceptual model for the contamination can be based on this information, as well as the design of the exploratory and main investigation; this includes issues of worker health and safety during intrusive investigations. The preliminary investigation should identify whether regulatory bodies should be informed prior to intrusive investigations. Sources of information on the history of a previously used site, described in Section 5.3, should shed light on the nature of contaminants likely to be found.

The resulting source-pathway-receptor scenarios and preliminary conceptual model will then be investigated. An exploratory investigation may help to confirm whether the initial conceptual model is correct and provide sufficient information to design the main investigation. This is likely to involve targeted collection and analysis of soil, surface and groundwater samples and, where appropriate, the use of non-intrusive techniques. On large sites, geophysical methods, particularly resistivity, can be useful in detecting anomalies that can then be investigated more thoroughly in a later stage of investigation.

The main investigation should provide sufficient information for the confirmation of a conceptual model for the site, the risk assessment and the design and specification of any remedial works. This is likely to involve collection and analysis of soil, gas, surface and groundwater samples and, where appropriate, the use of non-intrusive techniques. An investigation of the groundwater hydrology may also be required.

Sufficient, appropriate sampling points ensures that the information obtained is robust and enables the risk assessment and design of containment or remediation to be carried out. A large amount of material can be inspected relatively cheaply and quickly in trial pits.

In some cases, a supplementary investigation may be carried out to better define the areas for remediation, or may be required to confirm that the remediation has been successful.

Once drilling, sampling and laboratory testing of contaminated soils and groundwater are complete, the scale of any hazard should be assessed. Some substances are desirable in small quantities. For example, some of the metals that may be present on contaminated sites are essential in trace quantities for the health of crops and animals; they include zinc, copper, iron, molybdenum, manganese; other substances, such as lead, cadmium, mercury, nickel and arsenic, may be harmful, even in relatively low concentrations[23]. The determination of the level of hazard presented by a contaminated site is difficult. Appraisal of the health hazard due to contamination is fraught with uncertainty, and demands considerable judgement and experience, together with knowledge of ground chemistry.

8.6 Strategies for building on brownfield sites

In developing a risk management strategy, it is necessary to examine the potential interaction with regulations and legislation (described in Chapter 2) with due attention to guidance documents from authoritative sources. Risk management should include:
- identifying the likely hazards;
- evaluating the risks posed by these various hazards;
- assessing the acceptability of known risks;
- where necessary, implementing actions to reduce the probability or consequences of the occurrence of identified hazards.

Since there are many possible hazards on a brownfield site, the most significant problems should be identified at an early stage so that the risks that they pose can be evaluated. The key role of site investigation is described in Section 8.5. The standard land condition record (LCR) should provide a verifiable record[24] of the nature of any contamination and the previous use of the land (Box 8.12). It is also necessary to define what level of risk is acceptable. On housing developments, risks to human health from contamination can be a major issue but this should not distract attention from the hazards to the built environment.

Box 8.12 Standard land condition record (LCR)
The LCR is, in effect, a voluntary logbook for brownfield land which is kept and maintained by the landowner and transferred with the land. It uses standard forms for the collection and presentation of data in a technically robust and systematic way. It should provide a clear record of the physical and chemical nature of the land and give details of the use that the land has supported. Anyone with a vested interest in the site can view basic technical data, enabling them to make better informed decisions based on existing information. It should focus attention on the probable hazards of the site at the earliest possible stage of redevelopment.

When the likely hazards have been identified, and the risks that they pose for the proposed development have been assessed, it may be considered that the risks are unacceptable; appropriate action is then required to avoid, prevent, mitigate or transfer risk. Selecting an appropriate remedial strategy involves many factors related to technical adequacy, costs, environmental effects and perception; it should be kept in mind that a solution to one problem may cause another. For example, piles or vibrated stone columns installed to reduce settlement may provide a pathway for contaminants to move through the ground[25].

❐ Technical adequacy: it is essential that the problem has been diagnosed correctly and a technically adequate solution adopted. Any other factors should be subservient to this over-riding requirement.
❐ Costs: a strategy which involves relocating the development to another site may have large financial implications, whereas appropriate ground treatment may add relatively little to the cost of the development.
❐ Environmental effects: a solution that is technically adequate for dealing with the hazards identified could have harmful and unacceptable environmental effects.
❐ Perception: where a site is popularly thought to be hazardous, it is important to adopt a strategy that will not only solve the problem but will be clearly seen to have solved the problem; transparency is vital where popular perception is involved[26].

Some basic types of remedial strategy are:

(a) Avoid creating the hazard
It may be possible to avoid creating a particular hazard. For example, where the site reclamation process has not yet been carried out, the earthmoving operation can be controlled to prevent mixing fill materials which can cause expansive reactions and the generation of highly polluting leachates. The extra costs will be small and the gain considerable in terms of risk mitigation. Where a fill has to be placed as part of the site redevelopment, placement under controlled conditions of an adequately engineered fill can eliminate or greatly reduce the hazard of compressible fill. The extra costs involved in adequate compaction of the fill and sufficient supervision will not normally be great. The consequences of failure can be very great in terms of reconstruction of buildings damaged by settlement and expensive legal actions.

(b) Relocate the building development
Where major hazards are identified and the risks are great, abandoning the site or part of the site for the intended purpose should be considered. On large sites, it may be possible to avoid using the worst areas for the most sensitive uses. This type of approach can be helpful on large developments which include residential, retail and commercial buildings on sites with different types of brownfield land. There will be much less scope on smaller sites. An unwarranted over-sensitivity to risk will defeat the objective of locating building developments on brownfield sites.

(c) Segregate the buildings from ground hazards at the site
The foundations of a building can be isolated from the effects of a compressible fill by using piles to transmit the weight of the structure to a firm stratum underlying the fill. Where this approach has not been successful, it is because the piles have not been long enough to reach the firm stratum. Where expansive ferrous slag or some other type of expansive fill is present, it may be necessary to sleeve the piles to accommodate ground heave (Figure 8.6[27]). Even where the building is successfully isolated from a settling or heaving fill, services may be subjected to severe differential vertical movement where they enter the building as the fill moves relative to the building.

(d) Segregate the buildings from ground hazards adjacent to the site
Developing a contaminated site should be clearly differentiated from a site adjacent to contaminated land. Difficulties are often encountered at a site adjacent to contaminated land, not at the time of construction but some time later, when it is discovered that pollution is penetrating into the unpolluted land. The most common example is the migration of methane gas from adjacent landfill sites into housing estates (Sections 8.4 and 8.7). Containment of contamination has a vital role in the utilisation of many brownfield sites[28]. In this commonly used remediation method, the objective is to block or control

Figure 8.6 Piling system to isolate houses from heave of slag – *courtesy of Roger Bullivant Ltd*

the pathway between a hazard and potential receptors. Engineered cover layers[29] and in-ground vertical barriers, such as slurry trench cut-off walls[30] (Box 8.13), have been widely used as reliable means of containing contamination and isolating areas of pollution. Figure 8.7 shows a slurry trench cut-off wall being installed. Where the hazard is associated with the migration of gas, the flow of groundwater or the spreading of a subterranean fire, the site may be protected by a cut-off wall.

(e) Remove the poor ground
Problem soils are removed physically and, if necessary, replaced by an imported engineered fill. Depending on the quantity of soil to be moved and the nature of the contaminant, removal to another site may be prohibitively expensive and it may be difficult to find an appropriate licensed landfill to

Box 8.13 Slurry trench cut-off walls
Slurry trench cut-off walls using self-hardening cement-bentonite are the most common type of vertical barrier used to control lateral migration of pollution and gas from contaminated land and landfill sites in the UK. The walls are constructed by excavating a trench, usually 0.6 m wide, using a hydraulic backhoe type excavator. The slurry acts to support the sides of the excavation during its construction and then solidifies to form the permanent wall. To form a satisfactory barrier to leakage migration, a cut-off wall must have a low permeability and be continuous with an adequate toe-in to an underlying aquiclude. For control of gas migration, it is normally recommended that the slurry wall includes an HDPE membrane.

Figure 8.7 Installing a slurry trench cut-off wall

deposit the soil (Section 2.4). Alternatively, following excavation, the poor ground may be sorted to facilitate the removal or treatment of contaminated and other unacceptable material. The remaining acceptable material, together with treated material, can be placed as an engineered fill. This also can be expensive and there may still be a problem with disposal of the contaminants.

(f) Improve the poor ground in situ
In-situ ground treatment can be used to improve the physical or chemical properties of the ground so that the ground-related risks to the development are mitigated. Ground treatment is described in Chapter 9. Technologies for the in-place cleaning of contaminated soil have not yet been very widely used in the UK; they are described in Section 9.9.

8.7 Foundations and services
A foundation design must be robust and able to resist a variety of ground conditions which may be encountered on a brownfield site. Geotechnical and geoenvironmental specialists have to understand the requirements of the structural designer and the structural designer should have realistic expectations of ground behaviour. Foundations are primarily required to support the building but can also form a barrier to contaminants. Two particular hazards for foundations and services which may be encountered on brownfield sites, chemical attack and gas migration, are examined here; problems associated with ground movement are considered in Chapters 6 and 7.

Chemical attack
Interaction with contamination may result in loss of strength or change in volume of the foundation material; this can cause damage to a building or services as they enter the building. The form of construction of the building and its foundations affect the vulnerability of the building materials to chemical attack. Foundation elements, such as piles, strip footings and rafts, have different exposures to attack; slender sections are more at risk than massive foundations. The thicker the building element, the less likelihood there is of contaminant attack causing damage to the component or serious damage to the building. Because of the over-design of the element, thicker building elements will last longer but durability is also related to the rate of attack. Ground supported floor slabs are especially at risk where they can dry from the top, because this encourages the movement of contaminants into the concrete from the ground.

Strip, trench-fill and pad foundations are often less than 1m deep; on many sites contact with groundwater contamination is unlikely. Raft foundations are generally built onto an engineered sub-base of clean granular fill, close to ground surface and above the water-table, and the availability of contaminants to the foundation is considerably less than for deeper foundations. The granular fill and/or an impermeable membrane can be used as a capillary break within a clean cover system to control upward migration of contaminants.

Piling and penetrative ground treatment methods, such as vibrated stone columns and vibrated concrete columns, are commonly used for foundations on brownfield sites. These foundation systems and ground treatment processes are more likely to pass through any contamination on site and to come into contact with any contaminated groundwater. Therefore, the materials must be sufficiently durable to resist degradation[31].

Services in the ground include pipework carrying water or gas, electricity and telephone cables, drainage gullies, culverts, sewers and foul drainage, and soakaways for rainwater. A variety of materials are used for underground services and foundations of overground services; they include concrete, plastics, ceramic and metal pipes, coatings for pipes, sheathing and insulation of electric cables. Physical and chemical hazards should be considered in the design of the services in relation to such matters as protection against excessive differential settlement, the permeation of polyethylene pipes with organics, and the need to avoid existing buried obstructions.

Many brownfield sites result from building demolition and contain substantial foundations. These can be a problem for redevelopment, particularly as a cause of differential settlement, and it can be a considerable deterrent to developers to have to remove or bridge over old foundations, or to site buildings away from them. Re-using old foundations can deliver substantial benefits, but there may be difficulties:
❐ The proposed building layout may not be compatible with the old foundation layout.
❐ The new structure may impose loads of a magnitude for which the old foundation does not have sufficient capacity, and load distributions markedly different from the original; this could lead to the risk of differential movement and deformation of the new building.
❐ The remaining useful life of the old foundations may be inadequate; consideration needs to be given to such factors as the state of the materials, including any degradation due to sulfate attack and the position of reinforcement.

Precautions that may prevent or control chemical attack on foundations include:
❐ Appropriate design for the mortar and concrete, including the use of suitable materials, for example, appropriate choice of cement and aggregates[32].
❐ Appropriate control of production and placing that produces a dense concrete with adequate cover to reinforcement.
❐ Protection from ground conditions.

Gas migration
A commonly encountered problem on gas-contaminated land is the protection of building occupants from the ingress of hazardous gases. Figure 8.8 shows the locations at which gas can enter houses. Gas may accullulate in wall cavities and roof voids, beneath suspended floors, within voids caused by settlement of the ground and in drains and soakaways. Stringent precautions, such as the provision of high rates of sub-floor and

cavity ventilation and the installation of gas detectors and alarms, are required for many types of occupied building. The difficulty of ensuring that these precautions remain effective over the lifetime of a gassing site generally precludes low-rise housing development on recently placed domestic refuse, but problems can also arise with building developments on older, small, uncontrolled tips. Blocks of flats, offices and retail premises can be made sufficiently safe with the necessary precautions.

The hazards posed by gas migration should be taken into account in the design and construction of buildings[33], with the detailed assistance of appropriate specialists. Precautions should be taken to prevent the passage of gases into the building, either by placing barriers to their movement or by installing ventilation measures to remove the gases[34]. Arrangements should be made for maintenance and monitoring. There are two approaches to gas protection:
❏ prevent or regulate gas emissions and migration from the gassing source;
❏ prevent migration of gas into buildings and associated infrastructure.

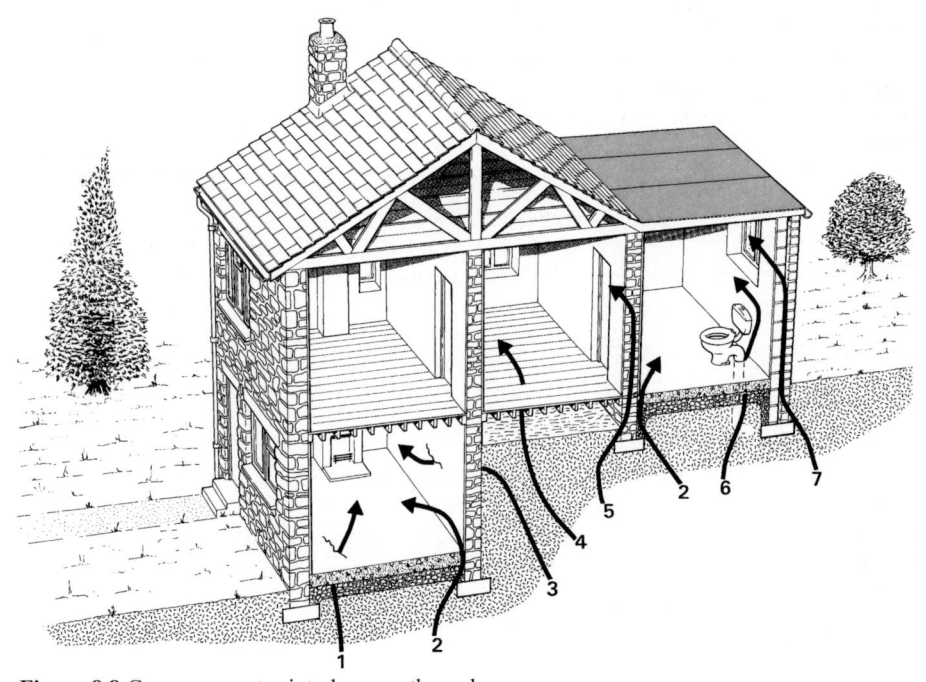

Figure 8.8 Gas entry routes into house – through:
1. cracks in solid floors
2. construction joints
3. cracks in walls below ground level
4. gaps in suspended floors
5. cracks in walls
6. gaps around service pipes
7. cavities in walls

BROWNFIELD LAND

Notes

1. A national target that, by 2008, 60% of additional housing should be on previously developed land (Department of the Environment, Transport and the Regions, 2000a) has been set against a background of social changes which could mean that up to 3.8 million extra households have to be accommodated by 2021 (Department of the Environment, Transport and the Regions, 2000b). The Urban Task Force (1999) considered that achieving the 60% target was fundamental to the health of society and recommended that ambitious targets should be set for the proportion of new housing to be developed on recycled land in urban areas of the UK where housing demand is currently low. At the time of writing it is believed that the 60% target is being met.

2. Previously developed land is land which has been affected by some human activity other than agriculture. In PPG3 (Department of the Environment, Transport and the Regions, 2000a), previously developed land is defined as land that is or was occupied by a permanent structure (excluding agricultural or forestry buildings), and associated fixed surface infrastructure. The definition covers the curtilage of the development and includes land used for mineral extraction and waste disposal where provision for restoration has not been made through development control procedures.

3. The issue of over-engineering brownfield sites has been debated by Wood and Griffiths (1994).

4. In ordinary speech, the terms *hazard* and *risk* are used more or less interchangeably, but in risk management they have acquired distinct technical meanings. Hazards are defined as properties or situations with the potential to cause harm, including human injury, damage to property, and environmental and economic loss. Risk is considered to be the combination of the probability or frequency of the occurrence of a hazard and some measure of the consequence of that occurrence.

5. BRE publications provide advice and useful background knowledge concerning building on brownfield sites. A general overview of the subject is provided by Good Building Guide GBG 59 *Building on brownfield sites*; Part 1 *Identifying the hazards* and Part 2 *Reducing the risks*. A number of publications deal with particular aspects of the subject: Digest 395 *Slurry trench cut-off walls to contain contamination*, Digest 427 *Low-rise buildings on fill* (in three parts), BRE reports BR 212 *Construction of new buildings on gas-contaminated land*, BR 255 *Performance of building materials in contaminated land*, BR 414 *Protective measures for housing on gas-contaminated land*, BR 424 *Building on fill: geotechnical aspects*, BR 447 *Brownfield sites: ground-related risks for buildings*.

6. The legacy of contamination from the defunct industrial process of gas manufacture merits particular attention. However, the rate of industrial change in the UK means that pollution from many other declining industrial processes can present obstacles to redevelopment and at the sites of derelict steelworks and obsolete chemical plants large quantities of substances harmful to the biosphere will have been used. Examples of sites likely to contain contaminants are given in Section 2 of Approved Document C of the Building Regulations.

7. A fuller description of gasworks sites is given by Leach and Goodger (1991) *Building on derelict land*.

8. During the reclamation of a 40 ha, 15 m-high steelworks waste tip at Hartlepool in 1980, some of the iron and steel slag was excavated and then refilled in thin layers with compaction to form a 4 m-deep engineered fill. The expansion of this layer was monitored between July 1980 and September 1983; it varied between 1.55% and 3.25%, with expansion still continuing. This instructive case history is described by Eakin and Crowther (1985) and is also summarised as Case history 13 in BRE Report BR 424.

9. The mechanism is described in BRE Report BR 332 *Floors and flooring*.

10. BRE Information Paper IP 18/01 *Blastfurnace slag and steel slag: their use as aggregates*.

11. See the warning in BRE Digest 276 *Hardcore*: '*Slags from steelmaking are not recommended for*

use as hardcore because they may contain phases which cause expansion on wetting, eg free lime, free magnesia or broken refractory bricks.'

12. Two examples of major building developments on brownfield sites are the Hampton township which is being built on a 1000 ha site at Peterborough, and the planned development of the 150 ha Barking Reach site north of the Thames in east London.

❐ At the Hampton site, clay was excavated for brick making and much of the site was derelict for decades. Between 1965 and 1991, large areas of land were restored to agriculture, which involved disposing of pulverised-fuel ash (pfa) as a slurry in the old clay pits. In addition to the pfa-filled pits, there were pits not filled with pfa, and also some ridge and furrow areas which required a major earthmoving operation. The presence of several thousand great crested newts, which had colonised parts of the old clay pits, led to the construction of a large newt reserve at a substantial cost. The township is planned to have more than 5000 houses.

❐ At Barking Reach, two coal-fired power stations, an oil-fired power station and a landfill had left a legacy of unsuitable materials and ground contamination which had to be dealt with. It is planned to build 10 000 new homes, schools and community facilities (*New Civil Engineer*, 22 April 2004, pp 12-13).

13. The term receptor is used to describe people, other forms of living organism, ecological systems, groundwater, or buildings and services which could be adversely affected by hazards; it forms part of the methodology for evaluating contamination problems using a source-pathway-receptor model. The term target has also been used. While buildings and people at risk from other forms of brownfield hazard can be described as receptors, many of these situations are better considered in terms of a hazard-event-consequence scenario.

14. Guidance on the risks to human health from land contamination is provided in a series of contaminated land reports (Department for Environment, Food and Rural Affairs and the Environment Agency):

CLR7 – *Assessment of risks to human health from land contamination: an overview of the development of soil guideline values and related research.*
CLR8 – *Priority contaminants for the assessment of land.*
CLR9 – *Contaminants in soil: collation of toxicological data and intake values for humans.*
CLR10 – *The Contaminated Land Exposure Assessment Model (CLEA): technical basis and algorithms.*

CLR8 deals with priority contaminants; these are identified on the basis that, at many of the sites affected by industrial or waste management activities, they are present in sufficient concentrations to cause harm and that they pose a risk to the human population, the built environment, water or ecosystems. CLR8 indicates those contaminants likely to be associated with particular industries.

15. General guidance on the performance of building materials used in contaminated land has been given in BRE Report BR 255 and by Garvin *et al* (1999).

16. Technical information on the impact of chemicals on building materials below ground is limited and largely confined to qualitative assessment based on professional judgement and experience. A notable exceptions concerns the impact of sulfates and acids on concrete which is described in BRE Special Digest SD1 *Concrete in aggressive ground*; five classes of soil contamination are defined and recommendations are made for the composition of concrete to resist sulfate and acid attack.

17. An explosion which completely destroyed a bungalow and badly injured three occupants at Loscoe in Derbyshire on 24 March 1986 is believed to have been caused by the lateral migration of methane gas from a landfill site 70 m from the bungalow (Williams and Aitkenhead, 1991). The worst disaster in the UK involving gas migration into a building occurred on 23 May 1984, when sixteen people were killed by an explosion in a water supply valve house at Abbeystead in Lancashire (Health and Safety Executive, 1985); although this accident was not associated with a brownfield site, it illustrates the major hazard posed by gas migration.

18. Two surveys of the incidence of subterranean fires in the UK were undertaken in the 1980s when it was found that the majority of fires occurred in sites containing domestic refuse and/or colliery waste.

19. British Standard BS 10175:2001, which forms a code of practice for the *Investigation of potentially contaminated sites* presents a risk-based approach to identify and quantify the hazards that may be present and the nature of the risk they pose. The Environment Agency and NHBC (2000) *Guidance for the safe development of housing on land affected by contamination* also follows a risk based approach. The stages of risk assessment are decsribed in Section 2 of Approved Document C of the Building Regulations. CIRIA Special Publication SP103 (Harris *et al*, 1995) gives guidance on site investigation and assessment of contaminated sites

20. A conceptual model of the ground conditions provides a means of assembling site data to better understand the implications of the likely behaviour of the ground for the proposed development. The principle of defining and revising a conceptual ground model for a brownfield site is a key theme and has six elements which are described in section 5.1. Four of the elements are as relevant to greenfield sites as they are to brownfield sites; the following two elements are relevant only to brownfield sites:
❐ Modifications caused by previous human activities. An understanding of the history of the site and of the ways in which it has been affected by various types of human activity is required.
❐ Soil contamination model. Source-pathway-receptor scenarios should be identified (Environment Agency and NHBC, 2000).

21. The Association of Geotechnical and Geoenvironmental Specialists (AGS) has published *Guidelines for combined geoenvironmental and geotechnical investigations*. The guidelines recognise that the activities carried out in each type of investigation can be similar, and seeks to ensure that the technical difficulties in carrying out combined investigations are recognised and overcome, and that the advantages of combining some of the activities involved in the investigations are realised. The stages of the investigation process are outlined, highlighting potential conflicts of requirements and technical difficulties at each stage and ensuring that the objectives of each type of investigation can be met by an integrated approach.

22. Chapter 4.1 of NHBC Standards deals with land quality. Documentation is required by the NHBC to verify that the site has been properly investigated and, where contamination is suspected, a form of validation has to be produced.

23. See CLR9 *Contaminants in soil: collation of toxicological data and intake values for humans*. (Department for Environment, Food and Rural Affairs and the Environment Agency)

24. The technical data in each LCR can be independently checked by engineers and other professionals accredited by the Institute of Environmental Management and Assessment as Specialist in Land Contamination (SILC).

25. Westcott *et al* (2001a & 2001b) give guidance on pollution prevention where piling and penetrative ground treatment techniques are used on land affected by contamination.

26. An example of the importance of perception has been given by Taunton and Adams (2001). It might have been expected that the removal of an old landfill in a semi-rural area of Cheshire would have been welcomed as the hazard of a contaminated site was being removed from the locality. However, the perception grew among some of the residents that there was a potential health problem related to the work and changing this perception was not easy!

27. New houses are being built at the site of Consett steelworks on ground containing slag which has the potential for major expansion. Foundations contractor Roger Bullivant Ltd has designed a foundation solution using grouted tubular steel piles, within an outer steel casing designed to isolate the loadbearing pile from any movement of the surrounding ground. The inner pile is anchored into mudstone at depths of 3 m to 7 m, while the outer casing can move to accommodate any ground

heave. The foundation scheme uses a precast concrete beam and slab house foundation system (*New Civil Engineer*, 11/25 December 2003, p37).

28. Engineering a new waste disposal site is quite different from containing an old site where filling began before the advent of properly designed lining systems. Modern disposal sites should have been constructed to contain or remove in a controlled manner any harmful gases or leachate, the liquid that accumulates in the refuse.

29. General guidance on the use of cover systems is given in CIRIA reports SP 124 (Privett *et al*, 1996) and SP 106 (Harris *et al*, 1995). BRE Report BR465 gives guidance on the thickness of cover systems.

30. A specification for slurry trench cut-off walls has been prepared by the Institution of Civil Engineers (1999). See also BRE Digest 395 *Slurry trench cut-off walls to contain contamination*.

31. BRE Digest 315 *Choosing piles for new construction* reviews the various types of piles and makes reference to maximum durability in aggressive soils and groundwater. Westcott *et al* (2001a and 2001b) have reviewed piling methods on land affected by contamination.

32. Concrete should be designed according to BRE Special Digest SD1 *Concrete in aggressive ground*. Approved Document A of the Building Regulation gives recommendations for the construction of foundations of plain concrete; it includes the design of concrete for use in chemically aggressive ground.

33. Approved Document C of the Building Regulations gives the general approach to building development on land with such gaseous contaminants as radon, and landfill gas and methane. A warning is given that where development is proposed within 250 m of the boundary of a landfill site, further investigation for hazardous soil gases may be required; the Environment Agency's policy on building development on or near to landfills should be followed.

34. BRE Reports BR 212 *Construction of new buildings on gas-contaminated land* and BR 414 *Protective measures for housing on gas-contaminated land* describe appropriate construction techniques on gas-contaminated land and describe protective measures. See also CIRIA Report 149 *Protecting development from methane* (Card, 1995).

CHAPTER 9: GROUND TREATMENT

Where the ground has poor load-carrying properties; significant differential movements could occur over the area of a building; improving the ground by appropriate ground treatment prior to development of the site can reduce the foundation problems. Before adopting a treatment technique, it is essential to define how the load-carrying characteristics of the ground are inadequate. The decision to use a particular type of ground treatment depends not only on technical suitability but also on the cost in relation to the cost of other technically adequate solutions. The treatment method most commonly used in the UK is the deep vibratory process described as vibrated stone columns. Proprietary techniques offered by specialist contractors are widely used but there are situations where methods involving bulk earthmoving may be preferable. This chapter is principally concerned with commonly used ground treatment methods designed to mitigate geotechnical problems; remediation techniques for contaminated land are also reviewed.

9.1 Introduction

Where site investigation has indicated that the ground has poor load-carrying properties, significant differential movements could occur over the area of a building; appropriate ground treatment prior to development of the site can improve the ground and form a major element in the provision of an economic and technically adequate solution to the foundation problems. The interest in ground treatment techniques has grown rapidly in the UK as the scarcity of good building land has led to construction on ground hitherto regarded as unsuitable for development.

The selected treatment technique must be able to remedy the inadequacies in the load-carrying characteristics of the ground. This means that the deficiencies in the ground properties must be correctly diagnosed before an appropriate treatment can be specified. This may seem self-evident, but in the past there has often been little engineering input to the design of ground treatment apart from that of the specialist contractor. Involving a suitably qualified geotechnical engineer in the initial investigation, design of ground treatment and site supervision, should ensure a continuity of design philosophy throughout the project. The responsibility of the different parties should be clearly defined.

Most ground treatment in the UK is carried out on non-engineered fill (Sections 3.9 and 7.4). The engineering properties are usually difficult to characterise so sophisticated design methods are of little practical use and empirical methods have to be applied. Where damaging settlements occur, they are usually due to causes not associated with the weight of a building (Chapter 7). Consequently, treatment which merely reduces settlement due to the weight of the building by stiffening a thin surface layer may be inadequate. Collapse compression on inundation is a particular hazard for loose, partially-saturated fill and may occur at depth within the fill (Sections 4.9 and 7.4). Collapse potential can be reduced or, in some cases, eliminated by densification of the fill. The more commonly encountered

techniques used to increase the density of non-engineered fill are:
- deep vibratory techniques (vibro compaction and vibrated stone columns: Section 9.2);
- compaction by surface impact loading (dynamic compaction, rapid impact compaction: Section 9.3);
- preloading (using a surcharge of fill or, occasionally, by lowering the water-table: Section 9.4);
- excavation and refilling in thin layers with adequate compaction as an engineered fill (using either excavated material or imported material: Section 9.5).

Soft saturated natural clay soil often requires treatment. Geotechnical problems with soft clay are associated with low undrained shear strength and high compressibility (Chapter 4). Inadequate bearing capacity and excessive settlement due to the weight of buildings founded on the clay are likely to be the main problems (Chapters 6 and 7). The very low permeability of the clay is one of the major factors which determine the usefulness of treatment methods on this type of ground.

Before adopting a ground treatment method on a particular site, it is essential to define the respects in which the load-carrying characteristics of the ground are inadequate; first diagnose the problem, then adopt a solution. Similar treatment methods may be used on different types of compressible ground but with different objectives. For example, some of the ground treatment techniques used on fill can also be used on soft clay (for example vibrated stone columns and preloading), but the purpose of the treatment and the mechanisms of soil behaviour which are utilised may be different.
- Where vibrated stone columns are used in a coarse non-engineered fill, it will often be possible to significantly densify the fill. However, the installation of stone columns in a soft saturated clay is unlikely to improve the clay and any benefit will be associated with the reinforcing effect of the relatively stiff stone columns.
- Preloading with a temporary surcharge of soil can be used on fill and soft clay. On fill, the density is increased by reducing the air voids and most of the compression occurs as the surcharge is placed. On soft clay, preloading consolidates the soil as water is slowly squeezed out of the pores. The speed at which consolidation occurs is controlled by the low permeability of the soft clay and it will usually be necessary to install vertical drains to achieve consolidation within a reasonable period.

The decision to use a particular type of ground treatment depends on technical suitability and cost in relation to the cost of other technically adequate solutions. Proprietary techniques have sometimes been selected because specialist contractors offer some form of 'guarantee' or insurance. There are important practical differences between the various treatment methods: some treat the whole site but others are used to treat only the ground immediately beneath the foundation. The distinction has both technical and economic consequences.

GROUND TREATMENT

The responsibility for investigation, design and construction should be clearly defined; this is particularly important where specialist treatment methods are used. Adequate in-situ testing of the treated ground is particularly important when non-engineered fill is treated. However, on small jobs the cost of testing can be a significant proportion of the total cost.

The primary purpose of ground treatment is to reduce differential settlement and create more uniform foundation conditions. Although the treatment should reduce ground movements, it is unlikely to eliminate them; ground treatment may not, therefore, be an adequate solution for structures that are sensitive to small ground movements. Very sensitive structures may have to be supported on piles.

CDM Regulations[1] apply to all stages of a construction project, including ground treatment; it is essential for safe and effective execution of treatment works that all relevant requirements are met, including provision of an appropriate risk assessment for each project as required by the Management of Health and Safety at Work Regulations[2].

This chapter is mainly concerned with commonly used ground treatment methods designed to mitigate geotechnical problems but Section 9.9 reviews remediation techniques for contaminated land. Only brief accounts, which contain general information and guidance, are given of some of the more common types of ground treatment; more comprehensive reviews of the subject can be found in a number of text books[3]. On sites where ground treatment may be required, the services of a suitably experienced geotechnical engineer will usually be needed.

9.2 Vibrated stone columns

The ground treatment technique most commonly used in the UK is the deep vibratory process described as vibrated stone columns or, more commonly, *vibro*[4]. The basic tool is a depth vibrator; this is in the form of a cylindrical poker with an eccentric weight in its bottom section. Rotation of the weight results in vibrations in a horizontal plane being transmitted to the soil as the vibrator penetrates the ground. Vibrators typically range in diameter from 0.3 to 0.45 m, and from 3 to 5 m in length. The long cylindrical hole formed by the depth vibrator is backfilled with stone, compacted in stages and tightly interlocked with the surrounding ground to form columns which stiffen the soil. The columns are usually between 0.6 and 0.8 m in diameter. Stone columns can be installed to considerable depths[5] using extension tubes, but much of the treatment for housing in the UK is carried out only to a shallow depth.

In the top-feed method, the vibrator is suspended from a crane and penetrates the ground under its own weight. Compressed air (dry method) or water (wet method) are used to remove any loose debris and assist penetration. The water is also used to support the sides of the hole in soft or unstable soils. The dry method is often used on small sites because there can be environmental objections to the quantity of water needed in the wet method and to its disposal.

The use of the bottom-feed method is increasing. The vibrator is mounted on a vertical slide feed mast on a crawler base unit; a tremie pipe delivers stone directly to the tip of the vibrator in the ground, allowing the vibrator to remain in the ground during column construction (Figure 9.1).

Two processes, vibro-compaction and vibrated stone columns, should be distinguished (Box 9.1). The distinction between the two is whether or not the soil can be effectively compacted. It has generally been considered that the soil can be effectively compacted if the percentage of silt-size particles is smaller than about 15%, but that otherwise the ground is improved principally by the presence of the relatively stiff stone columns[6].

Most of the applications of vibrated stone columns in the UK are in fill, with some in soft clay. The soil profile should control the required depth of treatment; in fill this is normally the full depth of the fill and any underlying soft natural soils. If it is not economic to provide full depth treatment, the use of partial depth treatment needs to be critically assessed before proceeding. Varying depths of fill, particularly at the boundary of the filled area, can give cause for concern.

Stone columns can form paths for water to penetrate into untreated fill at depth. In a loose partially saturated fill

> **Box 9.1 Vibro-compaction and vibrated stone columns**
> A loose sand can be compacted with no necessity to add any material, although the level of the ground surface will be lowered. This process, in which extra material may or may not be added, is known as vibro-compaction. It is rarely applicable in the United Kingdom. In most soils stone columns are installed. Vibrated stone columns are generally intended to stiffen and densify the ground, to identify weak spots and to reduce differential settlement.

Figure 9.1 Installing vibrated stone column

GROUND TREATMENT

susceptible to collapse compression on inundation, this might have serious consequences (Section 4.9). On a brownfield site, there could be concern that surface contaminants could pass down the column into an underlying aquifer; a system has been developed in which a concrete plug is installed at the base of the column.

For natural soft ground, very low undrained shear strength presents an installation problem. When using the dry top-feed process, a minimum undrained shear strength
$$c_u = 30 \text{ kPa}$$
may be required to ensure that the hole created by the depth vibrator stays open during backfilling. The dry bottom-feed process can be used in soil with lower strength and only a minimum undrained shear strength of
$$c_u = 20 \text{ kPa}$$
may be needed. A minimum value less than 20 kPa can be acceptable for the wet method. The minimum value should also be judged in relation to the required column load capacity and settlement criteria. The amount of improvement that can be achieved in natural soft ground may be quite limited. Typically, settlement might be reduced to half of the settlement of untreated ground[7]. This means that on this type of ground the method is often more appropriate to low-rise industrial buildings than settlement-sensitive housing.

Discrete zones of organic matter, or any substance that may decrease in volume with time owing to chemical reaction or solution, may have to be removed and replaced with coarse inert material. Significant volumes of such unacceptable material render a fill unsuitable for vibro treatment. The reduction in volume of the ground would lead to a loss of lateral support for the stone columns and consequent settlement.

Vibrated stone columns are well adapted to traditional housing and light industrial units where it is not necessary to treat the whole site in a uniform way. Treatment typically consists of placing lines of stone columns beneath loadbearing walls (Figure 9.2) or on a grid pattern beneath rafts. The geometry of the proposed structure constrains the spacing of the columns in plan. For small structures, such as low-rise housing, there is little scope for variation in treatment pattern; treatment points

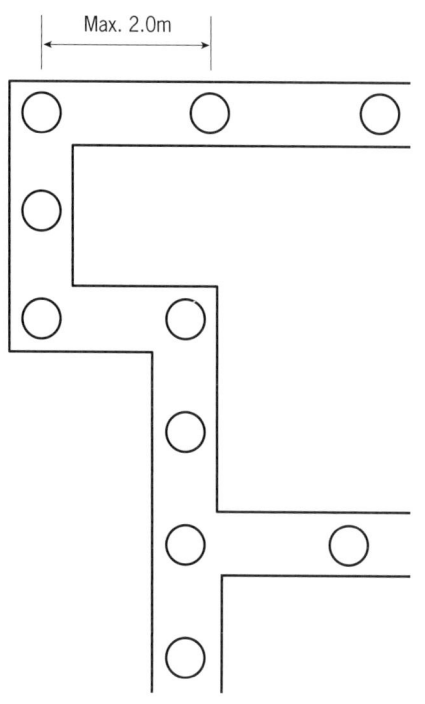

Figure 9.2 Location of stone columns under a strip footing

along each length of footing are generally at about 2 m centres with additional treatment points under the floor slab where a ground supported floor slab is considered suitable. In some cases, suspended floor slabs may be required.

Projects range in size from a few treatment points beneath strip footings for a pair of semi-detached houses, to the treatment of large areas with a uniform pattern of treatment points. The technique is particularly suited to site redevelopment and can be used relatively close to existing structures, much closer than dynamic compaction can be safely used.

There are practical problems which restrict the applicability of this type of treatment:
❐ Extensive buried obstructions can seriously impair a treatment programme if not identified and removed ahead of the works.
❐ A firm or desiccated crust may have to be pre-bored to avoid impeding the penetration of the depth vibrator.
❐ Vibration could cause settlement of adjacent structures and buried services; limiting distances of 2 m to 5 m have been quoted.
❐ Although the noise level is relatively low, there may still be environmental problems.

Vibrated concrete columns are a more recently developed treatment method[8]. The columns are installed using a modified guided bottom-feed vibro rig and concrete is pumped into the cylindrical void formed by the depth vibrator. The columns are effectively end-bearing piles, which are designed to transmit the structural loading to a suitable underlying bearing stratum.

9.3 Impact compaction

Dropping a heavy weight onto the ground surface is one of the simplest ways of compacting a soil and was used in antiquity. However, it was the advent of large crawler cranes which made the current high energy levels feasible[9]. There are two basic approaches to deep compaction by surface impacts[10]:
❐ dynamic compaction in which a heavy weight is dropped in 'free fall' from a considerable height;
❐ rapid impact compaction in which a modified piling hammer drives a steel compaction foot into the ground.

Dynamic compaction

In recent years, dynamic compaction has commonly been carried out using 8 to 10 tonne weights dropped by a crane from heights of around 10 to 15 m (Figure 9.3). The weight is usually a toughened steel plate. The treatment has mainly been used to compact fill[11].

A typical total energy input is of the order of 2000 kNm/m^2 but there are large variations depending on the type of ground and the degree of improvement required[12]. The weight is

GROUND TREATMENT

dropped a set number of times on a grid pattern to form a pass, treatment usually consisting of at least two passes; the first tamping consists of a number of high energy impacts at relatively widely spaced grid points and aims to produce compaction at depth. The second pass is at intermediate grid points with reduced energy per blow. The imprints formed at each drop position are infilled with granular material after each pass. In the final pass, the complete surface area is uniformly tamped with abutting compaction points using a low drop height to ensure compaction of the surface layers.

There are practical limitations on the use of dynamic compaction:
❐ High mobilisation costs associated with the large crane required to drop the weight usually mean that small areas cannot be treated economically. It is essentially a whole-site treatment method.
❐ It is inadvisable to use the method close to existing structures owing to the potentially damaging vibrations caused by the impact of the weight onto the ground surface; minimum distances of 30 m have been quoted, but much depends on circumstances.

Figure 9.3 Dynamic compaction

❐ Flying debris constitutes a hazard for personnel, vehicles and structures close to the impact point; some shielding of vulnerable targets may be required.
❐ With many fill sites, it will be necessary to provide a thick granular blanket to form a working platform for the crane; the cost of this granular material may be substantial in relation to the total cost of the treatment.

The use of specially shaped tampers, which are dropped in 'free fall' within 'leaders', enables a greater accuracy and potentially a higher proportion of the available energy is transmitted to the ground for a specific drop height. This development of the falling-weight method of dynamic compaction uses weights of between 4 and 10 tonnes and drop heights of up to 10 m.

Rapid impact compaction

The rapid impact compactor comprises a modified hydraulic piling hammer acting on an interchangeable articulating compacting foot[13]. A 7 tonne weight falls 1.2 m onto a 1.5 m-diameter steel articulated compaction foot at a frequency of about 40 blows per minute. The compaction process requires a minimum of two treatment phases, with high energy primary compaction using abutting compaction points to improve the ground at depth and a low-energy secondary treatment to ensure adequate compaction of the material up to the ground surface. The standard 1.5 m-diameter compaction foot is very effective at punching into the ground to give the maximum depth of treatment. A larger-square foot is often used for a lighter secondary treatment; alternatively, rolling with conventional compaction plant may be appropriate.

The technique produces compaction by impact loading and is similar in principle to dynamic compaction using a drop weight but there are differences:

❐ energy per impact is much lower;
❐ number of blows per unit time is much greater;
❐ foot that transmits the energy to the ground remains in contact with the ground.

The rapid impact compactor can improve the engineering properties of a range of fill and natural sandy soils to depths of about 4 m. Figure 9.4 shows the settlement versus depth profile for a building waste fill treated by the rapid impact compactor.

Vibrations are generally much smaller than with dynamic compaction, although the greater blow rate may lead to a less favourable perception. Measures have been taken to reduce noise emissions. The process produces virtually no flying debris and could work safely, with normal precautions, in close proximity to other site operations or structures, including roads and railways, where vibration and noise alone may not be of great importance. However, vibration and noise may make the method unsuitable for some urban sites.

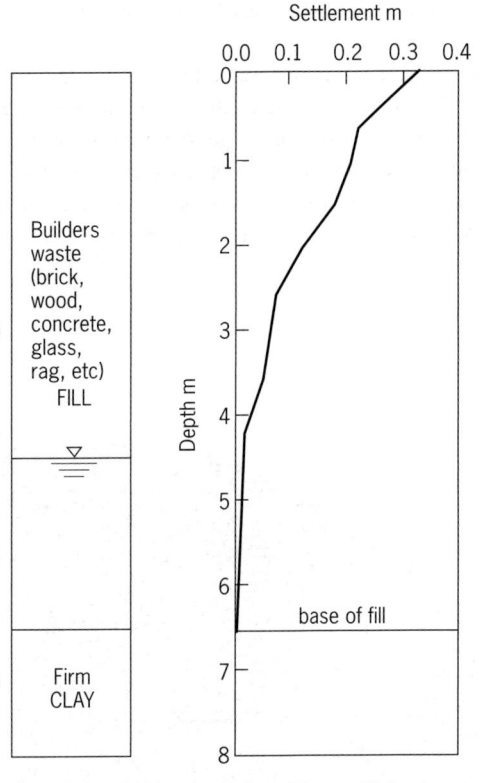

Figure 9.4 Settlement induced by rapid impact compaction of building waste
– *after Watts and Charles, 1993*

GROUND TREATMENT

9.4 Preloading

The superior load-carrying characteristics of many natural soils result from over-consolidation from preloading during their geological history (Section 3.2). Over-consolidation has made the soil stiffer than a comparable, normally consolidated soil. In a similar way, the load-carrying characteristics can be improved by temporary preloading with a surcharge of fill prior to construction. The method has been used on two very different types of soil:
- uncompacted fill with large air voids;
- soft saturated natural clay soil.

Where fill is in a loose unsaturated state, compression occurs mainly as the surcharge is placed and air voids are immediately compressed. Consequently, there is no need to leave the surcharge in position for an extended period. Figure 9.5 shows a 9 m-high surcharge used to improve a stiff clay fill prior to building on the site and Figure 9.6 shows the rapid response of the stiff clay fill to surcharge loading.

Consolidation of soft saturated natural clay soil may take a considerable time (Sections 4.5 and 6.8) because it is controlled by the low permeability of the clay[14]. The rate of placement of the surcharge may have to be controlled to ensure that the preloading does not cause instability owing to excess pore water pressures. It may be necessary to install vertical drains to speed up the rate of consolidation; relatively rapid drainage usually requires the installation of vertical drains as closely-spaced as practicable[15]. Prefabricated

Figure 9.5 Preloading with surcharge of fill

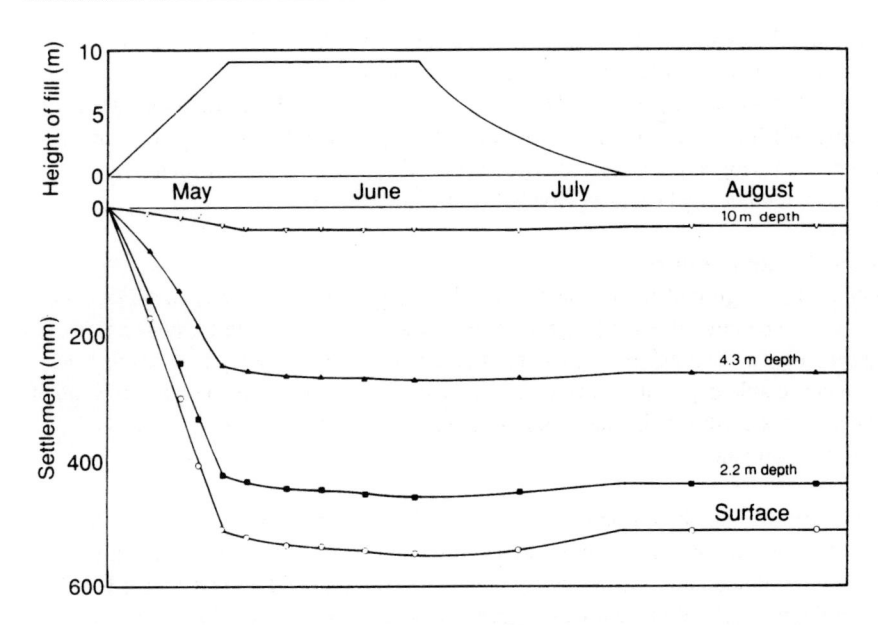

Figure 9.6 Settlement induced by preloading clay fill

band drains are most commonly used; typically they are 100 mm wide and a few mm thick, with a plastics core wrapped in a paper or fabric filter. Prefabricated sand wicks are also used.

The height of the surcharge should be calculated in relation to the properties of the ground that it is necessary to improve and the required depth of improvement. This is simple where settlement caused by the weight of buildings is the major problem, but more difficult where other causes of settlement are perceived to be major hazards[16].

Preloading with a surcharge of fill requires only normal earth-moving machines; the appropriate type of plant for a particular job depends on the quantity of surcharge fill to be moved, the haul distance and the trafficability of the surcharge fill. Treatment by preloading does not require specialist equipment or skills and the services of a specialist contractor are not needed unless vertical drains are required. Despite the simplicity of earthmoving methods, such as preloading with a surcharge of fill and excavation and refilling (Section 9.5), the lack of commercial interest in promoting them can result in these solutions being overlooked.

Factors which limit the practical usefulness of the method include:
❐ A relatively large site area is needed for preloading with a surcharge of fill to be feasible.

GROUND TREATMENT

❏ It is essentially a whole-site treatment method and cannot be readily adapted to the treatment of small areas where foundations are to be built.
❏ The cost of treatment depends on the haul distance for the surcharge fill; consequently a local supply of fill is usually required. However, on a large site the surcharge fill can be obtained by removing a relatively thin layer of material over an extensive area. The surcharge can then be applied successively to adjacent parts of the site.

9.5 Excavation and refilling

Excavation of poor ground and replacement with an engineered fill can be an appropriate solution where shallow mine workings or loose fills are present. Where the poor ground is deep, it may not be practical to excavate the full depth so it is necessary to determine the minimum acceptable depth of excavation. This method, like preloading with a surcharge of fill, involves bulk earth-moving and does not require specialist techniques; it is essentially a whole-site treatment.

The engineered fill may be formed from the original ground material, with unsuitable material identified and removed from the site during excavation. Alternatively, imported fill can be used. Foundations set partly on fill and partly on natural ground should be avoided and the site should be worked in such a way that structures are located either directly on natural ground or directly over a roughly equal thickness of fill. Some stepping of the natural ground may be necessary (Figure 9.7).

Compaction of the fill in thin layers should be carried out to an appropriate specification under controlled conditions[17]. For earthfills compacted to support low-rise structures, loose layer thicknesses greater than 250 mm are unlikely to be satisfactory, and layers of

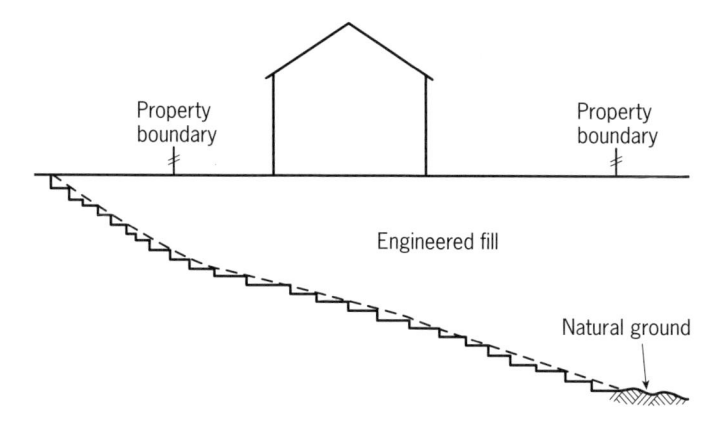

Figure 9.7 Benching of sloping natural ground for engineered fill
– *after Trenter and Charles, 1996*

200 mm or less may be necessary. With rockfills, thicker layers may be acceptable provided that adequately heavy compaction plant is available. Where landscape or other non-loadbearing areas form part of a development, they need not necessarily receive the same level of compaction as the loadbearing areas. Where a lower level of compaction is acceptable, there should be a transition zone around the loadbearing area (Figure 9.8). Sensitive structures may merit a capping layer formed from special fill.

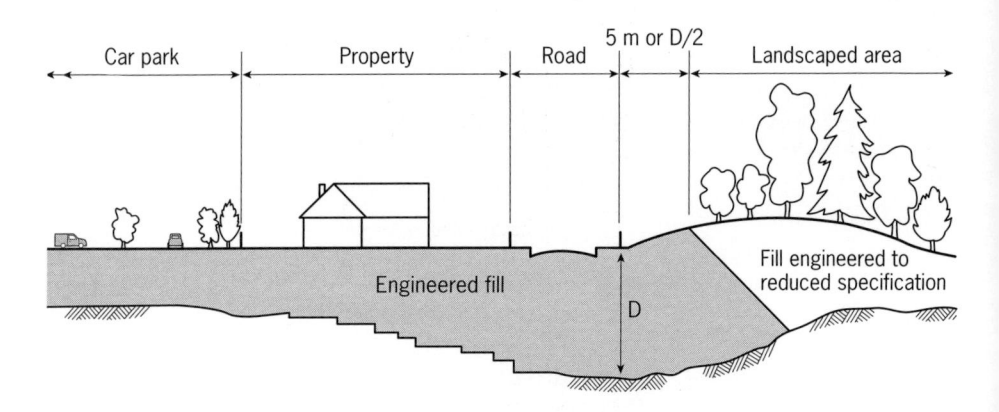

Figure 9.8 Fill beneath buildings, hard standing and landscaped areas
– *after Trenter and Charles, 1996*

The water content of earthfill at placement is fundamental to subsequent performance. If the fill is too dry, heave or collapse settlement could occur; if the fill is too wet, it may have insufficient strength and high compressibility. The significance of this is described in Box 9.2 in terms of laboratory compaction parameters (Section 4.10 and Figure 9.9).

Adequate site supervision and testing during the placement of engineered fill is vital (Section 9.10). The type and quantity of plant selected for a particular job depends on the area and depth of the excavation, the time available for earth-moving and the requirements of the specification. The behaviour of the engineered fill should be well defined and it should be possible to make reliable estimates of settlement under working loads, including long-term settlement and bearing capacity.

The method has wide application but there are sites where it would be difficult to use:
❒ On a small congested site, fill handling and storage is not easy; there may be insufficient space available for temporary storage of excavated material and it may be impracticable to excavate to any great depth.

GROUND TREATMENT

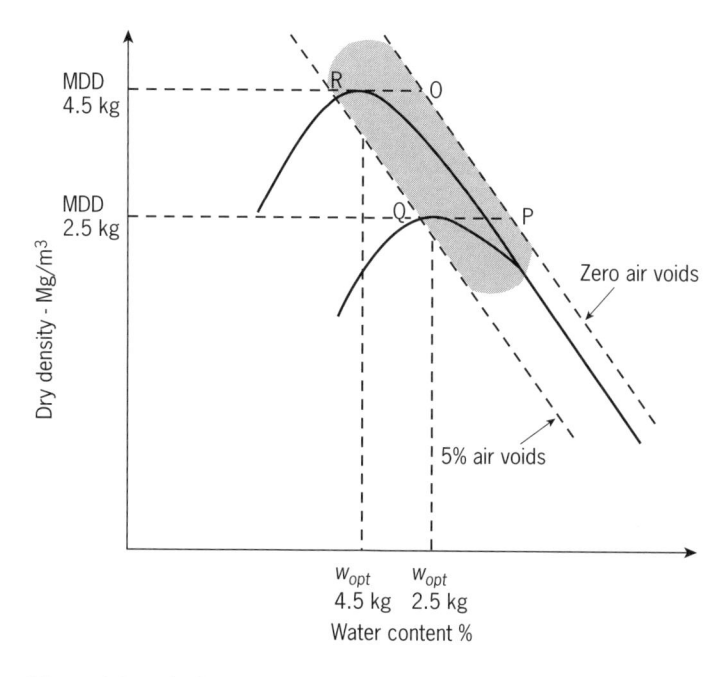

Figure 9.9 Basis for design of engineered fill

❐ Where there is a high groundwater level, excavation may necessitate dewatering the site; very wet fill is difficult to handle.
❐ The nature of the poor ground can present problems for excavation; special safety techniques and equipment may be required where the ground is contaminated or where combustion is occurring.
❐ The nature of the poor ground may be such that it is unsuitable for re-use on the site and difficult to find a site to which it can be taken; this may be the case where the ground is contaminated or biodegradable. A supply of fill from a near-by site is required.

Box 9.2 Placement water content for earthfill

Typical laboratory compaction curves for the standard 2.5 kg rammer Proctor compaction test and the 4.5 kg rammer heavy compaction test are shown diagrammatically in Figure 9.9. The significant parameters are the optimum water content w_{opt} and maximum dry density (MDD) for each test. The shaded area OPQR, which represents the range of water contents and densities expected to be acceptable for most earthfills designed to support low-rise structures, is bounded by the zero air voids line (OP), the 5% air voids lines (RQ), the maximum dry density for the 4.5 kg rammer method (RO), and the maximum dry density for the 2.5 kg rammer method (QP). For heavily loaded foundations or settlement sensitive structures, water content and density values should be sought in the upper part of the diagram, near line RO or above. Lightly loaded buildings which are not sensitive to settlement could remain in the lower part of the diagram, near line QP and below. It should be kept in mind that heavily compacted plastic clays are prone to heave.

9.6 Drainage

Water has a dominating effect on soil behaviour so control of water is a vital consideration at many sites. Ground treatment can involve the removal or addition of water to the soil and these methods include various types of drainage and inundation. Successful building development on some types of ground are largely influenced by, and contingent upon, appropriate drainage measures. While the natural processes of evaporation and transpiration can help, these processes may result in fine soils acquiring a crust of stronger material which is quite thin and which gives a misleading impression of the condition of the bulk of the soil.

Soil behaviour is controlled by the principle of effective stress and drainage methods can be used to reduce pore water pressures (Section 4.3). Deep drainage can control groundwater level, whereas surface drainage can control infiltration of surface water into the ground. Slope drainage can reduce pore pressures and hence improve slope stability and reduce the possibility of erosion. Although drainage can be a very effective method of stabilising a slope, as a long-term solution it suffers from the need for periodic maintenance to ensure continued functioning. The various functions of drains are described in Box 9.3.

> **Box 9.3 Functions of drains**
> Drains take several different forms and fulfil a wide variety of functions:
> ❑ Shallow drains can increase run-off and thereby reduce seepage forces and lower groundwater levels; they are relatively cheap and easy to maintain but are likely to fall rapidly into disrepair if neglected.
> ❑ Deep drains can modify the seepage pattern within the soil or rock mass, thereby removing water and decreasing pore water pressures in the vicinity of a potential slip surface; they are generally more expensive than shallow drains, but are likely to be more effective and to require less maintenance.
> ❑ Drains can dissipate excess pore-water pressures generated, for example, during the construction of an embankment. These drains may be shallow or deep, and are required to operate for only a limited period of time.

An open ditch is the simplest type of drain. Cheap to construct and capable of carrying high levels of discharge for short periods, ditches are commonly used on natural slopes to carry away discharge from springs, to lower the maximum level of ponds or to re-route streams. They are difficult to keep operational: high rates of discharge cause scour, low rates allow the growth of weeds. Scour can be reduced by lining the ditches with concrete but this increases cost and can be disrupted by small movements. The outfall of the ditch needs careful detailing to ensure that the discharge, together with any entrained material, does not cause problems.

Shallow gravel-filled trenches are often used in a herringbone or chevron pattern to intercept run-off on the face of a slope. To fulfil this function, they must be open at the top, not covered with earth. They need to be protected by a filter to prevent fines penetrating and blocking the drainage pathways. Where high rates of discharge are anticipated, it may be necessary to provide a perforated pipe to increase capacity. It may also be advisable to concrete the base of the trench to protect it from erosion.

Deeper rubble or gravel-filled trenches can be used to stabilise shallow slides by lowering groundwater levels. They are particularly effective where there is a layer of highly disturbed soil or debris overlying undisturbed soil of much lower permeability. Where the trench penetrates into the undisturbed soil it may provide some mechanical buttressing; this is described as a counterfort drain. The effectiveness of these drains depends on their spacing, depth and the ratio of the horizontal to vertical permeability of the ground[18].

Vertical drainage can be enhanced by the installation of drains and wells. Vertical drains can accelerate the consolidation of a preloaded saturated clay soil; band drains are commonly used (Section 9.4). Well points and deep vertical wells are similar in principle to vertical drains. Wells can be pumped out if necessary and used for in-situ permeability tests. Vacuum well points accelerate the consolidation of fine soils, but the cost of electrical power can make this expensive. Where it is necessary to modify seepage in the vicinity of a deep slip surface, vertical drains can be used to drain a perched water-table to an underlying permeable stratum.

Horizontal drainage layers, sometimes called drainage blankets, can be incorporated in an engineered clay fill during construction to accelerate consolidation. The drainage layers are installed between layers of fill or between the fill and the underlying soil; they allow lower permeability materials to be used as fill, which would otherwise be unacceptable because of the build-up of pore pressure during fill placement.

9.7 Soil mixing

The behaviour of certain types of soil, such as fill and soft natural clay, may be improved by treating them with an admixture. The stabilising agent, which can be a liquid, slurry or powder, is physically blended and mixed with the soil. While some admixtures simply act as a binder, active stabilisers, such as lime and cement, produce a chemical reaction with the soil with consequent desirable changes in the engineering properties. Chemical stabilisation may have one or more objectives:
- to improve load-carrying properties (for example improve bearing capacity, reduce compressibility, increase resistance to water softening);
- to reduce permeability;
- to stabilise chemical contaminants.

Additives may be mixed with the ground:
- during placement of fill in layers;
- by an in-situ deep mixing process[19];
- by injection into the ground (Section 9.8).

In-situ mixing of soils and stabilising such materials as cement, lime and bitumen has been used in highway works for many years. The surface soil layer is broken up, additives (and water when required) are mixed in and then the treated layer is compacted. The mixing has been achieved by earth-moving machines and only a very shallow depth of soil is affected. Quicklime is commonly used with fine soil, cement or pulverised-fuel ash with coarse soil.

Although the addition of lime and cement to clay soils to improve their engineering characteristics is well established, in a small number of cases the stabilised layer fails to meet the specification owing to sulfate attack[20]. Sulfates in the ground react with the lime or cement and water to form the highly expansive sulfate hydrates, ettringite and thaumasite. Types of construction which have been affected include capping layers for roads, hardstanding for vehicles and stabilised materials under the groundbearing floors of warehouses and retail units.

In-situ deep mixing to form stabilised soil columns is a more recent development[21]. Dry mixing methods are commonly used in Scandinavia in soft clay to form lime, lime/cement and cement columns. In the deep soil mixing method developed in Japan, a stabilising agent such as cement slurry is forced into the ground under pressure and mechanically mixed with soft clay; the method has been used to improve the ground under the foundations of detached houses.

9.8 Grouting

Fluid grout is injected into the ground to reduce the permeability or to improve the load-carrying properties. Successful grouting requires a high level of specialist knowledge and expertise. There is much literature on the subject[22].

Care is needed to ensure that grouting operations do not inadvertently cause uplift of surrounding structures. Where there is a risk of this, or where grouting is being undertaken to counteract the settlement of a building, optical levelling can be used to monitor settlement. More sophisticated systems based on the use of electro-level allow the movements to be monitored continuously (Section 10.9).

Grouting techniques
There are four basic types:
- permeation grouting;
- hydrofracture grouting;
- compaction grouting;
- jet grouting.

In permeation grouting, the grout is injected into coarse soil or fissured, jointed or fractured rock to produce a solidified mass to carry increased load or to fill voids and fissures to control water flow. The grout is introduced into the soil pores without any essential change to the soil structure; a wide range of cement-based and chemical grouts are used. With the tube-à-manchette method, sleeved port pipes are installed in a predetermined pattern and permeation of the ground is by injection through discrete ports at specified intervals, rates and pressures. Re-injection via adjacent or previously injected ports is possible.

In hydrofracture grouting, the ground is deliberately fractured and the grout is forced into the fractures so that lenses and sheets of grout are formed, causing an expansion. The process can be used to control or reverse the settlement of structures. Grout injection tubes are installed to a predetermined pattern and grout is injected through sleeves with careful process control to induce compensating movements. The control of settlement is carried out from outside the building so there is no disruption to the occupants. The process can be repeated allowing continuing settlement to be controlled. Control can be very selective, inducing small level changes over the space of several metres. Because the process requires that the soil is fractured and not permeated, hydrofracture grouting may be used in most soil types ranging from gravels to clays or weak rocks.

In compaction grouting, stiff, mortar-type grout is injected under relatively high pressures to displace and compact soils. A dense coherent bulb of grout is formed around the injection point and the surrounding soil is compressed. Compaction grouting is most effective in coarse soils where disturbance has occurred. Normally sand:cement grouts are used with additives. Grout pipes are installed in a specified pattern to the required depth. Grout is injected until a predetermined criterion is met, such as a maximum grout volume or until ground or structure heave is observed.

Jet grouting uses high pressure erosive jets of grout to break down the soil structure and partially replace the soil, with the remainder mixed with the ground in situ. Soil is loosened by the jetting action and the loosened soil is partially removed to the surface via air-lift pressure as the remaining soil is simultaneously mixed with grout. A wide range of soil types can be treated, from gravels to clays above and below groundwater level. Jet grouting can produce solidified ground of predetermined shape, size and depth, to a characteristic strength, permeability or flexibility.

Old mine workings
One particularly important use of grouting techniques is in the mitigation of the hazard posed by old mine workings (Section 7.7). They are usually stabilised by injecting grout. The efficiency of the grouting can be tested by drilling holes to intercept the treated workings and carrying out injection tests. Testing cores can be misleading in view of the variability of the material encountered by the grout. The grouting operations are normally undertaken by geotechnical or mining engineers with appropriate knowledge and

experience. Box 9.4 describes the treatment of old mine workings. Old mine shafts can be a major hazard for buildings. The treatment of old shafts, which depends on the extent to which they have been backfilled, is described in Box 9.5.

9.9 Remediation of contaminated land

With the current emphasis on the redevelopment of brownfield sites, the remediation of contaminated land has become a substantial business in the UK[23]. The approaches to risk management of contamination and the types of remedial action can be categorised in terms of the source-pathway-receptor methodology[24].

Source
❐ excavate and remove the contaminant;
❐ treat the contaminant by physical, chemical or biological processes to neutralise or immobilise the source.

Pathway
❐ block the pathway by isolating the contaminant beneath protective layers;
❐ prevent migration by installing barriers.

Receptor
❐ remove receptor by changing layout of the building development;
❐ protect receptor by selection of resistant building materials.

While there is substantial experience in treating the physical problems encountered on brownfield sites, remediation techniques for contamination have not been so extensively used and, consequently, there is less understanding of what can be achieved in respect of remediation. There can be unrealistic expectations of what is likely to be

---

Box 9.4 Treatment of old mine workings
Old mine workings are usually stabilised by the injection of cementitious grout mixed with pulverised-fuel ash (pfa). Where large cavities are to be filled, sand or pea-gravel may also be added. For collapsed or partially collapsed workings, primary grouting typically is carried out through 50 to 75 mm-diameter holes on a grid at 6 m centres, followed by secondary grouting at 3 m centres. Grouting is normally continued until the grout returns to the surface or a limiting pressure is reached. Where open workings are to be filled, a perimeter grout curtain is essential to minimise the grout wastage. This is normally formed by using a larger diameter hole (75 to 100 mm) with sand or fine gravel added to the grout as it is fed into the hole under low pressure.

Box 9.5 Treatment of old mine shafts
The treatment of old mine shafts depends on the extent to which they have been backfilled (whether they are empty, partially backfilled or completely backfilled).
Empty shaft A shaft that is open or just capped at the surface is normally consolidated by filling from the base of the shaft; sand is satisfactory for dry shafts but coarser material, such as gravel or hardcore, is preferred where there is water. Colliery spoil can be used provided it is relatively incombustible and chemically inert.
Partially backfilled shaft Where the backfill is resting on a platform or obstruction, the best solution is to drill through the barrier down to the base of the shaft and backfill with sand or gravel to within a few shaft diameters of the barrier. The remaining void is filled with a cement grout and the fill above the barrier is treated if necessary.
Completely backfilled shaft A shaft containing loose backfill can be injected with cement grout to fill any major voids and prevent collapse. A hole is drilled and cased to the bottom of the shaft, then a cement-based grout is injected under pressure as the casing is gradually withdrawn up the shaft. Large-diameter shafts are likely to need an array of holes to ensure that all the fill is properly consolidated.

GROUND TREATMENT

achieved by particular techniques and an unwarranted enthusiasm to apply new techniques at considerable cost without first establishing applicability and efficiency[25]. A more realistic balance is needed between target clean-up criteria and what the available technology can achieve[26]. The interaction of chemical and biological processes with physical processes requires careful consideration.

A large number of remediation processes are currently available[27]; some of the more commonly used are listed in Table 9.1. Some of these have been used for some time; others are dependent on new technologies and are less well established.

The most common solution has been to excavate and remove the contaminated material to a licensed site. The main advantage is that future liability for any harm caused by the contamination is removed. However, this an increasingly expensive option (Section 2.4).

Material removed from a site can be treated to remove contaminants, then returned to the site for placement as an engineered fill (Section 9.5). The volume removed permanently from the site may need to be replaced. Where this is the case, suitable imported material should be compacted to an appropriate engineering specification to address any problems of residual contamination and to provide suitable load-carrying characteristics.

Physical removal by excavation is the most basic solution to a contamination problem but, as Table 9.1 indicates, there are a number of more sophisticated ways of removing or reducing the risk posed by the contamination. The pathway between the source and the receptor can be physically blocked by civil engineering works, such as slurry trench cut-off walls, cover layers, various types of grouting, and interception wells. The receptor can be removed to some other site or may be provided with some form of protection.

Treatment of groundwater can be very important. In-situ treatment involves the flow of groundwater through a treatment zone and generally requires prior removal of the source contaminant. Thereafter, the natural groundwater flow, or a modification due to injection or pumping, can be used to cause groundwater to flow through the treatment zone, the treatment itself being achieved through chemical reaction, volatilisation or degradation. Ex-situ treatment requires extraction wells, monitoring wells and above-ground treatment plant. Changes to groundwater levels can have an impact on the physical behaviour of the ground; rising groundwater levels may cause settlement of fill or heave of natural soil, changes in water content may initiate degradation or chemical processes; lowering or cycling of groundwater may result in consolidation and improvements in soil properties.

Chemicals that can destroy or detoxify contaminants, or reduce their availability to the environment, are generally introduced to the soil by application to the surface layers via trenches or, for deeper treatment, pumped into wells. Chemical treatment can be combined

Table 9.1 Techniques used in remediation of contaminated land

Technique	Description	Objective	Application	Limitation
Excavation and refilling	Ground is excavated and suitable material compacted in thin layers	Unsuitable material is replaced with engineered fill	Very wide	Cost of removal of contaminated soil
Stabilisation/ solidification	Binding agent put into ground by soil mixing augers or pressure injection	Immobilise contaminants by solidifying or binding into a matrix resistant to leaching	Heavy metals and some organic contaminants	Effectiveness depends on physical properties
Grouting	Pumpable materials injected into ground	Increase strength, stiffness and density, reduce permeability, infill cavities	Very wide	Difficult to control where grout goes
Leaching and washing	Contaminants removed in solution by leaching, and in suspension or an emulsion by washing	Remove contaminants that are carried or mobilised by liquids	Wide range of contaminants	Applicable only above water-table
Bioremediation	Suitable bacteria or fungi introduced into ground for in-situ treatment; ex-situ methods employ treatment beds or bioreactor systems	Transform, destroy or fix organic contaminants by natural processes of micro-organisms	Organic contaminants	Cannot degrade inorganic contaminants
Soil vapour extraction/air sparging	Gaseous contaminants removed in vapour phase; contaminated fluid then treated	Remove contaminants that can be carried by air	Volatile and semi-volatile contaminants	May be limited to sandy soils
Groundwater treatment	Ex-situ treatment involves extraction of groundwater and treatment above ground	Remove contaminants from groundwater	Contaminated groundwater	Depends on hydrological and geological conditions
Chemical treatment	Chemicals introduced into ground via trenches, or pumped into wells for deeper treatment	Destroy/detoxify contaminants	Pesticides, solvents and fuels	May affect flora and fauna
Thermal processes	Heat to 150°C to volatise water and organic contaminants; vitrification needs higher temperatures	Remove or immobilise contaminants by heating	Vitrification has been used on VOCs, PCBs and heavy metals	Dewatering may be necessary
Cover layers	Single or multiple layers of engineered fill and geosynthetic materials placed on ground surface	Reduce movement of contaminant	Forms element of a containment system	Surface clay layer may dry out, greatly increasing permeability
Vertical containment barriers	Formed by sheet pile wall, grouting, soil mixing or slurry-trench wall	Reduce movement of contaminant by encapsulation	Forms an element of a containment system	Difficult to achieve very low permeability

GROUND TREATMENT

with deep soil mixing to ensure good coverage of the chemical.

Contaminant toxicity, mobility and concentration may be reduced in some situations by naturally occurring processes, without human intervention. Contaminant plumes that can be dealt with in this way are likely to be above the water-table. Contamination will remain on site, and should therefore be considered in the geotechnical design and construction sequencing since piling, excavations, ground treatment or any hydrological changes could affect the state of the contamination. Monitoring should be carried out during construction and in the longer term, and issues of long-term liabilities should be addressed.

9.10 Specification and quality management

It is important that ground treatment and remedial measures are carried out to an appropriate specification under controlled conditions, with adequate testing to ensure that the required quality is achieved. For some of the more common types of ground treatment, specifications and codes of practice have been published[28]. This section is mainly concerned with the widely used geotechnical treatment processes described in Sections 9.2 to 9.5, which aim to improve the load carrying properties of the ground; it is also relevant to the types of treatment described in Sections 9.6 to 9.9.

Specification

Three basic approaches to specification can be identified:
- method;
- end product;
- performance.

The specification may, therefore, describe the method of treatment in detail, specify some required test results or define the required performance of the fill. The relative merits of these approaches have been widely debated and there are problems with all three.

The precise treatment procedure is stated in a method specification. Site control involves inspection of the works to ensure compliance. However, the ground may be heterogeneous and it may be necessary to modify substantially the treatment as the work progresses and the extent of the required improvement becomes clearer.

In an end product specification, the treatment is specified in terms of a required value for some property or properties of the ground following treatment. Control by on-site testing should be rigorous. However, test results may be difficult to relate to field performance.

In a performance specification, some facet of the behaviour of the treated ground is specified. A performance specification will be satisfactory only if a significant aspect of field performance can be measured shortly after treatment. This might be in terms of, for example, a maximum permissible settlement over a specified period or a required result from a zone loading test. The performance specification may appear attractive to a client as it provides a direct link with performance requirements and places responsibility for performance on the contractor. However, although compliance with a performance requirement may be checked, it may be difficult at such a late stage to obtain any adequate redress where non-compliance with the specification is established.

Quality management
Supervisory personnel should be on site during ground treatment, not just the operatives. This will increase costs but is considered essential to ensure adequate quality control and quality assurance. Appropriate testing is required and depends on the type of ground treatment used. This may have some or all of the following objectives:
- to investigate the properties of the ground prior to treatment;
- to confirm that a specified improvement has been achieved, testing may be carried out after treatment;
- to assess the degree of improvement effected by ground treatment, by comparing properties measured before and after treatment;
- to assess the effectiveness of the treatment as treatment progresses, geotechnical testing may be needed during treatment;
- to confirm that, for example, vibration and noise levels are acceptable, environmental testing may be needed during treatment;
- to determine the load-carrying characteristics of the treated ground by field load tests;
- to monitor long-term movement subsequent to treatment.

Various types of testing are described in Box 9.6. The most common form of testing for vibrated stone columns is the plate loading test. A 600 mm-diameter plate is placed on top of a stone column and the load versus deformation behaviour is determined during a relatively quick loading and unloading cycle. The load is applied by a hydraulic jack using the weight of a vehicle or crane as reaction. This type of test is normally carried out as a routine control procedure. It may give some indication of workmanship and uniformity but the results of the test cannot be used for design or to predict the long term movements of structures which stress a large number of columns and the intervening ground.

Box 9.6 Testing ground treatment
Many different forms of testing may find application where ground treatment is used on a site:
- laboratory tests at the design stage to predict field behaviour (Sections 4.5 and 4.6);
- control testing during the treatment process;
- load tests to confirm predicted behaviour;
- in-situ penetration tests, such as SPT, CPT, DP (Section 5.7) to compare treated and untreated ground;
- geophysical measurements (Section 5.8) sometimes prove useful in indicating improvement;
- long-term settlement of buildings built on the treated ground should be monitored where possible.

GROUND TREATMENT

Notes

1. Construction (Design and Management) Regulations 1994.

2. Health and Safety Commission (1992).

3. Detailed accounts of ground treatment can be found in the specialist text books compiled by Van Impe (1989), Moseley and Kirsch (2004) and Xanthakos *et al* (1994). CIRIA have published two useful reports on the subject: C573 *A guide to ground treatment* (Mitchell and Jardine, 2002) and C572 *Treated ground - engineering properties and performance* (Charles and Watts, 2002). A number of conference proceedings, such as the proceedings of the Third International Conference on Ground Improvement Systems (Davies and Schlosser, 1997), contain much valuable information.

4. Helpful guidance on this ground treatment method is in BRE Report BR391 *Specifying vibro stone columns*. A European standard *Execution of special geotechnical works – Ground treatment by deep vibration* is being prepared.

5. The vibrocompaction of a sand fill to a depth of 56 m in Germany has been reported by Degen (1997). Such a treatment depth is very unusual.

6. It has been affirmed that the development of new vibrators and modified construction techniques has enabled sands with significantly higher fines content to be compacted (Slocombe *et al*, 2000).

7. Numerous analytical, laboratory and field studies have been carried out to investigate the relationship between the spacing and size of stone columns and the reduction of settlement that installation of the columns has effected. Some of these data are summarised in Figure 9.2 of Charles and Watts (2002).

8. Vibrated concrete columns were developed in Germany and introduced into the UK in 1991.

9. Dynamic compaction has been used in the UK from the early 1970s.

10. Both approaches to surface impact compaction are covered in BRE Report BR458 *Specifying dynamic compaction*.

11. It has been claimed that the method can be successfully applied to soft saturated clay, where it has been termed 'dynamic consolidation'. However, this type of application is not usually recommended.

12. There is a considerable literature on the depth and degree of improvement achieved by dynamic compaction. Mayne *et al* (1984) presented a large quantity of field data. Charles and Watts (2002) have provided an introductory summary of this information.

13. The mobile rapid impact ground compactor was initially developed for repairing explosion damage to military airfield runways and has been adapted for use in building and civil engineering applications. Field trials carried out by BRE have confirmed that this is a promising technique for the improvement of miscellaneous fill of a generally granular nature up to depths of about 4 m (Watts and Charles, 1993).

14. Consolidation times are a function of the square of the length of the drainage path and horizontal permeability is usually much greater than vertical permeability.

15. It was once common to install vertical sand drains, which were 200 mm to 500 mm in diameter, and spaced from 1.5 m to 6 m centres, using either displacement or non-displacement techniques. Prefabricated band drains are now generally used.

16. Charles *et al* (1986) and Charles (1996).

17. The best known earthworks specification in the UK is *Specification for highway works* (Highways Agency, 1998). However, this specification is applicable to fills placed to form highway embankments. Fills which are placed for building on are likely to require a higher standard and a

model specification for engineered fill for building purposes has been developed by BRE (Trenter and Charles, 1996). Some pitfalls have been identified by Charles *et al* (1998). See also BRE Digest 427 Part 3 *Low-rise buildings on fill: engineered fill* and Parsons (1992).

18. Bromhead (1992) has provided design charts to determine the required spacing of drains to achieve a given reduction in groundwater level.

19. A classification of deep mixing methods has been presented by Bruce *et al* (1999):
❐ binder – slurry (W) or dry (D)
❐ mixing mechanism – rotary (R) or jet assisted (J)
❐ location of mixing action – over length of shaft (S) or at end of mixing tool (E).

20. Longworth (2004).

21. Deep foundation stabilisation of soft clay was first used in Sweden in 1967 and the method has been widely used in Scandinavia since 1975. From the early 1970s, deep soil mixing has been developed in Japan to improve the properties of cohesive soils to considerable depths; cement or lime have been used and depths of as much as 50 m have been treated; cement is now the primary agent.

22. A helpful account of *Grouting for ground engineering* is in CIRIA report C514 (Rawlings *et al*, 2000). Moseley and Kirsch (2004) and Xanthakos *et al* (1994) have several chapters covering various types of grouting.

23. Kwan *et al* (2001) stated that over £500 million is spent every year in the UK on removing contaminants from polluted sites.

24. The source-pathway-receptor methodology provides a helpful way of assessing the hazard posed by contamination.

25. The requirements for low-rise housing are particularly onerous and reference should be made to *Guidance for the safe development of housing on land affected by contamination* (Environment Agency and NHBC, 2000). The latest revision of Approved Document C of the Building Regulations gives much emphasis on contamination.

26. *Realism in remediation* by Clark (1998) is a helpful guide to what can be achieved.

27. Useful reviews of remedial processes for contaminated land have been provided by Clark (1998) and Loxham *et al* (1998). CIRIA have published many informative reports including SP101 – SP112 *Remedial treatment for contaminated land* in 12 volumes, the most relevant of which are listed in the *References* under Harris *et al* (1995, 1998), and C548 *Remedial processes for contaminated land – principles and practices* (Evans *et al*, 2001).

28. In 1987, the Institution of Civil Engineers published *Specification for ground treatment* with *Notes for guidance*. These two documents dealt with the general requirements for ground treatment work and specialist aspects of vibrated stone columns, dynamic compaction and deep drains. *Specifying vibro stone columns* (BR391) was published by BRE in 2000 and *Specifying dynamic compaction* (BR458) in 2003. The Institution of Civil Engineers published *Specification for cement-bentonite cut-off walls* in 1999. European Standards on ground treatment are listed in the *References*.

CHAPTER 10: FOUNDATION MOVEMENT AND DAMAGE

While foundation failure rarely causes a building to collapse, it is not uncommon for foundations to fail to fulfil their primary function of carrying the superstructure loads safely to the ground without any unacceptable movements or damaging deformation of the building. However, foundation movement is just one of the many processes that cause cracking in buildings. When building damage can be attributed to ground movement, the cause of the movement should be identified and the likelihood assessed of further movement causing more damage. Ground movement can take the form of horizontal strain or heave but settlement is usually the main concern. In this chapter, the nature of structural damage, the processes that can give rise to damaging ground movements, and techniques to diagnose the cause of damage in existing buildings are described. The role of monitoring is reviewed. Remedial treatment and underpinning are considered in Chapter 11.

10.1 Introduction

Foundations rarely fail in the sense that their inadequacy causes a building to collapse or parts of a building to fall down. However, foundation movement can seriously damage a building. Also, foundation concrete may be attacked by chemicals in the groundwater and steel may corrode, but it is rare that these processes degrade the foundations to the stage where they are no longer capable of supporting the building. When foundations are described as having failed, it usually means that the foundations have not met their primary function, which is to carry the superstructure loads safely to the ground, without unacceptable movements or damaging deformation of the building. Foundation failure in this sense is by no means uncommon.

Except in rare circumstances, it is both uneconomic and unnecessary to design foundations to be totally immobile throughout the life of the building. As explained in Chapter 6, the essence of successful foundation design is to ensure that the movements transmitted to the superstructure never exceed acceptable levels.

When a damaged building is investigated to determine if the foundations are defective the following questions should be addressed:
- Have the foundation materials degraded?
- Have the foundations moved?
- What has caused the ground movements?
- What potential is there for further movement to occur?
- If further movements are likely, what can be done to reduce their potential for causing further damage?

To answer these questions the cause of the damage has to be identified. However, the scope of the investigation will be constrained by considerations of cost. It is important at an early stage to determine whether or not remedial works are required so that decisions can be made regarding the cost-effectiveness of further investigations. This chapter describes the nature of structural damage and the processes that can give rise to damaging ground movements. The techniques that can be used to diagnose damage in existing buildings and the role of monitoring are reviewed.

10.2 Causes of building damage

Foundation movement is just one of many processes that can cause distortion and cracking in buildings. The first step in investigating a damaged building is to identify any damage that can be attributed to causes other than foundation movement[1]. Some common processes that affect buildings are as follows:

Moisture changes

Many building materials shrink as they lose moisture; examples are: lightweight concrete products, plain concrete slabs, some plasters, rendering and timber. This shrinkage normally occurs during the first few years after construction but can be triggered in older buildings by the installation of central heating. In contrast, new, ordinary clay bricks expand as they absorb moisture. The damage associated with these processes is usually minor and is unlikely to be a continuing problem.

Thermal effects

Varying amounts of expansion and contraction in different building materials produce strains that can cause cracking; flat roofs and long, south-facing walls without expansion joints are particularly vulnerable. Dramatic thermal effects can be generated by fires; for example, an expanding steel roof truss can punch through the masonry supporting its ends.

Overstressing

Local crushing of masonry may occur where loading is increased, for example by structural alterations. Increased floor loadings, for example storage of heavy items in a loft, can cause ceilings to sag and crack.

Frost attack

Freezing and thawing of absorbed water can cause cracking and spalling in porous materials, such as fired clay products, natural stones, and weaker concrete materials.

Chemical attack

Aqueous solutions of sulfates can attack cement and hydraulic lime mortars and concrete blocks causing them to expand. The problem is restricted to exposed walls and slabs that remain wet for prolonged periods. Some concrete aggregates may react with alkali present in the hardened cement causing the concrete to expand and crack, a process known as alkali aggregate reaction.

FOUNDATION MOVEMENT AND DAMAGE

Corrosion
The expansion associated with the oxidation of steel fixings and reinforcement buried in porous building materials can cause movement, spalling or cracking of the materials in contact with the metal. Corroded wall ties can lead to the failure of cavity brickwork.

Creep
Over long periods of time, some building materials deform slowly under load; these movements may eventually lead to cracking. An extreme example of creep is roof spread, where prolonged loading of inadequately tied pitched roofs can result in appreciable lateral loading being transferred to the walls.

10.3 Types of foundation movement

Ground movements in both the vertical and horizontal directions can occur beneath the foundations of a building. Settlement is usually the major concern, although horizontal strain and heave can be important in some circumstances.

Most of the processes listed in Section 10.2 tend to produce a profusion of small, widely distributed cracks. Although shrinking and swelling of bricks can cause local cracking around windows and at junctions of walls, it is rare for the cracks to exceed 5 mm and they seldom extend through the damp-proof course (DPC). In contrast, foundation movement tends to produce a few isolated cracks, often continuous through the DPC and sometimes more than 5 mm wide. Foundation movements tend to affect large sections of wall, so the cracking usually manifests itself at weak points, such as window openings and doors. A building may deform in sagging or hogging mode and the pattern and taper of the cracking can be associated with a particular mode of distortion (Figure 10.1).

Other indications of foundation movement are sticking doors and windows, displaced roof joists, gaps in roof tiles and disrupted services. However, the best way of confirming that the foundations have moved is to measure the out-of-level along a course of bricks or to measure the verticality of external walls[2]. Damage is unlikely to have occurred unless there has been at least 25 mm of differential movement[3] and this should be distinguishable from any construction tolerances, which should be less than about 20 mm for a typical pair of semi-detached houses. Building errors are usually randomly distributed whereas foundation movement is indicated by a trend in level change. Indicators of foundation movement are summarised in Box 10.1.

The most acute foundation problems on poor ground are often associated with small buildings for which deep foundations were not an economically viable solution. Where a stiff raft foundation has been provided, it should prevent distortion of the building due to differential settlement (Figure 10.2) and should resist horizontal tensile forces; the remaining concern is whether the tolerable tilt will be exceeded (Section 6.3).

GEOTECHNICS FOR BUILDING PROFESSIONALS

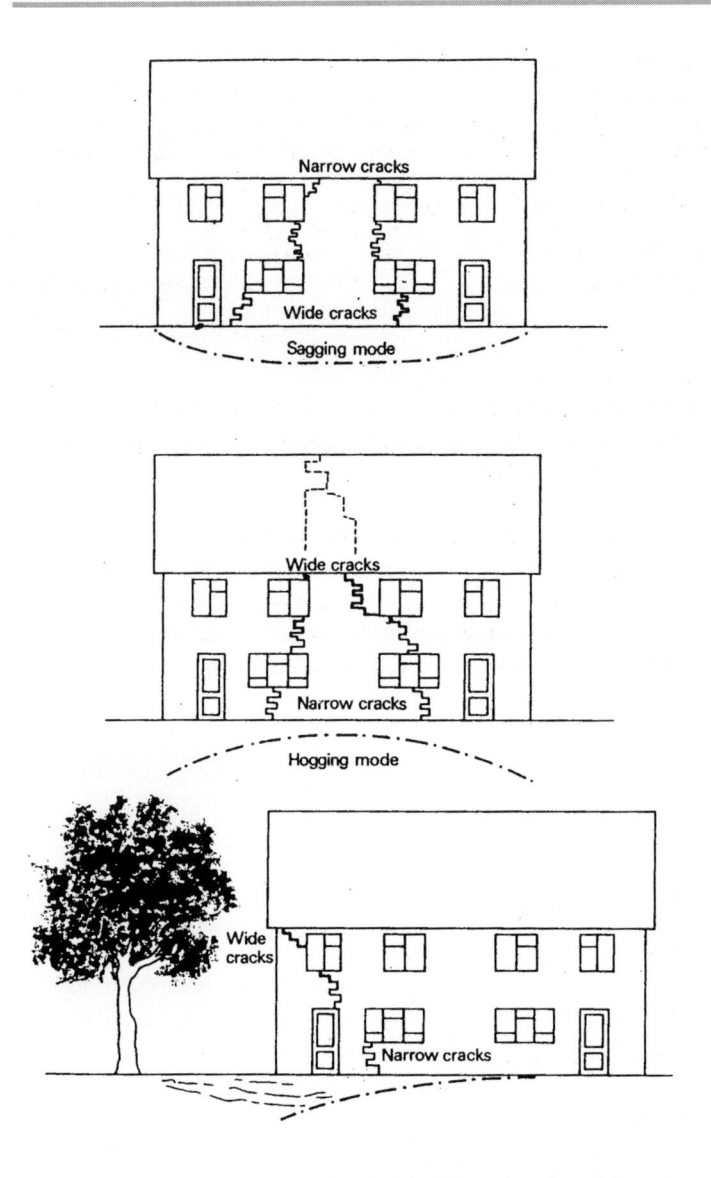

Figure 10.1 Crack patterns associated with different modes of distortion

FOUNDATION MOVEMENT AND DAMAGE

> **Box 10.1 Indicators of foundation movement**
>
> There are a number of features which may indicate that foundation movement has occured and may be continuing:
> - isolated cracks at weak points in structure;
> - cracks taper from top to bottom or vice versa;
> - cracking is continuous through DPC;
> - crack widths may exceed 5 mm;
> - both external and internal cracks which are usually at the same location
> - pattern of cracks which is compatible with a credible mechanism of foundation movement;
> - sticking doors and windows;
> - disrupted drains and services;
> - measurable out-of-level or out-of-plumb of walls.

10.4 Causes of foundation movement

In order to assess the seriousness of foundation movement and the likelihood of future ground movement, it is necessary to identify the principal causes of ground movement[4]. Some of the more common are discussed here[5].

In the early 1970s, damage caused by subsidence was added to the cover offered by most UK domestic building insurance policies. It is necessary, therefore, when discussing damage caused by foundation movement, to take note of the terminology employed by insurers[6]. Foundations can move downwards, upwards or sideways and these modes are described in British insurance policies as subsidence, heave and landslip.

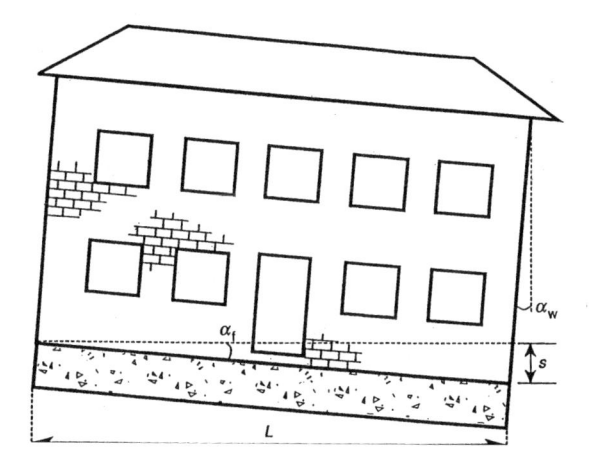

Figure 10.2 Tilting of low-rise house (settlement exaggerated!)

Weight of building

All materials deform under applied loading. On coarse soil, such as sand and gravel, the deformation is almost instantaneous. On clay, there is a long-term component associated with the gradual expulsion of pore water and in a low permeability clay this consolidation process can take several years. Highly compressible soil, such as soft clay or fill, tends to creep under the applied load; this process, which is sometimes termed secondary compression, can continue throughout the life of the building, albeit at an ever-decreasing rate (Section 4.5).

Different parts of the building may settle by varying amounts owing to differences in the applied foundation loading and variations in the properties of the underlying soil. It is this differential settlement that can lead to cracking. The settlement should be small, often no more than a few millimetres, provided that the allowable bearing pressure of the soil is not exceeded – Section 6.5; any associated damage should manifest itself during the first few decades after construction and should be easily repairable. Initial settlement tends to be a serious problem only where the allowable bearing pressure is exceeded and this can normally be attributed to inappropriate foundation design or the existence of an unidentified soft layer (peat, for example) at shallow depth.

The removal of load due to the excavation for a basement or the general lowering of the ground level will produce some heave in the soil. This will be small in a coarse soil and will occur instantly; in a clay soil it may be more significant and take place over many years[7]. However, unless large quantities of soil are removed, the movements are likely to be no more than a few millimetres. In some situations it may be appropriate to design foundations to allow the underlying soil to swell without lifting the building, for example by providing a void under the floor slab of a piled building.

Clay shrinkage or desiccation

The effects of evaporation and moisture extraction by vegetation on clay soil have been described in Section 7.5. These processes can damage buildings that are not founded at a sufficient depth to protect them from the volume changes that occur in the surface soil (Figure 10.3); the damage is most likely to occur during prolonged periods of dry weather. The analysis of insurance claims indicates that clay shrinkage is the commonest cause of damage to low-rise buildings in the UK and that the extent of damage correlates with dry weather[8].

Even during prolonged periods of dry weather, the effects of evaporation are largely confined to the surface metre of soil. Hence, where there are no trees or other large vegetation in the vicinity, it is mainly buildings whose foundations do not conform to the requirement of a minimum foundation depth (0.9 m on clay soils with medium volume change potential) that are susceptible to movement as a result of clay shrinkage. Where

FOUNDATION MOVEMENT AND DAMAGE

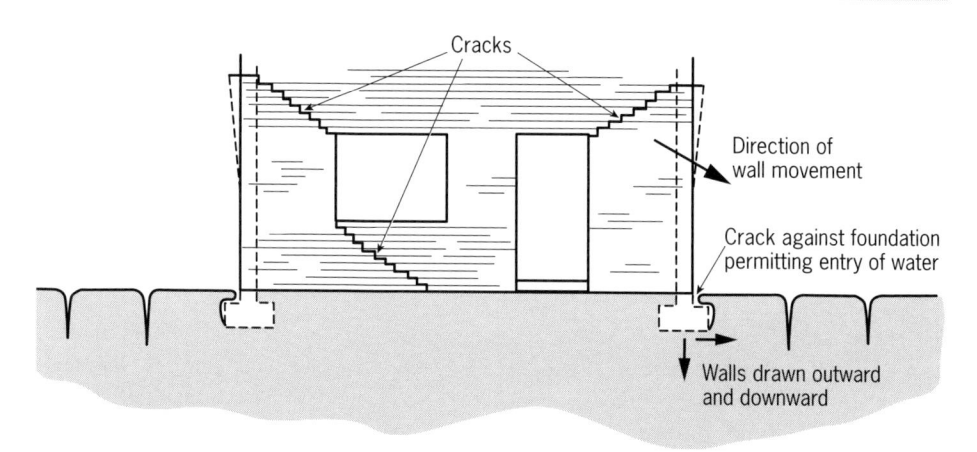

Figure 10.3 Cracking associated with shallow foundations on shrinkable clay

trees or other large vegetation are present, the foundation depth may be inadequate because the building predates the current recommendations (Section 7.5), or because circumstances have changed since the building was constructed owing to potentially large trees being planted too near to the foundations. An extension may have been built on shallow foundations too close to large vegetation.

Desiccation is largely a reversible process; if a tree is removed, water will return slowly to the soil causing it to swell (Figure 10.4) and the process will continue until the pore water suctions that desiccation induces have dissipated to equilibrium values[9]. The time-scale over which these changes take place depends on the permeability of the soil, although the extent and magnitude of the desiccation and the availability of free water are also important factors. A well-documented case history[10] has shown that measurable movements were continuing 25 years after the removal of some large elm trees, although heave was 90% complete within 10 to 15 years of tree removal. The effect of tree removal on a building depends on the relative ages of the tree and the building (Box 10.2).

Erosion and soil softening

While surface erosion by wind and rain can remove soil from around buildings, this process is seldom allowed to continue to the stage where it has an adverse effect on foundations. However, erosion within the soil can cause damage to buildings. Percolation of water through granular soils can wash out fine particles or dissolve soluble minerals, creating voids or reducing the density of the remaining soil and producing settlement of the ground surface[11]. Most problems are caused by fractured drains or water pipes.

Where the lateral stresses acting on clay soil are reduced, for example adjacent to a cutting, excavation or steep slope, the clay will tend to absorb water and soften. This

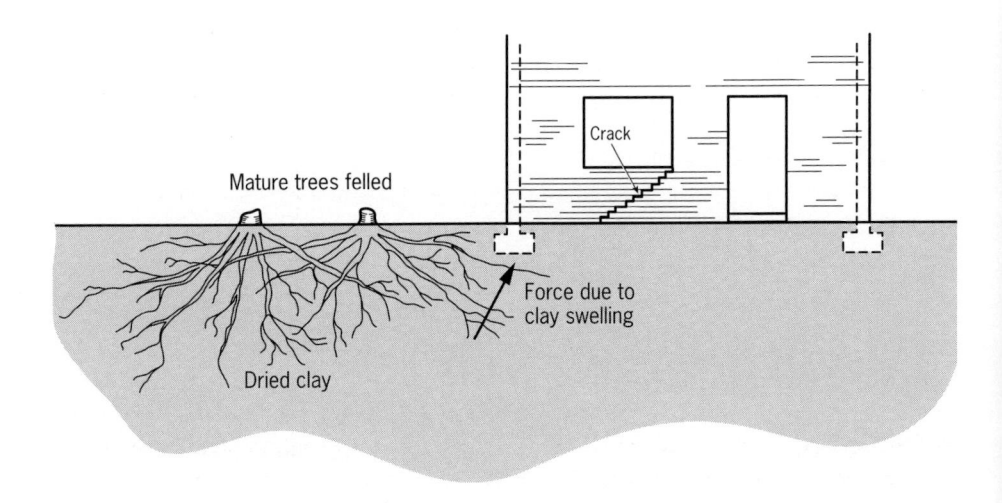

Figure 10.4 Consequences of tree removal for building with foundation on shrinkable clay

process can reduce the bearing resistance of the soil and could, in extreme cases, lead to the foundation loads inducing a shear failure in the soil. More commonly this process produces some limited movement as it allows soil to move towards a loosely backfilled trench, such as a deep drain. This softening will occur wherever the soil is below the water-table for at least part of the year; in the UK, this means that many excavations that go below foundation level are likely to induce some limited ground movement. Having softened, the soil should be largely unaffected by a sudden influx of water, for example from a leaking drain. However, it is a common practice to lay drains in a bed of granular material (Section 12.3) and, in such circumstances, leaking water may wash out fine particles allowing further movement of the already softened clay.

Box 10.2 Effect of tree removal on buildings founded on shrinkable clay

The effect of tree removal depends on whether the tree is younger or older than the building:
- If a tree is younger than the building, its removal can, at worst, only return the foundations to their original position and should not, therefore, cause significant damage.
- If a tree is older than the building, or the property has been recently extended, there is a danger of damage resulting from the removal of the tree.

A more common cause of damage is the removal of trees prior to construction; this can lead to problems unless appropriate measures are taken to isolate the building from the effects of the swelling soil (Section 7.5). This type of problem normally manifests itself within ten years of construction, and the past presence of trees can usually be confirmed by reference to local records, building control submissions and aerial photographs.

A leaking drain can influence foundation movements in clay soil where there are trees nearby. A long-term leak could reduce the state of desiccation in the soil, producing local heave, or supply sustenance to a tree; repairing the leak could then trigger dessication.

Change in groundwater level
Water abstraction or general changes in the groundwater regime can cause certain soils, such as loose sand, silt and peat, to reduce in volume and hence to subside. This is a common problem in fen areas where lightweight buildings on rafts have been adversely affected by drainage schemes. In such situations, roads may also subside.

Rising groundwater levels are a recognised problem in parts of the UK. In some cities, notably London, Birmingham and Liverpool, cessation of pumping from underlying aquifers, where water was formerly extracted for industrial use, is allowing groundwater to rise to levels not seen for a century[12]. In clay strata overlying aquifers, a rise in the water-table causes swelling which may result in differential vertical heave in buildings, particularly those founded partly on deep and partly on shallow foundations. Long piles may fail in tension. In both clay and coarse soil there may be a reduction in bearing resistance of end-bearing piles. There may be a risk of seepage into basements and attack on foundations and buried services from corrosive (for example sulfate-rich) groundwater.

The cessation of pumping for regional drainage in former coal mining areas may lead to a rapid rise in groundwater levels[13]. The rise of groundwater into unstable room-and-pillar mine workings may trigger local surface subsidence (Section 7.7) and the rise of groundwater into opencast backfill may cause collapse compression (Sections 4.9 and 7.4). There may be a temporary hazard from increasing methane gas emissions.

Compression of fill
Unless it is placed under carefully controlled conditions, fill tends to be more compressible than most natural soil (Section 7.4). Hence, where the presence of a small amount of fill goes undetected during construction (for example when building on the edge of an old backfilled pit), there is an increased likelihood of damage as a result of excessive differential settlement.

Even where the presence of fill is recognised, special care is needed because of the tendency of fill to compress slowly with time, either as a result of self-weight consolidation or degradation of any organic component. Fill may also be sensitive to a change in the groundwater level and may settle dramatically due to collapse compression the first time it is inundated (Section 4.9).

Collapse of mines and natural cavities
Shallow mine workings are prone to collapse causing localised subsidence of the overlying ground surface as the void migrates upwards (Section 7.7). A similar phenomenon can occur in limestone and chalk deposits that contain natural cavities or solution features formed by the percolation of water. Deeper, abandoned mines can cause subsidence over wider areas as groups of pillars, left to support the roof, collapse progressively.

Modern longwall mining, where deliberate roof collapse follows coal extraction, can produce a wave of subsidence that may distort and crack any buildings in its path. However, the movements are often of relatively short duration and in many cases the buildings will be left in a stable and serviceable condition. Problems are most likely to occur towards the periphery of the subsidence area where permanent differential movements are greatest and in areas containing faulted rock; the faults may be reactivated by the mining subsidence, locally causing large differential movements at the surface.

Construction processes
Construction work, particularly deep excavation and tunnelling, generates deformations in the ground that can affect nearby buildings. There is a tendency for the surrounding soil to move towards the excavation or tunnel; the magnitude of the movements depend largely on the method of construction and the properties of the soil. It is commonly assumed that no movement occurs at distances further than three times the depth of the excavation. The effects of excavation are well appreciated in urban redevelopment schemes. Careful monitoring of adjacent buildings is frequently employed to give warning of undue disturbance and to provide data for controlling the excavation or tunnelling process[14].

Vibration
Although vibrations induced in buildings by ground-borne excitations are often noticeable, there is little evidence that they cause damage to buildings[15], other than in extreme cases such as driving piles close to the building, explosions or earthquakes. However, vibration may exacerbate existing damage (encourage cracks to propagate) and could result in compaction of soils such as loose sand or poorly compacted fill.

Frost heave
Prolonged periods of freezing weather can cause surface layers of chalky and silty soils to expand, resulting in heave damage to buildings on very shallow foundations (Section 6.2). However, in most parts of the UK it is generally accepted that this process is unlikely to affect buildings founded below about 0.5 m[16]. Where damage has occurred, it has frequently been to new, part-completed buildings in highly exposed conditions. Special consideration may need to be given to refrigerated buildings used for cold storage[17].

FOUNDATION MOVEMENT AND DAMAGE

Chemical reaction
Most chemical reactions that affect foundations are expansive and, therefore, can result in heave (Section 6.11). The commonest form is sulfate attack of concrete[18]. Lifting of floor slabs supported on sulfate-rich fill material has been a quite common problem. Fill materials that may be particularly high in sulfates include burnt colliery spoil, brick rubble with adhering plaster, and some industrial wastes. Rarely, sulfate attack can weaken foundation elements, such as piles, and result in failure.

Sulfate attack occurs only where there is water. A high or fluctuating water-table will normally saturate the whole of the foundations and cause general uplift; however, where the source of the water is internal leakage or external run-off, the damage may be concentrated in the wettest areas.

Landslip
Landslip can take the form of a slope failure, in which there is a sudden, bodily movement of a large mass of soil sliding on a shear plane, or of slope creep, in which there is a gradual movement downhill of a surface layer. Many incidents of landsliding are slow movements of old, reactivated slope failures (Section 7.6).

It is uncommon in the UK for buildings to be sited on slopes that are naturally unstable. But there are exceptions such as on the Isle of Wight, where hundreds of houses are situated on the Under Cliff; this is a complex of old landslides that are slowly moving[19]. Evidence of previous movements should normally be detected during the site investigation (Chapter 5) and the location avoided or steps taken to improve its stability (Section 7.6). The stability of existing slopes may be adversely affected by excavating near the toe of the slope, or by loading the crest with a building or fill material, or by making local alterations to the natural slope or the groundwater regime.

There is little information on the potential for soil creep in relatively shallow clay slopes (say, $7°$ to $12°$), which can migrate slowly downhill. Damage caused by slope creep may be wrongly diagnosed as clay shrinkage, but it is possible that the two phenomena are related. Desiccation cracks that open up during periods of exceptionally dry weather may not close completely, causing a net movement down the hill. Similarly, slopes can exacerbate clay shrinkage by increasing run-off, thereby reducing the amount of rainfall infiltration into the ground.

10.5 Investigating damage

There has been a substantial increase in the level of insurance claims for heave and subsidence damage to domestic properties in recent years. However, until recently there has been a dearth of guidance on how best to investigate claims and select appropriate remedial measures. This section outlines the key elements of a cost-effective investigation[20]. Chapter 5, which deals with ground investigation for new construction, and Section 6.3, which deals with acceptable ground movements, are relevant.

Damage assessment
The initial survey of the damaged property[21] should include a sketch showing the position, width and taper of all internal and external cracks[22]. This provides an objective record against which future damage can be compared and facilitates the identification of the mode of distortion (the way in which the foundation movements are affecting the building as a whole). It may highlight damage that is being caused by, for example, rotation of a wall that is well removed from the area that is subsiding or heaving. The location of the damage does not necessarily correspond to the location of the foundation movement. The severity of the damage should be evaluated according to Table 10.1.

The history of the damage should be established and, wherever possible, information supplied by the owner or occupier should be corroborated by examining the surfaces of cracks for dirt deposits indicative of age, and by looking for evidence of past repairs. Where the building is founded on a clay soil, the size, position and species of trees close enough to have an effect on the foundations (say within 1.0H where H is the tree height) should also be noted.

Distortion survey
Where there is exposed brickwork on the outside of the house, a simple survey should be performed using a portable water level or an optical level to determine whether there are any significant level changes along brick courses. Alternatively, where the external walls are rendered, an indication of previous movement may be obtained by measuring the verticality of walls[26].

In some cases, other types of distortion survey are appropriate[27]. Where heave is suspected, it may be appropriate to measure levels on the floor slabs as well as the external walls. Ideally the distortion survey should form part of the initial inspection; it should take about one or two hours for one person to check brick course levels around a detached house using a water level.

Desk study
The local geology should be checked by reference to the relevant 1:50,000 scale drift (or solid & drift) geological map or, in some areas, larger scale maps (Section 5.3). In some cases it may be appropriate to check old maps, records and aerial photographs for the location of pits, streams and trees. It may be possible to obtain copies of the plans for the original building work and any extensions; these can be used to identify the depth of foundations, heave precautions and floor slab details. There is always a possibility of significant differences between the plans and the actual construction.

FOUNDATION MOVEMENT AND DAMAGE

Table 10.1 Classification of visible damage to walls with particular reference to ease of repair of plaster and brickwork or masonry[23]

Category of damage	Description of typical damage *Ease of repair in italic*	Typical crack widths[24]
0	Hairline cracks of less than about 0.1mm which are classed as negligible. *No action required*	Up to 0.1 mm
1	*Fine cracks which can easily be treated using normal decoration.* Damage generally restricted to internal wall finishes; cracks rarely visible in external brickwork.	Up to 1 mm
2	*Cracks easily filled. Recurrent cracks can be masked by suitable linings.* Cracks not necessarily visible externally; *some external repointing may be required to ensure weather-tightness.* Doors and windows may stick slightly and *require easing and adjusting.*	Up to 5 mm
3	*Cracks which require some opening up and can be patched by a mason. Repointing of external brickwork and possibly a small amount of brickwork to be replaced.* Doors and windows sticking. Service pipes may fracture. Weather-tightness often impaired.	5 to 15 mm or several of, say, 3 mm
4	*Extensive damage which requires breaking-out and replacing sections of walls, especially over doors and windows.* Windows and door frames distorted, floor sloping noticeably[25]. Walls leaning or bulging noticeably; some loss of bearing in beams. Service pipes disrupted.	15 to 25 mm, but also depends on number of cracks
5	*Structural damage which requires a major repair job involving partial or complete rebuilding.* Beams lose bearing, walls lean badly and require shoring. Windows broken with distortion. Danger of instability.	Greater than 25 mm, but depends on number of cracks

Trial pits

Where some foundation movement appears to have occurred, it is generally desirable to confirm this with a direct investigation (Section 5.5). This is relatively expensive, so it is important that it is properly targeted. At least one trial pit should be excavated early in the investigation and, in straightforward cases, may be the only intrusive ground investigation that is required. Trial pits are needed to confirm the depth and condition of the foundations and can also be used to obtain soil samples for index property tests, which can then be used to establish the volume change potential of a clay soil (Sections 4.8 and 7.5). Deeper samples, which can give an indication of the depth of any desiccation, can be obtained by hand augering below the base of the pit.

Boreholes

There are a number of methods of forming boreholes for investigating building damage[28]. Light cable percussion boring (Figure 5.3) is occasionally used. It is more expensive than trial pitting but its use may be warranted in some situations; examples are where removal of a large, mature tree is being considered and it is important to determine accurately the depth and degree of desiccation, or where a house is suspected of being on a backfilled pit and it is important to confirm the ground conditions. When investigating shrinkable clay, samples of undesiccated clay should be obtained from a control borehole for comparison (Figure 4.11) wherever possible[29].

Where only disturbed samples are required to a maximum depth of about 6 m, a hand auger can be used. Driven tube window samplers, which are being used increasingly in clay soil, can form boreholes to a depth of about 8 m relatively rapidly (Section 5.5).

10.6 Diagnosis

A correct diagnosis of the nature of the problem is crucial but it is sometimes difficult to identify the cause of damage with confidence, especially where the damage is relatively minor. However, in most cases an investigation of the type outlined in Section 10.5, together with the application of a basic knowledge of how buildings and the ground behave, should make it possible to determine the likely causes. Where the damage is severe or the case complex, further closely targeted investigations may be needed. An exhaustive account of all the deductive processes that can be applied to the information obtained during an investigation is beyond the scope of this chapter, but the following comments illustrate the basic principles.

Type of damage

Some possible causes can be eliminated by establishing whether the damage has been caused by settlement, heave or landslip, since certain types of damage are characteristic of particular causes. For example, under-sailing of brickwork below DPC indicates heave associated with swelling soil or landslip where the building is on a slope. The pattern of damage, aided where necessary by simple measurements of changes in level along brick courses and the verticality of walls, should be examined.

Soil type

The nature of the ground is one of the most important factors in determining the cause of foundation movement. For example, if the soil is identified as being predominantly coarse (for example, sand, gravel, chalk, limestone), seasonal shrinkage and swelling of the surface soil can be eliminated since such behaviour is restricted to clay. On the other hand, if the soil is found to be a clay, the possibility of erosion can usually be eliminated, although care is needed since many clay soils contain lenses of coarse soil that could be affected; the existence of internal erosion can usually be identified by its effect on general ground levels and the presence of unusual quantities of water.

If the soil is a soft clay or fill, there is a strong possibility of the damage being caused by excessive settlement, particularly where the thickness of the fill or soft layer varies across the site. On fill, there is the added possibility of movement as a result of self-weight or collapse compression; this can often be recognised by the effect on ground levels generally

(Section 7.4). The likelihood of a building having been constructed on soft clay or fill can often be determined from local knowledge, geological maps and records and plans showing, for example, the position of old pits. Failing this, a ground investigation with several boreholes may be needed to confirm the nature and variability of underlying soil.

Foundations

Foundations that are more than 0.5 m deep are unlikely to have been adversely affected by frost heave and any that are more than 1.0 m deep are unlikely to have been affected by clay shrinkage, unless there are trees or excavations nearby. Where the foundations are significantly deeper than 1 m and the soil is a shrinkable clay, clay heave may be a cause of damage if trees have been cleared from the site prior to construction. Evidence of this may be provided by deep roots in boreholes and aerial photographs that pre-date construction. The use of piles, pads or raft foundations for low-rise, lightweight buildings could indicate that the builder or designer was anticipating difficult ground conditions (Section 6.2).

Poor workmanship or an inadequate specification may have contributed to the damage. For example, the forces generated by swelling clay can break inadequately reinforced piles or displace insufficiently protected trench-fill foundations, and sulfate attack can weaken ground beams and other foundation elements not constructed with sulfate-resisting concrete. Poorly prepared trenches can cause strip and trench-fill foundations to settle excessively and sometimes this movement will occur during a dry period as soil shrinks away from the side of the concrete.

In some circumstances the design of the floor slab may also have an important influence. A ground supported slab significantly increases the building's susceptibility to ground heave, particularly where the internal leaf of cavity walls, as well as internal walls, rest on the floor slab.

Environment

Local topography can be important. In clay soil, some form of landslip is possible wherever the slope of the ground exceeds about 7° (Section 7.6). Localised movements are possible near cuttings and excavations; these features may also have an influence on groundwater levels and hence on the degree and extent of any desiccation. Small movements are possible wherever there are poorly backfilled trenches at or below foundation level, especially where there is a flow of water to wash out any coarse backfill material. The existence of current or old mining activity and the likelihood of the ground containing natural cavities should be checked by reference to relevant geological maps and local records (Section 7.7).

Water
The presence or absence of water can provide important evidence. Groundwater running freely into a trial pit or borehole excavated in a clay soil indicates the presence of layers of coarse soil, which can have a major effect on the permeability of the soil mass; desiccation levels are likely to change far more quickly than they would in a homogeneous clay. Where there is free water for any length of time there can be no desiccation. The discovery of water in a borehole should, therefore, be considered carefully in determining the likelihood of clay shrinkage or swelling as explanations of building damage.

Chemical attack is most likely where the foundations are inundated and the possibility of chemical processes having contributed to movement can usually be confirmed or denied by testing the groundwater for such aggressive compounds as sulfates. Chemical testing can help to identify the source of the water; for example high levels of organic matter may indicate leakage from foul water drainage (Section 6.11).

The possibility that changes in the groundwater regime have induced ground movement should be considered. A rising water-table can cause increased loading on retaining walls, basements, and other watertight elements, and may cause collapse compression in fill; a general lowering of the water-table can lead to settlement of loose sand, silt, peat and soft clay with a relatively high organic content. The pumping and de-watering operations that often accompany large excavations can have a local effect on groundwater levels.

Trees
Trees and other large vegetation play a key role in influencing ground movement on shrinkable clay (Section 7.5 and Figure 10.5). A large broad-leaf tree, such as an oak, can have a significant effect on desiccation levels to a depth of 6 m and over an area of more than 1000 m^2. Wherever damage is attributable to clay shrinkage, the position of nearby trees (say within a distance from the building of $1.0H$ where H is the tree height) should be noted.

Although cases of trees causing damage in non-shrinkable soil, such as sand, gravel and chalk, are very rare, the possibility should be considered where large trees are extremely close to the foundations.

Composite effects
In many cases damage cannot be attributed to a single cause and there may be several contributory factors. For example, a building may be more susceptible to damage as a result of clay shrinkage because it settled differentially during the first few years following construction. Similarly, damage to a house built on an unstable clay slope may be exacerbated as a result of seasonal volume changes in the surface soil. In making a correct diagnosis it is important to gauge whether the manifest damage and distortion are

FOUNDATION MOVEMENT AND DAMAGE

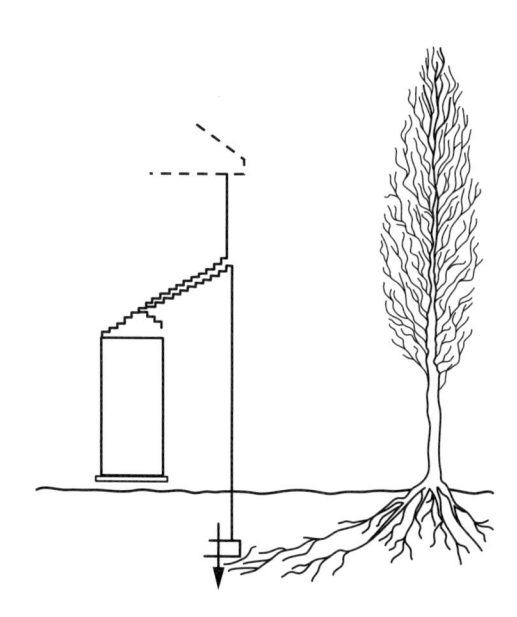

Figure 10.5 Effect of tree on shallow foundation in shrinkable clay

consistent with the movement attributed to a particular cause. Where a house is thought to have been damaged by clay shrinkage, it should be questioned whether the measured pattern and magnitude of the foundation movements are compatible with the depth of the foundations, the properties of the soil, and the size, type and proximity of trees.

10.7 Uses of monitoring

Where a building is thought to have been damaged by ground movement, a decision may be made to monitor the damage, or to monitor the movement of the building. It is sometimes desirable to measure the movement of the ground; for example, where landslip is thought to be a contributory factor, measurements of lateral ground movement are likely to be helpful. Monitoring can help to confirm the cause of damage and assist in determining an appropriate course of action but it is not a substitute for a proper and rational investigation. The various techniques that can be applied to monitoring damaged buildings are described briefly in Sections 10.8 and 10.9.

There are three ways in which monitoring can be used to aid the investigation of a damaged property:

❐ *Monitoring to establish cause of damage.* Where the cause of the damage is not obvious, monitoring can be a cost-effective diagnostic tool. For example, where a house is founded on a slope, monitoring vertical foundation movements can help distinguish clay shrinkage from landslip. Measurements are likely to be needed for at least a year before any long-term settlement or heave can be distinguished from seasonal movements. This may rule out monitoring as a diagnostic tool if the damage is serious enough to make it unreasonable to postpone repairs for at least twelve months.

❑ *Monitoring to measure rate of movement.* Where the cause of the damage is obvious, monitoring can be used to establish whether the damage is progressive and, if so, the rate of progression. It can help to determine an appropriate course of remedial action. This type of monitoring can be useful where it is known that the damage has been caused by a process that has a limited duration, such as heave following removal of a tree, but in cases where the damage is attributable to a persistent effect, such as seasonal shrinkage and swelling of a clay soil, it is likely to be of limited benefit.

❑ *Monitoring to check effectiveness of remedial action.* Where sensible action has been taken to mitigate the cause of the damage, such as removing or pruning nearby trees, monitoring can be used to gauge the effectiveness of the remedy and assess whether it is likely to provide a satisfactory long-term solution. At the end of the monitoring period, a decision can be taken about whether further measures, such as underpinning, are required to stabilise the foundations. This form of monitoring is particularly useful because it does not unnecessarily delay remedial action.

10.8 Structural monitoring

The most common and simplest way of monitoring damage caused by ground movement is to measure changes in the width of existing cracks; there are several ways [30]:

Steel ruler

Provided sufficient care is taken, crack widths can be measured to the nearest 0.5 mm using a steel ruler. Readings tend to be subjective and it is difficult to ensure that the crack is measured at the same point each time. Consequently, this form of measurement is normally used only for recording the state of damage during the initial investigation.

Magnified graticule

Internal cracks in plaster or other smooth finishes can be monitored by measuring the offset between two pencil marks using a magnifying glass fitted with a graticule. With care, movements can be measured to a resolution of 0.1 mm.

Plastic tell-tales

The most popular system[31] consists of two over-lapping plates screwed to the wall; one is marked with a cursor, the other with a scale marked in millimetre intervals (Figure 10.6). The two plates are mounted on opposite sides of the crack so that the cursor is initially in line with the centre of the scale; any subsequent shear or normal movement of the crack can then be resolved to the nearest millimetre by recording the position of the cursor with respect to the scale. Readings can be taken at any time by anyone, including the occupiers

FOUNDATION MOVEMENT AND DAMAGE

of the building, without the need for additional measuring equipment or an original zero reading. However, the plates are relatively obtrusive, vulnerable to vandalism, and have a limited accuracy.

Glass tell-tales

Cementing glass strips across cracks used to be a popular method of detecting progressive movement. However, the strips are vulnerable to vandalism and, when they crack, they give no indication of the magnitude and direction of the movement.

Brass screws

Small brass screws are installed on either side of the crack and the distance between them is measured using a calliper (Figure 10.7). The system is simple, robust, relatively unobtrusive and, by using the callipers in different modes, capable of measuring cracks in corners and other awkward positions. Moreover, if three screws are arranged in a right-angle triangle, it is possible to resolve both normal and shear crack movements. Using a digital calliper, a resolution of 0.02 mm can be obtained and an overall accuracy better than ±0.1 mm should be achieved.

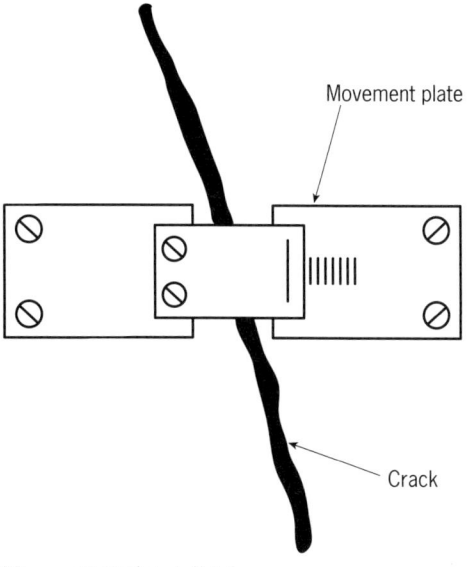

Figure 10.6 Plate tell-tale

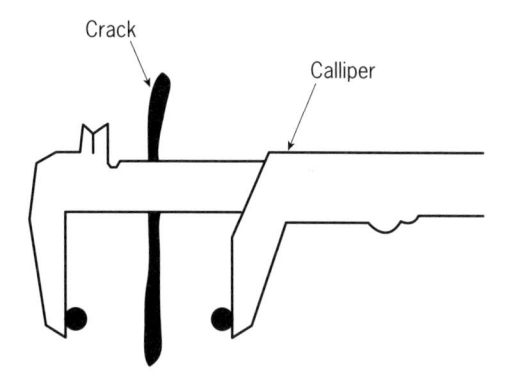

Figure 10.7 Pin tell-tale

Displacement transducer

There may be circumstances when more sophisticated and relatively expensive techniques of monitoring crack widths are justified. It may be necessary to monitor continuously in order to activate an alarm in the event of sudden movement, or the crack may be in an inaccessible place, such as a railway tunnel. Two commonly used devices are Linear Variable Displacement Trans-formers (LVDTs) and potentiometric displacement transducers. Both these devices can be read either manually, using a hand-held unit, or automatically, using a data logger, and can easily provide an accuracy of ± 0.1 mm.

10.9 Foundation and ground movement monitoring

The results of monitoring crack widths can be ambiguous; a crack may form for one reason and progress for another. Previous distortions that are unconnected with the cause of the cracking may lead to further movements as locked-in stresses re-distribute themselves throughout the building, while the foundations remain stationary. In effect, only the symptoms of ground movement are being monitored, not the cause. Generally it is preferable to measure the movements of the foundations in addition to crack widths.

Foundation monitoring

Vertical movements are usually monitored using precise levelling[32] because rates of movement need to be determined and these may be only a few millimetres a month. Consequently, the portable water level gauge is not sufficiently accurate nor, in most circumstances, is normal site surveying equipment. In rare cases where a continuous check on building movements is needed, for example where an historic building is being affected by nearby excavation work, electronic tilt sensors or precision water gauges with electronic pressure sensors may be needed.

A precision optical level has a resolution of 0.01 mm and, for a single house, it is normally possible to achieve an overall accuracy of better than 0.5 mm (the closing error of the survey should be not more than 0.5 mm). In order for measurements of this accuracy to be meaningful, it is essential that monitoring points of some form, such as the BRE levelling station[33], are attached to the building (Figure 10.8).

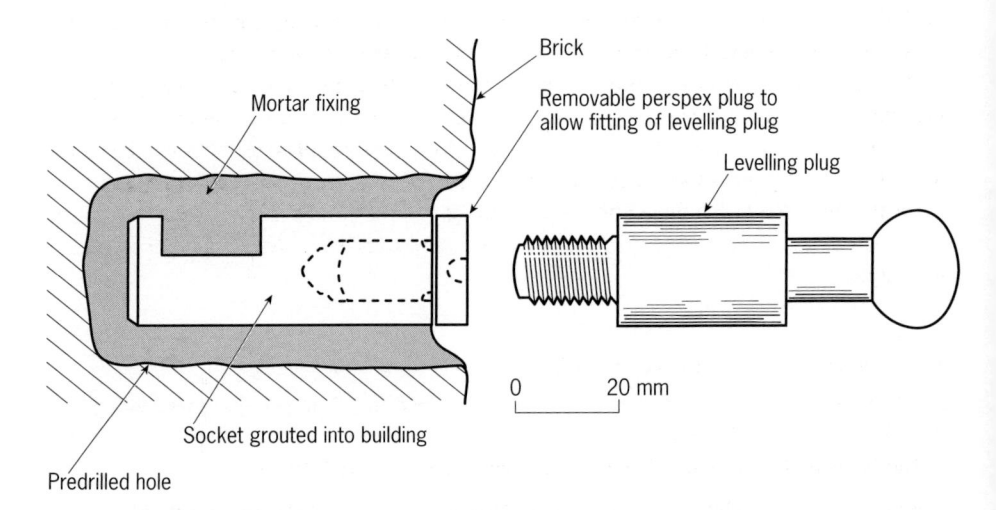

Figure 10.8 BRE levelling station

FOUNDATION MOVEMENT AND DAMAGE

Wherever possible, level measurements should be referred to a fixed datum. For most domestic applications, a storm-water drain or similar feature is usually sufficiently stable for this purpose. However, where there are no deep drains or where absolute accuracy is essential, it is necessary to install a deep datum to a suitable depth[34].

Monitoring is normally required for a period of at least 12 months to distinguish seasonal movements from long-term settlement or heave; ideally, readings should be taken every month but in practice every two months is usually adequate.

Foundation movement can be monitored by measuring changes in the inclination of elements of the structure. This technique is limited to the detection of differential movements but it is more amenable to the use of electronic sensors and a check can be kept on absolute movement by periodic surveys using a precision level.

Although sensitive electronic tilt-meters or inclinometers have been available for years, cost has restricted their use to demountable devices. However, the development of cheaper electro-levels has meant that it is now practical to semi-permanently install enough sensors on a structure to allow its movement to be monitored continuously. They are sensitive enough to measure extremely small movements; typically, an electro-level with a range of ± 3° mounted on a 1.5 m-long beam can resolve relative movements of the ends of the beam to ± 0.1 mm. Precision water gauges can also be used[35].

Ground movement monitoring

Techniques similar to those described for measuring foundation movement can be used to measure ground movements. The vertical movement of the ground surface or of rods anchored at various depths can be measured by optical levelling. Where relatively large movements are taking place, extensometers can measure movements at various depths in the ground. Demountable inclinometers lowered down flexible plastics tubes can be used to profile lateral ground movements and trains of electro-levels can be deployed in either vertical, horizontal or inclined tubes, to monitor profiles of ground movement. Some of the grouting operations described in Sections 9.8 and 11.3 can be monitored using this latter technique.

Notes

1. More information is in BRE Report BR 292 *Cracking in buildings*.

2. BRE Digests 343 and 344 provide guidance on simple methods of measuring and monitoring movement in low-rise buildings.

3. This might be 40 mm in older buildings (Skempton and Macdonald, 1956).

4. More detailed descriptions of the causes of damage, and how they can be recognised and remedied, are given in BRE Report 184 *Foundation movement and remedial underpinning in low-rise buildings* and Digest 352 *Underpinning*. See also Atkinson (2000).

5. BS 8004: 1986 Code of practice for foundations lists more than 13 possible causes of foundation movement.

6. The terms settlement and subsidence are both widely used to describe downward movement of the ground and of buildings. In non-technical usage, the two terms are virtually interchangeable; insofar as there is any distinction, settlement tends to imply a more gradual movement and subsidence is associated with the ground caving-in. Two examples of dictionary definitions demonstrate this:

- In the 5th edition of the Concise Oxford Dictionary of Current English, published in 1964, the following descriptions are given:
settle - subside, soil, house, foundation settles - comes gradually to lower level
subside - (of ground) cave in, sink, (of building) settle down lower in ground.
- In the 1952 edition of Chambers's Etymological English Dictionary, the meaning of 'to settle' is given as 'to subside' and the meaning of 'to subside' is given as 'to sink in level'. Subsidence is the act or process of subsiding, settling, or sinking, especially the sinking or caving-in of the ground or the settling down of a structure to a lower level.

In geotechnical engineering, the usual term for downward movement of buildings or the ground on which they are founded is settlement; it has, therefore, been used in this book. Subsidence is sometimes used to describe widespread movements, particularly mining subsidence.

British house insurers provide cover for subsidence rather than settlement. For insurance purposes, settlement is regarded as movement due to the distribution or redistribution of loading and stresses within the various elements of a building which normally occurs in the early stages of the life of a building; subsidence is defined as the downward movement of a building and its foundation caused by loss of support of the ground beneath the foundations. *Subsidence of low-rise buildings* (Institution of Structural Engineers, 2000) gives a detailed account of this distinction.

7. Construction of the Shell building on the South Bank of the Thames in 1953 involved a 12 m-deep basement excavation that came to within 1 m of the roof of the Bakerloo line tunnels of the London Underground. The bottom of the excavation and the tunnels are within the London Clay. Relief of overburden pressure has caused long-term heave of the London Clay, which has been monitored by BRE. The continuing upward movement of the southbound tunnel beneath the Shell Centre basement had reached 50 mm by 1986 and has provided a well documented example of heave due to unloading (Burford, 1988).

8. There have been several unusually dry periods in south-east England since domestic buildings insurance was extended to include damage caused by subsidence; 1975/6, 1989/90 and 1995/6. There were sharp rises in insurance claims following these dry periods, particularly following 1989/90 when the effects of two consecutive dry summers were exacerbated by a dry intervening winter. This is illustrated in Figure 2.1. A dry period in 2003 also led to a surge in claims.

9. BRE Digest 412 *Desiccation in clay soils*.

10. Cheney (1988) presented this case history of long-term monitoring of some cottages at Windsor built on London Clay.

11. Where earth embankments act as water retaining dams, internal erosion can be a serious problem. The subject, on which there is extensive literature, is outside the scope of this book.

12. The engineering implications of rising groundwater levels in London and Birmingham have been reviewed by Simpson *et al* (1989) and Knipe *et al* (1993), respectively. Long-term plans to lower the water-table in London were reported in *New Civil Engineer* (25 March 1999, p6): '*Rising groundwater levels under London which threaten tube lines and deep basements will be halted by a long term programme of borehole extraction, a London meeting was told this week. The General Aquifer Research, Development and Investigation Team (GARDIT) – an unofficial body made up of water companies, infrastructure owners and insurers – presented its strategy to local authorities and property owners at the Guildhall in the city of London.*'

13. Younger (1993).

14. See, for example, Burland and Hancock (1977), Price *et al* (1994), Burland *et al* (2001).

15. BRE Digest 403 *Damage to structures from ground-borne vibration*.

16. See BS 8004: 1986 clause 3.2.9.1 and clause 2E4 of Approved Document A of the Building Regulations.

17. Cooling and Ward (1948).

18. BRE Special Digest 1 *Concrete in aggressive ground*.

19. An international conference on slope stability engineering was held on the Isle of Wight in April 1991 (Chandler, 1991). A session of this conference was devoted to landslides on the island.

20. This section is based on BRE observations of foundation behaviour over the last 50 years and BRE involvement in a number of recent claims.

21. More information is in BRE Digest 251 *Assessment of damage in low-rise buildings* and BRE Report BR 292 *Cracking in buildings*.

22. BRE Digest 343 *Simple measuring and monitoring of movement in low-rise buildings* Part 1: *cracks*.

23. The table is taken from BRE Digest 251 *Assessment of damage in low-rise buildings*.

24. Crack width is one factor in assessing category of damage, but should not be used on its own.

25. Local deviation of slope, from the horizontal or vertical, of more than 1/100 will normally be clearly visible. Overall deviations in excess of 1/150 are undesirable.

26. BRE Digest 344 *Simple measuring and monitoring of movement in low-rise buildings* Part 2: *settlement, heave and out-of-plumb*.

27. Robson (1990).

28. These drilling techniques are described in more detail in BRE Digest 411 *Site investigation for low-rise building: direct investigations*.

29. BRE Digest 412 *Desiccation in clay soils* gives advice on appropriate sampling and physical investigation techniques for the identification of desiccation.

30. BRE Digest 343 *Simple measuring and monitoring of movement in low-rise buildings* Part 1: *cracks*.

31. This type of equipment is sold by Avongard Ltd.

32. BRE Digest 386 *Monitoring building and ground movement by precise levelling*.

33. The BRE levelling station is described in BRE Digest 386 *Monitoring building and ground movement by precise levelling*.

34. A deep datum designed by BRE is described in BRE Digest 386 *Monitoring building and ground movement by precise levelling*.

35. Automatic instrumentation systems were designed and installed to monitor the movement of the Mansion House in the City of London during tunnel construction for the Docklands Light Railway. Instrumentation included both electro-levels and a water levelling system (Price *et al*, 1994).

CHAPTER 11: REMEDIAL TREATMENT AND UNDERPINNING

Where foundation movement has caused building damage (Chapter 10), some form of remedial treatment may be required. Underpinning is often disruptive and expensive and should be pursued only after due consideration has been given to such alternatives as mitigating the cause of the movements or repairing the superstructure. Alternatives to underpinning are dealt with at the beginning of this chapter; the remainder of the chapter is concerned with different forms of underpinning: mass concrete, pier-and-beam, pile-and-beam and piled raft, and mini-piling. Some general design concepts of underpinning are outlined and particular concerns with partial underpinning are described.

11.1 Introduction

Figure 2.1 shows the growth of insurance claims in Great Britain for subsidence; in most years the total value of claims exceeds £100 million[1]. The types of ground movement that can cause damage to buildings are described in Chapter 10. Where building damage has occurred, and it has been confirmed that the damage is attributable to foundation movement, and that there is a likelihood of further movements occurring, it has to be decided what action is needed to prevent further damage.

One commonly adopted solution to foundation damage and failure is to underpin the foundations. But underpinning is often disruptive and expensive so it is important to give due consideration to alternatives; they include:
❐ mitigating the cause of the movement;
❐ stabilising the soil;
❐ repairing and/or strengthening the superstructure.

The next three sections of the chapter deal with these approaches; the rest of the chapter is concerned with different forms of underpinning:
❐ mass concrete;
❐ pier-and-beam;
❐ pile-and-beam and piled raft;
❐ mini-piling;
❐ partial underpinning.

11.2 Mitigating the cause of ground movement

Having identified the cause of the movement, it may be possible to remove it or to mitigate its influence on the foundations of the damaged building. Some examples of this approach to foundation damage are as follows:
❐ repair of a leaking drain;
❐ removal of a tree causing clay shrinkage;
❐ installation of a retaining wall to prevent movement towards an open excavation;
❐ installation of gabion walls to stabilise a slope.

REMEDIAL TREATMENT AND UNDERPINNING

Unfortunately, this type of approach cannot always be relied upon to prevent further movement since a change may have occurred in the ground that is not reversed by removing the cause. In the case of leaking drains, for example, cavities may have formed in the ground which may take some years of collapse and consolidation before their effect on the foundations is complete. Nevertheless, in many cases it may be possible to mitigate the cause of ground movement at modest cost and then monitor the damage to assess whether this action has provided a satisfactory long-term solution.

Many cases of damage are attributable to clay shrinkage and where there is reason to believe the shrinkage has been exacerbated by the presence of trees, there are essentially four ways in which the effect of the tree can be mitigated and which can provide a cost-effective remedy:

- tree removal;
- tree reduction;
- root pruning;
- root barriers.

Removing the tree is likely to have the greatest and most immediate effect on the level of desiccation in the soil. If the trees are no older than any part of the house, this should provide a safe solution, since any consequent heave can do no more than return the foundations to their original level. In such circumstances there is no advantage in a staged reduction in the size of the tree: the tree should be completely removed at the earliest opportunity. However the time taken for the soil to recover depends largely on the permeability of the soil and in a plastic clay, such as London clay, it may take tens of years for the ground to come to equilibrium[2], even though most of the heave occurs during the first few years. In more permeable soils, or soils containing permeable layers, full recovery may be achieved in one winter. Hence, depending on the soil conditions, the degree of damage and the circumstances of the owners, tree removal may or may not be an acceptable solution. If the tree is older than the house, or there are newer additions to the house, it may not be advisable to remove the tree altogether because of the danger of inducing damaging heave. The heave potential in the soil should be estimated as described in Section 7.5, before deciding on whether or not the tree should be removed.

Where it is unsafe to remove the tree altogether and the damage is relatively minor, some form of pruning can be considered. This may reduce its water extraction, but different trees react in different ways to pruning and arboricultural advice should be sought[3].

Roots can be pruned by excavating a trench between the tree and the damaged property to a depth sufficient to intercept and cut most of the roots. The trench should be positioned far enough away from the tree to ensure that its stability is not jeopardised[4]. The tree will almost certainly grow new roots to replace those that are cut but, in the short term, there will be some recovery as the degree of desiccation in the soil under the foundations reduces; where the damage appeared in a period of exceptionally dry weather, a return to the normal weather pattern may prevent further damage occurring.

Root barriers are a variant of root pruning where, instead of backfilling the trench with soil, the trench is filled with concrete or lined with an impermeable layer to form a permanent barrier to the roots. There is considerable uncertainty about how extensive a root barrier should be, both laterally and in depth[5]. Where a barrier is effective, the outcome will be the same as if the tree were removed. Barriers should, therefore, be considered only where tree removal will not cause unacceptable heave. Root barriers require careful consideration and detailing at the ground surface, since roots could potentially grow over the barrier and around services passing through the barrier. Inserting root barriers close to trees can be dangerous; if too much of the root system is severed, the tree may lose lateral stability and fall. Even if rapid instability is not caused, the tree may slowly die and become unstable later. The potential for undermining the building foundations while constructing the barrier should also be considered.

11.3 Soil stabilisation
Certain types of unstable ground can be improved by injecting grout to fill voids, reduce permeability and increase strength (Section 9.8). Cement grouts can permeate silty fine sands with permeabilities down to about 5×10^{-4} m/s, while chemical grouts can permeate finer soils with permeabilities as low as 1×10^{-6} m/s; they become more expensive as higher permeation is required. The various grouting techniques that can be used include permeation grouting, hydrofracture grouting, compaction grouting and jet grouting.

Compaction and hydrofracture grouting can be used to correct or prevent differential displacements of structures. These processes are referred to as grout jacking when carried out after the displacement has occurred, or compensation grouting when used to prevent differential displacements occurring, especially during tunnelling operations[6].

Certain chemicals can be used to reduce the shrinkage potential of clay soils. However, owing to the extremely low permeability of most shrinkable clays, mechanical mixing is normally required to introduce the chemicals effectively into the soil mass. This limits the usefulness of chemical additives, although some proprietary treatments are available.

11.4 Repairing or strengthening the superstructure
Where damage has been caused by a process that is now largely finished, such as initial settlement due to the weight of the building, or is likely to occur only rarely, such as clay shrinkage during a period of exceptionally dry weather, it should be possible to prevent further damage by repairing and/or strengthening the superstructure. The techniques that can be used include:

REMEDIAL TREATMENT AND UNDERPINNING

❐ tie rods;
❐ strapping;
❐ resin bonding of brickwork and cavities;
❐ brick stitching;
❐ mortar bed reinforcement.
Details of most of these techniques can be found in textbooks on structural repair[7].

Corseting is a relatively new technique; it consists of casting a reinforced concrete beam around the perimeter of the building, usually at or below ground level and connecting it to the brickwork with dowels[8]. This stiffens the building at foundation level and helps it bridge local areas of subsidence. It can, therefore, reduce the differential settlement associated with variations in the compressibility of underlying fill or soft soil. Corseting has also been applied to buildings suffering damage from slope creep movements.

11.5 Remedial underpinning

Where existing foundations are inadequate, they can be improved by *underpinning*, which involves either providing new foundations or, more usually, extending the existing foundations downwards to reach stiffer or more stable ground. Underpinning has a variety of applications, including increasing the bearing resistance of foundations prior to refurbishment work and preventing foundations being undermined by nearby excavations. The most common use is as a remedy for damage caused by foundation movement.

The use of underpinning as a remedy for subsidence and heave damage has increased significantly since the early 1970s[9]. Where there is relatively severe damage and little else that can be done to mitigate the cause of the damage, underpinning is often the correct solution and should be specified without delay. In other cases, where the cause of the damage can be mitigated, alternative measures, such as those described in the preceding sections, may be more desirable and more cost-effective.

The four main underpinning techniques that can be applied to low-rise buildings are described in the following sections[10]:
❐ mass concrete (Section 11.6);
❐ pier-and-beam (Section 11.7);
❐ pile-and-beam and piled-raft (Section 11.8);
❐ mini-piling (Section 11.9).
The choice of a cost-effective underpinning system is largely governed by the ground conditions and the required foundation depth. Underpinning must be appropriately specified, since poorly designed underpinning may create more problems than it solves. Although partial underpinning is often acceptable (Section 11.11), special care is needed where the underpinning is terminated because of the danger of creating hard spots. Similar care is needed where a semi-detached or terraced property is being underpinned

11.6 Mass concrete underpinning

Traditional underpinning was constructed in brickwork, but with high labour costs and the availability of ready-mixed concrete, mass concrete is now invariably employed. In mass concrete underpinning, an existing strip or pad foundation is extended downwards to reach a stratum that is stable and of sufficient strength (Figure 11.1). The underpinning is carried out in a series of bays. The length of each bay is determined by the ability of the walls to span the gap created; for most low-rise buildings with competent brick or stonework, this is likely to be in the range 1.0 to 1.4 m. Where there is any doubt about the ability of the wall to span the bay, the wall should be needled[11] and the load transferred to temporary supports bearing on the ground.

The bays are excavated so that no more than 20 to 25% of the wall is left unsupported at any one time. The mass concrete is cast into the bay to leave a small gap between the concrete and the underside of the existing foundation; when the concrete has been allowed to cure for a minimum of 24 hours, this gap is then pinned-up by ramming in suitable concrete. Having completed a series of bays, work can begin on the next series of bays, although it is normal practice to allow at least 24 hours between pinning-up and the excavation of an adjacent bay. This procedure is repeated until the bays become contiguous to form a strip under the section of the building that is being underpinned.

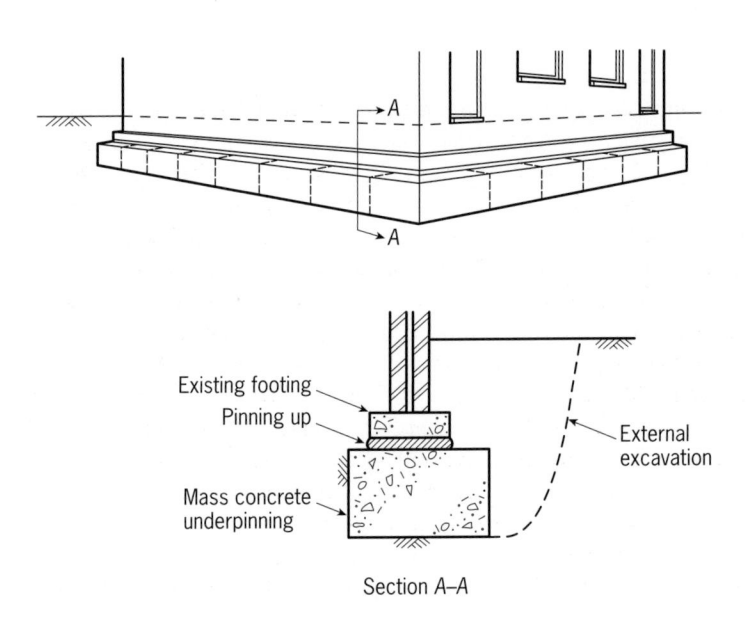

Figure 11.1 Mass-concrete underpinning

REMEDIAL TREATMENT AND UNDERPINNING

Where existing footings can span a short distance and loadings allow, the underpinning can take the form of discrete bays or piers instead of forming a continuous strip. The span between piers is determined by the strength of the existing footings. This form of mass underpinning should not normally be used where foundation loads are high or where footings are shallow, insubstantial or cracked as a result of excessive ground movement. It can be particularly suitable where existing foundations are reinforced.

Mass concrete underpinning can be installed without specialist equipment and is cost-effective for shallow depths. However, the cost increases rapidly as the required depth increases, because of the increasing cost of materials, soil removal and hand excavation. The presence of a high water-table in the ground may preclude its use, particularly if the ground through which excavation is to be made is loose or otherwise unstable. Where continuous underpinning is founded on a sloping impermeable stratum, it may be prudent to provide weep-holes to allow groundwater to pass through the foundations.

For low-rise buildings, the use of mass concrete underpinning tends to become uneconomical at depths between 2 and 2.5 m because of the high labour costs involved in hand excavation. For larger buildings, where the applied foundation pressures are higher, the greater costs of excavating to depths of considerably more than 2.5 m can be justified. Mass concrete is also used to greater depths in refurbishment and redevelopment works, where the underpinning may act as a retaining wall; in these circumstances, it should be not less than 0.6 m thick.

11.7 Pier-and-beam underpinning

This widely used system of underpinning was introduced shortly after the Second World War. A reinforced concrete beam is formed at or above footing level and spans between mass concrete piers founded at appropriate depths (Figure 11.2). The beam is usually constructed first. Small areas of brickwork below DPC are broken out and sacrificial props, known as stools, are fixed in the wall. The stools are usually manufactured in steel and have precast concrete top and bottom plates. They are normally supported on the existing footing or are inserted in the wall above the footing. If the footing is too weak or too shallow, individual footing pads resting directly on the ground may be used for support. Once the stools are in place, the remaining brickwork, or existing footing, is removed. Reinforcement is threaded through the stools and formwork fixed to the sides to allow the beam to be cast in situ; the stools are thus incorporated into the beam. Where the underpinning has been provided to arrest heave in desiccated clays, the underside of the beam will require protection from the heaving soil; this is usually provided by a proprietary compressible or collapsible board material. After a suitable period of time, typically 24 hours, the beam is pinned-up to the existing brickwork above it. Isolated piers are then excavated to an appropriate depth, concreted and pinned up to the underside of the beam or original footing.

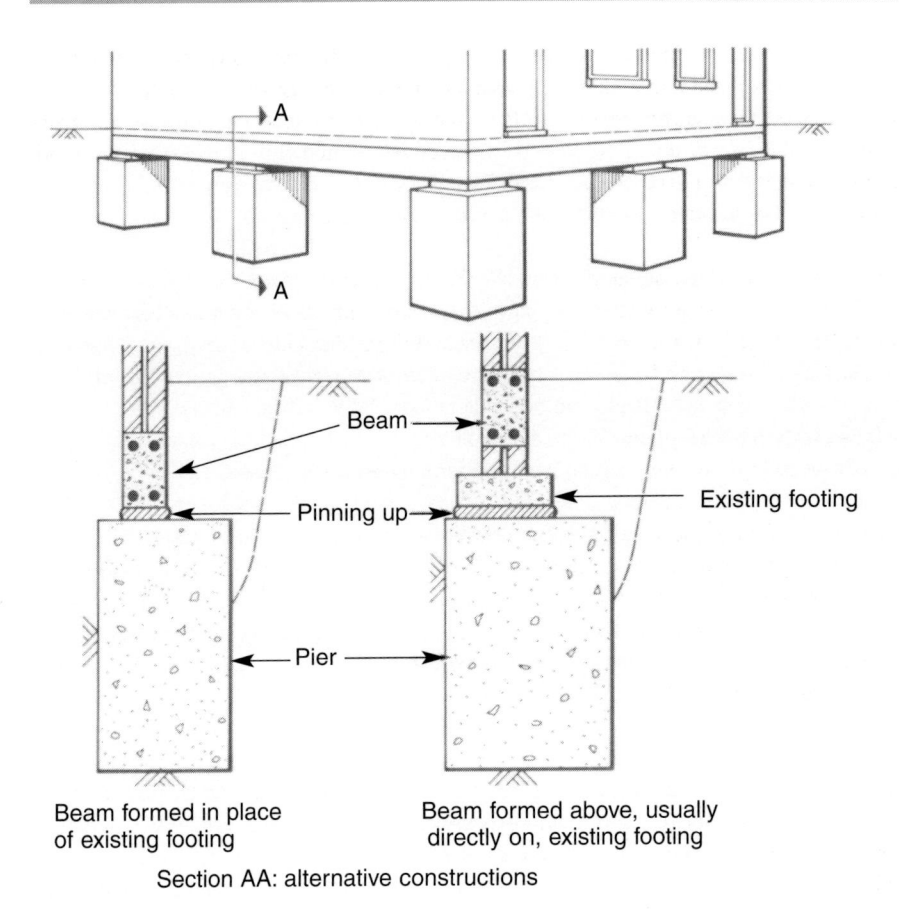

[Figure showing pier-and-beam underpinning with Section AA labels: Beam, Pinning up, Existing footing, Pier; captions "Beam formed in place of existing footing" and "Beam formed above, usually directly on, existing footing"; "Section AA: alternative constructions"]

Figure 11.2 Pier-and-beam underpinning

The beam, which is installed prior to excavation for the piers, serves to strengthen the building early in the underpinning process. Because the beam is constructed first, the building and the operatives are at less risk than with mass concrete underpinning.
The method is particularly appropriate where significant lateral movement of the ground is anticipated. Being relatively massive in comparison with piles, piers are able to withstand large lateral forces and, in the event of their movement, lateral forces are not necessarily transferred to the building because there is no structural connection between the piers and the footing or beam.

Pier-and-beam underpinning is feasible in most ground conditions. However, running sand and silt may prove to be impossible to excavate without endangering the structure so alternative methods of underpinning should be considered. While it is possible to construct piers to depths of 8 m or more, the high labour costs associated with hand excavation tend to restrict the maximum depth to about 5 m. Excavations can be carried out below the water-table by using shields, but this adds considerably to the cost.

The method is particularly suitable for use in clay that is likely to exhibit shrinkage or heave. The piers can be excavated to depths at which the effects of shrinkage and heave are minimal and, provided the sides of the piers and the under-sides of the beams are protected, the building will be isolated from the effects of volume changes in the soil.

11.8 Pile-and-beam and piled-raft underpinning

These pile-based systems are similar in concept to pier-and-beam underpinning but have advantages where there is no suitable bearing stratum available within a depth economical for hand excavation of piers (say 4 or 5 m) or where it is necessary for the underpinning to pass through loose or water-bearing strata. Beams are normally constructed in a manner similar to that for pier-and-beam underpinning, except that the beams are extended at corners and intersections to form caps for attaching to the pile heads (Figure 11.3). Intermediate supports are formed by pairs of piles using needle capping beams or, where internal access is restricted, cantilever pile caps.

Because of access limitations and the proximity of vulnerable structures, the piles are usually augered or bored and cast in situ; smaller piles may be driven. Alternatively, where external access is restricted, the existing floor slab can be removed and the piles installed inside the property; the piles are then connected using a reinforced concrete raft that is keyed to the external walls below ground level by removing small sections of brickwork.

Pile-and-beam underpinning can be used satisfactorily in most circumstances and becomes more economical than pier-and-beam when founding depths are greater than about 5 m or where ground conditions make hand excavation difficult. It is preferable to install piles after the construction of the beams, since vibration and loss of ground during piling may have serious effects on the foundations of a severely damaged building. This makes detailing and construction of pile caps difficult. Piled-raft systems are attractive where access to the outside of the building is restricted, or where a ground supported floor slab has to be replaced to avoid damage as a result of clay heave. The slender nature of the piles used in domestic underpinning, generally 125 to 400 mm diameter, means that they are less stable than piers under lateral loading.

11.9 Mini-piling underpinning

In this method of underpinning, loads are transferred directly from the structure to the

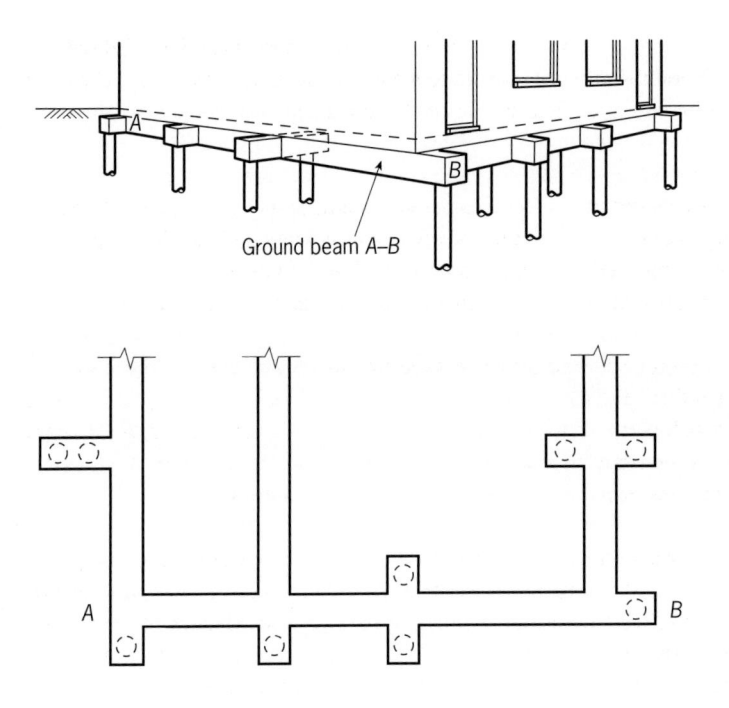

Figure 11.3 Pile-and-beam underpinning

piles, either by needles or cantilevered pile caps, or by placing the piles directly through the existing footing. In low-rise buildings the piles are normally at 1 m centres, or less. Because of the small spans between piles, pile loads are small and small diameter piles or mini-piles are invariably used. Typically, these piles have diameters ranging from 75 to 150 mm; they are formed by either filling bottom-driven steel or plastics casings with cement grout, or are augured and cast in situ[12].

The technique adopted depends largely on the ground conditions. Mini-piling systems can be categorised as follows[13]:

❐ *Needle piling.* The wall to be underpinned is supported by short needles at approximately 1 m centres carried on pairs of piles installed either both sides of the wall (Figure 11.4) or on one side only in a cantilever configuration (Figure 11.5).

❐ *Inclined piling.* Holes are cored at an angle through the existing foundations and an inclined or raking pile is formed and cast directly into the footing (Figure 11.6). Piles are generally inserted alternately from either side of the footing at 1 m centres or less, depending on the condition of the existing footing and the soil. The success of the system depends on the ability of the existing footings to withstand the lateral, vertical and torsional forces imposed by the alternately positioned piles.

REMEDIAL TREATMENT AND UNDERPINNING

❏ *Floor slab piling*. Subsiding ground supported floor slabs and rafts can be stabilised by coring through the slab and inserting mini-piles to a suitable depth (Figure 11.7). The head of the pile is then cast into the slab. The spacing of the piles depends on the strength of the floor slab.

Mini-piling tends to be cheaper than other pile-based systems and can compete with mass concrete underpinning in terms of cost. It has a low labour content and has the advantage, where driven piles are being used, of being a relatively clean and disturbance-free operation. Mini-piles can also be used to stabilise ground supported floor slabs.

Mini-pile systems rely on the strength of the existing foundations to transmit the wall loadings; they are, therefore, distinct from pile-and-beam and piled-raft underpinning. This form of underpinning differs from most others in the degree of dependence on the strength and integrity of the existing foundations and structure, which may be significant when considering older properties or properties in a poor structural condition.

Mini-piles are particularly suitable for underpinning buildings on uniform, relatively shallow thicknesses of fill and natural soil not susceptible to shrinkage and heave. Because of their slenderness, mini-piles are unsuitable where significant lateral loads are envisaged; their use to remedy clay shrinkage or heave movements is not normally recommended since horizontal forces may be transmitted unless precautionary measures are taken. Their main applications are where destabilising forces are acting vertically and adequate lateral support is afforded to the slender pile by the surrounding ground.

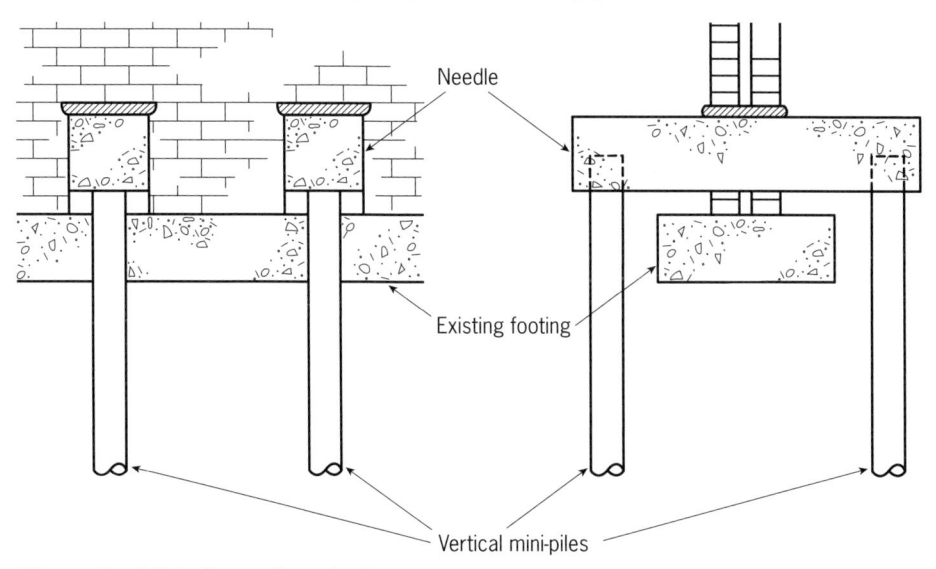

Figure 11.4 Mini-pile needle underpinnng

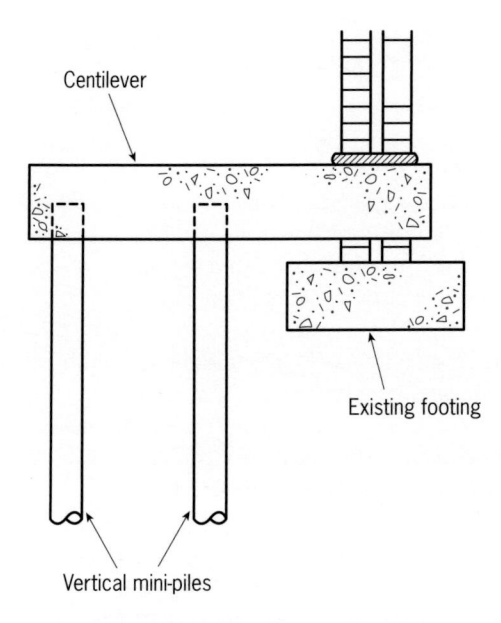

Figure 11.5 Mini-pile cantilever needle underpinning

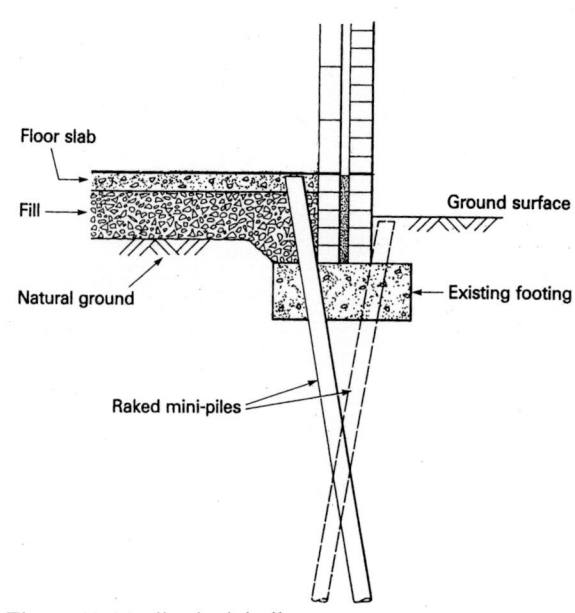

Figure 11.6 Inclined mini-piles

REMEDIAL TREATMENT AND UNDERPINNING

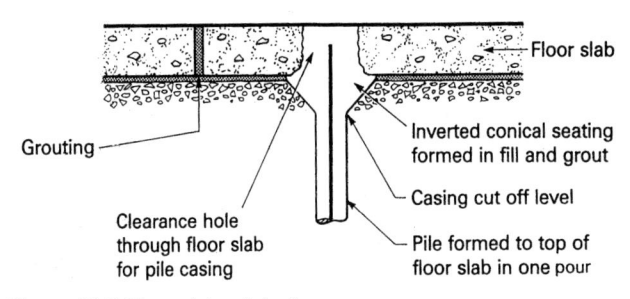

Figure 11.7 Floor slab mini-piles

11.10 Design of underpinning

As with foundations for new buildings (Chapter 6), it is generally uneconomical, and often unwise, to design underpinning not to move at all; the objective should be to limit movements to an acceptable, non-damaging level. However, since a damaged building may be more vulnerable to further foundation movement than a new building, the required depth for the underpinning may be determined on a more conservative basis than would be required for the foundation of a new building in similar circumstances.

The load applied to the soil by mass concrete underpinning and the piers of pier-and-beam systems must not exceed the bearing resistance of the soil (Chapter 6). Where significant amounts of future settlement are anticipated, jacks can be inserted between the underpinning and the original footings or ground beam. These can then compensate for future movements. However, it is rare for jacking to be used in domestic applications.

When installing mass concrete or pier-and-beam underpinning it is usually necessary to excavate an access pit on the outside of the wall before removing the soil from beneath the original footing. Consequently the line of action of the wall loading will normally be eccentric to the centre-line of the finished underpin, inducing an overturning moment. In more heavily loaded applications, this eccentric loading may lead to the bearing resistance of the soil being exceeded on the inside edge of the underpinning. This may be avoided by adopting a width of underpinning which is sufficient to ensure that the line of action of the resultant load passes through the middle third of the base.

Pile design criteria, described in Section 6.6, should be applied to the design of pile-and-beam, piled-raft and mini-piling systems. Appropriate soil parameters can be determined from a ground investigation, usually incorporating boreholes. Where driven mini-piles are used, they are usually driven to a set[14] and it is sometimes argued that a detailed ground investigation is not, therefore, needed. However, the set can be achieved on a buried obstruction and this can be a major problem when using this type of pile on fill sites.

Particular attention should be paid to the uplift, downdrag and lateral loading[15] that may be induced by future movements in the unstable ground. In pile-and-beam systems it may be possible to improve lateral restraint further by extending transitional beams into the unaffected parts of the property.

Where underpinning is being specified as a remedy for ground movements associated with fill, a layer of soft soil, internal erosion, collapse of an underground cavity, or landslip, the selection of a suitable depth for the underpinning is fairly straightforward; the underpinning should be taken to a depth where the ground is stiff enough and stable enough to provide a foundation that will be unaffected by these processes. For piled systems, this implies that the length of the pile may have to be approximately twice the depth of the unstable soil to ensure that there is sufficient resistance to downdrag and/or uplift forces. Similarly, where underpinning is being specified as a remedy for clay swelling following tree removal, the foundation should be taken below the depth of desiccated soil for a sufficient distance to provide adequate resistance to uplift forces.

Where large, mature trees are present, it is not uncommon for a shrinkable clay soil to be significantly desiccated to a depth of 6 m or more. However, it is likely that the levels of desiccation below about 3 m will remain at an approximately constant level as long as the tree remains in place (Section 7.5). The design of an underpinning system to withstand the ground movements associated with the removal of the tree will, therefore, be radically different from one designed simply to provide an adequate foundation while the tree remains as described in Box 11.1.

If a tree is to remain, the depth for underpinning is sometimes based on guidelines for new buildings[16]. Profiles of water content or suction can be used to quantify the extent and the magnitude of the desiccation (Sections 4.8 and 7.5). Some judgement is required to distinguish areas where the level of desiccation varies seasonally from those where it is relatively constant. In a soil with high volume change potential, it is often assumed that the depth of seasonal variation is about 1m plus half the remaining depth of desiccation; in a low volume change potential soil, movements may be almost entirely seasonal[17].

Alternatively it may be required that the excavation for the underpinning extends a distance, say 1 m, below the depth of the last visible root. This can lead to difficulties if the required depth is significantly greater than anticipated. There are three potential problems with this approach:

❐ trees can extract water through very fine fibrous roots which may be difficult to identify in an excavation;
❐ the zone of desiccation below a large tree may extend several metres beyond the last visible root for the reasons given in Section 7.5;

REMEDIAL TREATMENT AND UNDERPINNING

> **Box 11.1 Underpinning where large trees are present on shrinkable clay**
>
> There are two different situations that should be clearly distinguished:
> - an underpinning system designed to withstand the ground movements associated with the removal of the tree;
> - an underpinning system designed simply to provide an adequate foundation while the tree remains.
>
> The underpinning required in the first situation is likely to be radically different from that required for the second. For example, where the soil is desiccated to 6 m, the former requirement might necessitate a pile-and-beam system with pile lengths in excess of 12 m and extensive heave protection; the latter could probably be achieved using mass concrete to a depth of 2.5 m under external walls. There is a substantial cost differential between these two options, which for a large house could be in the order of £50,000.
>
> The lifespan of most common, large deciduous trees, such as oak, willow, ash, sycamore, horse chestnut is more than 100 years; it may not be necessary, therefore, to design against the eventuality of the tree being removed, unless there is good reason to believe the tree is diseased or nearing the end of its life. Wherever possible, nearby trees should be preserved and particular attention should be paid to avoiding damage to the trees and their roots during the underpinning operations.

- the presence of roots does not necessarily indicate that the ground in desiccated. Nevertheless, this approach may help identify situations where root activity is unusually vigorous[18].

Where the tree responsible for the damage is not yet fully mature, there are two options:
- remove the tree and design the underpinning to withstand the consequent heave;
- do nothing to the tree and make provision in the design of the underpinning for the increasing level of desiccation produced as the tree grows; consideration should also be given to the appropriateness of designing for the tree's future removal or death.

Special precautions are needed where the underpinning is being designed to protect the structure from the effects of soil swelling following tree removal. The underpinning has to be founded at a depth where the movements produced by the change in desiccation are tolerable[19]. Where large deciduous trees are involved, a ground investigation incorporating boreholes is likely to be needed to provide the necessary information. The procedure is essentially similar to that for new foundations (Section 7.5). In less extreme cases, for example where there are small deciduous trees or conifers, it may be cost-effective to make conservative assumptions regarding the depth of desiccation and to use this as the required depth for the underpinning.

With a pile-based system, due consideration should be given to the expected uplift forces generated by the swelling soil (Section 7.5). The required length of pile is likely to be in the order of twice the depth required for mass concrete or pier-and-beam underpinning.

In addition to selecting a depth to ensure adequate stability, it will be necessary to protect the underpinning and the existing foundations from the lateral and uplift forces generated by the swelling soil. The vertical faces of mass concrete, piers, ground beams and original footings can be protected by layers of low-density expanded polystyrene or other suitably compressible materials. This material should be placed on the inside of the foundations, where the soil is likely to be more confined and, therefore, capable of generating greater pressures. Even where trees are left in place, the underpinning is likely to cut through roots and cause the soil under the building to swell.

With a pile-and-beam or pier-and-beam system, uplift forces can be reduced by providing a compressible layer on the underside of the beam, where the beam is replacing the original footing. Alternatively, where the beam has been built into the wall, the original footing can be removed creating a void under the beam. Uplift forces on piles can be reduced by using various proprietary sleeving materials; however, it is not practicable to remove the uplift forces altogether and the friction acting on the sleeve has to be included in the design of the pile (Section 6.6). Swelling will also affect ground supported floor slabs and, where significant movements are anticipated, it may be necessary to replace any ground supported floor slabs with suspended floors.

11.11 Partial underpinning
Where only part of the building has been adversely affected by ground movement, it is generally inappropriate to underpin the whole structure. An underpinning scheme that does not include all loadbearing walls is normally referred to as partial underpinning. The underpinning may be restricted to one side of the property or it may be the internal walls that are left in their original condition. Similar considerations apply to the underpinning of a semi-detached or terraced house where it may not be practicable to extend the remedial scheme to the neighbouring properties.

When properly engineered, there is no reason why partial underpinning should not be satisfactory. However, special care is needed where the affected property is founded on shrinkable clay. All buildings founded on clay soils, even those with foundations complying with current guidelines, will move up and down as a result of seasonal moisture changes in the surface soil. The occupants are unlikely to be aware of these movements because the whole house is moving as a unit and differential foundation movements are small. However, there is a danger that, where part of the building is underpinned, the levels of differential movement in the remaining part will be increased, because the stability of the underpinned section will have been greatly improved.

One way of avoiding damage as a result of partial underpinning is to extend the underpinning under the unaffected part of the building and reduce the depth of the underpinning in steps. With mass concrete underpinning, the depth of each bay can easily

REMEDIAL TREATMENT AND UNDERPINNING

be varied and the usual practice is to decrease the depth of underpinning in 0.3 m steps over bays of 1 m length until it merges with the original foundations. This approach can also be applied to pier-and-beam and pile-and-beam systems, although coarser steps are likely to be needed because of the greater distance between piles or piers. Moreover, there is a practical minimum depth to which piles can be installed. For mini-piling systems, it should be possible to use progressively shorter piles at the extremities of the underpinning, although this is not common practice.

The likelihood of damage can be reduced by ensuring that the depth of the underpinning is not too deep and that the underpinned section of the building continues to move in sympathy with the rest of the structure. This may increase the risk of existing cracks re-opening, but may provide a more cost-effective solution than, for example, having to underpin all internal load-bearing walls.

Notes

1. BRE Digest 352 *Underpinning*, published in 1990, states that '*Every year in Britain about £80 million is spent on the repair of cracks and distortion in homes. Much of the damage arises from foundation movement caused by problems such as subsidence and heave of the ground, and landslip. About half the money is spent on underpinning foundations of houses and blocks of flats.*'

2. An illustration of this has been provided by the monitoring over a 25-year period of the heave of a single-storey building, which was constructed on clay following tree removal, (Cheney, 1988).

3. Inappropriate pruning can irreparably damage some trees and insufficient pruning may have little effect on water extraction. Pollarding is often mistakenly specified, because the height of the tree is the most commonly quoted measure of its likelihood to cause damage. However, since the vast majority of the moisture extracted by the roots passes straight up through the tree and is transpired, the leaf area is a more relevant parameter. Pollarding can severely damage the tree and pruning to thin or reduce the crown is generally preferable. The need for specialist advice is emphasised.

4. BS 5837: 1991 *Code of practice for trees in relation to construction.*

5. Marshall *et al* (1997) give general information on barrier positioning and materials.

6. Mair (1994) provides a useful review of these processes. See also Greenwood (1987).

7. For example, Brown (1992). See also BRE Good Repair Guides GRG 1 *Cracks caused by foundation movements*, GRG 2 *Damage to buildings caused by trees*, GRG 3 *Repairing damage to brick and block walls*.

8. A patented system *Hoopsafe* has been developed utilising a post-tensioned beam. This system does not use dowels; prior to concreting, the beams are tensioned to compress the structure contained within them. Burford *et al* (1999) have described the application of the *Hoopsafe* system to a damaged building on very high plasticity clay.

9. The increased use of underpinning for low-rise housing in the UK can be related to the perils of subsidence and heave damage being added to domestic building insurance policies. Remedial underpinning work may not always have been technically justified, and this could be at least partially responsible for a reluctance amongst some insurers to accept underpinning solutions.

10. More detailed guidance is in BRE Digests 352 *Underpinning* and 313 *Mini-piling for low-rise buildings*. Greenwood (1987) has described underpinning by grouting. See also Atkinson (2003).

11. A needle is a beam placed through a hole in the wall to transfer the weight of the wall above onto supports under the ends of the beam.

12. Guidance on the design, supervision and approval of remedial works and new foundations for low-rise buildings using mini-piles is given in BRE Digest 313. The term *mini-piles* is widely used for small diameter piles, but the European Standard prEN 14199:2001 refers to *micropiles*, defined as small piles which can be installed by means of a small rig: bored piles have a diameter smaller than 300 mm and displacement piles have a diameter smaller than 150 mm.

13. Root piles (*pali radice*) are a proprietary form of mini-piling in which a high-strength grout is injected into small-diameter holes to form the piles. The grout is usually placed by means of a tremie pipe and compressed air forces the grout into intimate contact with the soil as the tube is withdrawn. Small-diameter (100 mm) root piles are reinforced with a single bar, larger diameters (250 - 300 mm) use a cage or tube. Special care is needed when forming root piles in loose soils or below the water-table to ensure that the grout is continuous and sound.

14. A set is the penetration of a driven pile for each blow of the drop hammer which is driving the pile into the ground.

15. Guidance on the design of laterally loaded piles is given by Elson (1984); most building foundation piles can provide considerable resistance to lateral forces.

16. Chapter 4.2 of the NHBC Standards gives recommended depths for new house foundations when building near trees. These depths can be considerably less than the depth of desiccation as it is generally accepted that significant desiccation is often found at depths of 6 m and more near large trees (Chandler *et al*, 1992). Designing underpinning to these NHBC guidelines for new foundations could, therefore, produce a foundation that continues to move. The level of movement should be within the bounds that are tolerable by new buildings, say a maximum of 20 mm, and this may be acceptable for the underpinned building provided the structure has not been unduly weakened by the previous foundation movements.

17. When using this approach, it would not be prudent to specify depths that are shallower than those recommended in Chapter 4.2 of the NHBC Standards.

18. It would not be prudent to specify any underpinning depths based on the last visible root that are less than those recommended in Chapter 4.2 of the NHBC Standards.

19. It may be acceptable to use an allowable movement of 20 mm where damage has been caused by subsidence prior to the removal of the tree, on the basis that any future movement is tending to reverse existing distortions. Where existing damage is attributable to ongoing heave, it would be prudent to use a smaller value for the allowable movement, say 10 mm.

CHAPTER 12: ANCILLARY WORKS

Chapters 1 to 11 have been primarily concerned with the geotechnical aspects of foundation design and behaviour. This final chapter addresses some other facets of construction which commonly occur in building developments and which require geotechnical input. These are: hardcore, soakaways, drains and sewers, small earth embankments, small retaining walls and free-standing walls. The design and construction of embankments and retaining walls are major subjects; only the small structures typically found on building sites are considered here.

12.1 Hardcore

The principal uses of hardcore are as a make-up material to provide a level base on which to cast a floor slab, to raise levels and provide a dry, firm base on which work can proceed, or to carry construction traffic. For economic reasons, it is desirable to use readily available local material and a wide variety of materials have been used satisfactorily. Concerns over hardcore mainly occur where it used as infill within the foundations of a house to provide a level base on which to cast a ground supported floor slab[1]. Unsatisfactory behaviour of hardcore can result in gaps between floors and skirtings of perimeter walls, cracking of floor-mounted walls and disruption of services.

Problems with hardcore can be avoided by ensuring that:
- the site is properly prepared;
- appropriate types of material are selected;
- the materials receive adequate compaction.

Before hardcore is placed, the site should be adequately prepared. Unsuitable materials, such as topsoil, very soft clay and waste left from previous use of the site, should be removed. The ground on which the hardcore is to be placed should be able to support the hardcore and the concrete slab without undue movement.

Acceptable materials

A variety of materials have been used satisfactorily as hardcore, including natural soil and rock and various types of waste material. However, unsatisfactory behaviour can be experienced and a number of factors must be taken into account in the selection of materials. Materials which are to be used as hardcore should be:
- free-draining;
- not affected by water and not subject to volume change with change in water content;
- chemically inert;
- free from vegetable matter.

However, few of the materials available at reasonable cost satisfy these requirements completely. Some common materials are described in Table 12.1.

Table 12.1 Suitability of materials for use as hardcore

Gravel and crushed hard rock
Well graded materials are particularly suitable as they compact more easily. Graded supplies of gravel or crushed rock tend to be relatively expensive and an economic option may be to use coarse ungraded gravel and rock as a base layer covered by a layer of well graded material.

Quarry waste
Generally clean, hard and safe to use, but may be poorly graded and difficult to compact to provide a good working surface. Crusher-run stone should be economical near to the quarry. Gypsum waste should not be used as it frequently contains a mixture of limestone and gypsum which can attack concrete to form thaumasite[2].

Chalk
Widely and successfully used in some areas, but susceptible to frost heave (it has a tendency to expand when subjected to prolonged periods of sub-zero temperature). Problems are most likely to occur during construction while the floor slab is exposed to the elements and protection or heating may be needed during long, cold periods. Difficulties are likely to be encountered only after a few days of continuously freezing weather. Occupied buildings are unlikely to be affected but unheated buildings and cold stores merit special consideration.

Concrete rubble
Forms a good material when clean and suitably graded. Care should be taken with demolition rubble which may contain mixtures of materials, including gypsum plaster which could cause sulfate attack if in close proximity to concrete or brickwork in wet conditions. Dry rot in new buildings has been attributed to the use of hardcore containing pieces of timber from demolished buildings where dry rot was established.

Brick and tile rubble
Can be a useful material when taken from old buildings. Unless the bricks are soft and crumble easily, there may be a lack of fine material and it may be difficult to compact. Many clay bricks contain sulfate so a water barrier between the fill and a concrete floor is a wise precaution, particularly if new bricks, or old bricks with adhering gypsum plaster, are used. Refractory bricks, such as those from chimneys and furnaces, should be avoided as some types can expand when exposed to moisture.

Slag (Box 8.7)
Blastfurnace slag is a strong material that is often used as hardcore. It may contain sulfates but these are predominantly calcium sulfate which has a limited solubility in water. Slag is a free-draining material which will not retain water in contact with the concrete or brickwork. Large volumes of slag placed in wet stagnant conditions have occasionally caused water pollution problems by sulfur species leached from the slag. Old slag banks should be thoroughly sampled and tested as they may contain mixtures of materials. Old, partially vitrified slag may swell owing to attack by sulfate solutions arising from the groundwater or other components of the fill. Slag from steelmaking is not recommended because it may contain phases which cause expansion on wetting, such as free lime, free magnesia or broken refractory bricks.

Colliery spoil (Box 8.6)
Burnt colliery spoil, which results from combustion of residual coal and carbonaceous matter in old, poorly consolidated tips, has caused the failure of ground floor slabs by sulfate attack on wet sites. Although burnt spoil is a useful free-draining material, it can be variable because of incomplete combustion and tends to have a higher soluble sulfate content than unburnt spoil because of oxidation of sulfides (mainly pyrites) during burning. Unburnt colliery spoil is normally preferred; sulfate content is generally lower and good compaction is easier to achieve. When well compacted, access of oxygen is prevented and combustion should not take place. There is no convincing evidence that colliery spoil has caused failures by swelling.

ANCILLARY WORKS

Oil shale residue
Spent shale from the extraction of oil from oil shale is available in the Lothians area of Scotland. It is a tough material which resists disintegration by weather and has been extensively used. Its soluble sulfate content can, however, be appreciable and variable; on wet sites, precautions should be taken against sulfate attack on concrete or mortar.

Pulverised-fuel ash (pfa)
Fine waste material from precipitators of coal burning power stations can be used as hardcore. When conditioned with water, pfa (flyash) produces a lightweight material with self-hardening properties. It is generally low in sulfate but when obtained from a power station mixed with furnace bottom ash (the other major waste from the burning process), the total sulfate content may be increased.

Particle size is a key factor in achieving acceptable performance and excessive fines (particles smaller than 0.06 mm) can lead to the development of excess pore water pressure. Although it is not possible to give a definitive range of acceptable particle size distribution, the following general guidelines should be helpful. It is preferable that the hardcore should be a well-graded coarse material with, ideally, no more than about 10% fines. The maximum permissible particle size is related to layer thickness and with a layer thickness of 200 mm, a maximum allowable particle size of 150 mm is appropriate.

Hardcore should be placed in thin layers and each layer adequately compacted. The maximum allowable layer thickness depends on the compaction equipment that is used. Only light compaction equipment, such as a vibrating plate compactor, vibro-tamper or power rammer, is likely to be available when a small floor slab is constructed, and consequently the maximum permissible layer thickness after compaction is unlikely to be larger than 200 mm. The hardcore should be placed in a moist condition. While adequately compacted hardcore normally functions satisfactorily, difficulties can arise, in relation to house construction, associated with:
❐ compaction in a small confined area;
❐ inadequate supervision and control of compaction.

Common problems
Failures of hardcore are commonly associated with inadequate preparation of the site, poor compaction, excessive fines and the presence of water. Main hazards to be avoided are:
❐ chemical attack on concrete and mortar;
❐ swelling due to chemical changes;
❐ settlement of hardcore.

Chemical attack on concrete and mortar. Concrete floor slabs are vulnerable to attack by water soluble sulfates in hardcore and there have been many failures. In damp conditions sulfates can migrate into the underside of the slab and react with constituents of the cement. This sulfate attack produces a gradual breakdown of the concrete and causes it to expand and lift. Colliery spoil has been a particular problem but other materials containing sulfates have also been responsible, such as industrial waste and demolition rubble that has contained substantial quantities of gypsum plaster. The effects of the expansion are usually noticed first. It causes distortion and cracking of the relatively thin floor slab and consequent disturbance and movement of any partition wall built off it and can, in some circumstances, exert sufficient pressure on the enclosing walls to cause them to move outwards. The movements are quite slow and usually unnoticed for several years. In general, hardcore material containing water-soluble sulfates should be avoided; alternatively, a concrete quality should be chosen that will resist the effects of the sulfates.

Swelling due to chemical changes. The number of cases where movement in buildings has been caused by hardcore materials swelling is relatively small. Although swelling may have been suspected initially, the movement has usually proved to be due to sulfate attack and consequent volume change of concrete in contact with the hardcore. Material that contains a significant proportion of clay (for example some colliery spoil) can swell if placed in a dry condition but subsequently becomes wet. However, most cases where hardcore swells are caused by chemical reactions.

❒ Calcium oxide (free lime) found in steel slags (Box 8.7) reacts with moisture to form calcium hydroxide, with a large increase in volume and, subsequently, carbonates.

❒ Material which contains broken concrete or old, partially vitrified slag from old slag dumps may expand when exposed to sulfate solutions which originated in either the groundwater or other components of the hardcore. Modern, air-cooled blastfurnace slag is not susceptible to attack by sulfates.

❒ Sulfides in the form of pyrite can oxidise to form soluble sulfate in the presence of air, moisture and, possibly, bacterial action. They can therefore contribute to problems of sulfate attack on concrete. In addition, if pyrite and calcite are present together there is danger of expansion due to the growth of gypsum crystals formed by reaction between calcite and sulfuric acid from the oxidation of pyrite[3].

Settlement of hardcore. Hardcore may settle after building operations are complete owing to the following:
❒ inadequate compaction;
❒ depth too great;
❒ degradation of materials in wet conditions.

When settlement of the hardcore occurs, the ground supported floor slab loses its support over part or much of its area, gaps appear between the floor and skirting board, the slab can crack and any partitions and features built off it can be disrupted (Figure 12.1). There have been many cases where this has occurred and sites are particularly vulnerable where a considerable depth of hardcore is needed. It is, therefore, sensible to restrict the maximum permissible depth of hardcore[4]. It should be noted that this type of movement does not affect the foundations. Even on sites where the average depth of hardcore over the site is relatively small, difficulties can still be experienced at the edges if deep trenches have been formed to facilitate the construction of the foundations.

ANCILLARY WORKS

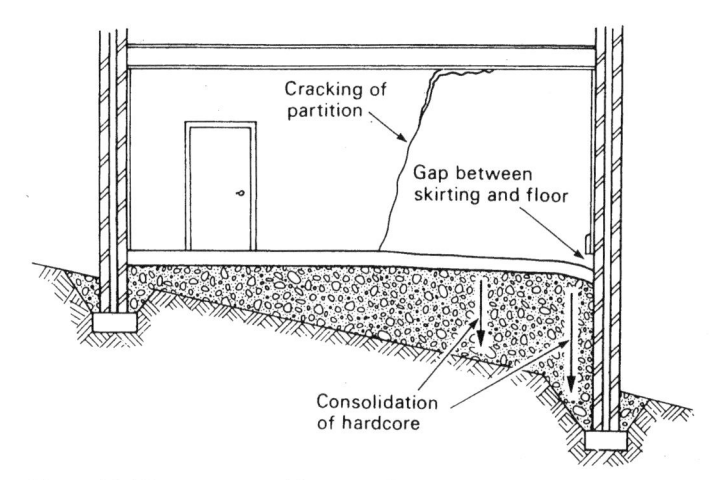

Figure 12.1 Damage caused by consolidation of hardcore

12.2 Soakaways

Soakaways are pits or trenches which store stormwater from buildings and paved areas and allow it to percolate gradually into the ground, so preventing flooding. Traditionally their use was restricted to areas remote from a public sewer or watercourse. However, they have become increasingly common in urban areas to limit the impact of new developments on existing sewer systems.

Soakaways can be constructed in many different forms and from a range of materials[5]. To function effectively, they should have an adequate capacity to store immediate stormwater run-off and should discharge the stored water sufficiently quickly to provide the capacity required to receive run-off from a subsequent storm. The time taken for discharge depends upon the size and shape of the soakaway and the infiltration characteristics of the surrounding soil.

Site investigations should be undertaken thoroughly and competently so that all aspects of soil properties and hydrogeology are adequately reviewed alongside the hydraulic design of soakaways. Natural groundwater should not rise to the level of the base of the soakaway during annual variations in the water-table. A soakaway should not normally be constructed closer than 5 m to building foundations[6]. In chalk, or other soil and fill material subject to modification or instability, the advice of a geotechnical specialist should be sought as to the advisability and siting of a soakaway. The risk of polluting the groundwater is discussed in Box 12.1

With run-off areas of less then 100 m^2, soakaways usually have been built as square or circular pits, either filled with rubble or lined with dry-jointed brickwork or precast

> **Box 12 1 Soakaways and the pollution of groundwater**
> The risk of pollution reducing the quality of groundwater should be considered. Roof surface run-off does not seem to cause damage to groundwater quality and may be discharged directly to soakaways. Pollutants entering the soakaway from roofs tend to remain in the soakaway or in its immediate environs, attached to soil particles. However, paved surface run-off should be passed through a suitable form of oil interception device prior to discharge to soakaways. Maintenance of silt traps, gully pots and interceptors will improve the long-term performance, and the use of wet well chambers within the soakaway system can further assist in trapping pollutants and extending operating life.

perforated concrete ring units surrounded by suitable coarse backfill. An alternative is to use trenches which have larger surface areas for a given stored volume compared with square or circular shapes. The merits of the more compact square and circular forms of soakaway have to be weighed against the better rate of discharge from the trench for the soil type, available space, layout and site topography.

For drained areas in excess of 100 m², soakaways can either be formed of precast concrete rings or be of the trench type. They should not be substantially deeper than soakaways that serve small areas: a depth of 3 to 4 m is adequate if ground conditions allow. Limiting the depth means that the length of trench soakaways has to be increased, but they are cheaper to dig with readily available excavating equipment. Trenches have a marked advantage where there is a high water-table.

Care should be taken that the introduction of large volumes of surface run-off into the soil does not disrupt the existing sub-surface drainage patterns; it may be advantageous to use extended trench soakaway systems. The effect of ground slope should be considered when siting soakaways to avoid water-logging of downhill areas.

Soakaways can provide a long-term, effective method of disposal of stormwater from impermeable areas of several hundred square metres. Long-term maintenance and inspection should be considered during the design and construction phase. Access for inspection is desirable with all types of soakaway but with wet-well soakaways it is particularly important that there should be suitable access to inspection covers so that vehicle mounted suction emptying and jetting equipment can be used.

Design
The required storage volume for a soakaway, S, is the maximum anticipated inflow of run-off from the catchment area during a storm, less the seepage into the ground that occurs during the storm. Hence, S, which is defined as the volume over the effective depth of the soakaway (the depth between the base of the soakaway and the invert of the drain discharging into it), can be derived from:

$$S = (A\ R) - (a_{50} f D) \qquad (12.1)$$

where:

> A is the impermeable area drained to the soakaway
> R is the total rainfall in a design storm
> a_{50} is the internal surface area of the soakaway below 50% of the effective depth and excluding the base area, which is assumed to clog with fine particles and become ineffective in the long term
> f is the infiltration rate of the surrounding soil
> D is the storm duration.

There are four steps in the design procedure[7]:
❐ carry out a site investigation to determine the soil infiltration rate;
❐ decide on a likely construction type;
❐ calculate the required storage volume S using equation 12.1;
❐ review the design to ensure its overall suitability considering space requirements, site layout and time for emptying; the time taken for the soakaway to discharge from full to half-full should not exceed 24 hours so that the soakaway is ready to receive the run-off from a subsequent storm.

This approach to design contains assumptions that generally combine to produce an adequate factor of safety against surface flooding, because:
❐ inflow into the soakaway is over-estimated as no account is taken of the losses due to surface wetting and filling of puddles;
❐ flow of water into the soil is under-estimated as higher rates of infiltration occur at greater depths of storage;
❐ required storage volume is over-estimated as no allowance is made for the time taken for the run-off to reach the soakaway; in practice, the time period over which water enters the soakaway will be considerably greater than the duration of the storm allowing a greater volume of water to seep away.

Soakaways are commonly designed for a 15-minute or 30-minute duration storm. It has been suggested that the value of R should be based on the 10-year return period storm (the maximum amount of rainfall that can be expected in a storm of given duration once every 10 years[8]. However, it is becoming increasingly common to specify longer return periods.

Infiltration characteristics

The infiltration characteristics of the soil are determined by excavating a trial pit, filling it with water and measuring the time taken for it to empty:
❐ the pit should be of sufficient size to represent a section of the design soakaway;
❐ the pit should be filled several times in quick succession whilst monitoring the rate of seepage to ensure that the water content of the surrounding soil is typical of the site when the soakaway becomes operative;

❒ existing site data should be examined to assess whether there are any geotechnical or geological factors likely to affect the long-term percolation and stability of the area surrounding the soakaway, such as variations in soil conditions, areas of filled land, preferential underground seepage routes, and variations in groundwater level.

The soakage trial pit should be excavated to the depth anticipated for the full-size soakaway; for areas of no more than 100 m^2, this will be 1 to 1.5 m below the invert level of the drain discharging to the soakaway, or a typical overall excavation depth of 1.5 to 2.5 m below ground level. The trial pit should be 0.3 to 1 m wide and 1 to 3 m long. It should have vertical sides trimmed square. A suitably dimensioned pit can be excavated with a backhoe or mini-excavator. Where necessary for stability, it should be filled with coarse granular material; in this case a full-height perforated vertical observation tube should be positioned in the pit to allow the water levels to be monitored using a dipmeter. It is necessary to know the porosity of the granular material, so that the water storage volume can be calculated. The dimensions of the pit should be carefully measured before trials begin and, for safety reasons, no one should enter the pit.

Determination of the soil infiltration rate will require a great deal of water, so a bowser may be needed. The inflow should be rapid so that the pit can be filled to its maximum effective depth in a short time (to the proposed invert level of the drain to the soakaway). Care should be taken to ensure the inflow does not cause the walls of the pit to collapse. The pit should be filled and allowed to drain three times to near empty; each time, the water level and time from filling should be recorded at sufficiently close intervals to define clearly the drop in water level as a function of time. The three fillings should be on the same or consecutive days.

The soil infiltration rate f is calculated from the time taken for the water level to fall from 75 % to 25 % of the effective storage depth in the pit:

$$f = \frac{V}{a_{50} t} \quad (12.2)$$

where:
    V is the effective storage volume of water in the trial pit between 75% and 25% of the effective depth
    a_{50} is the internal surface area of the trial pit up to 50% effective depth and including the base area
    t is the time for the water level to fall from 75 % to 25 % of the effective depth.

For design purposes, the lowest value of f obtained in the three tests is used.

ANCILLARY WORKS

Where a test pit is deep, it may be difficult to supply sufficient water for a full-depth soakage test. In such circumstances, tests may be conducted at less than full water depth, but the soil infiltration rates determined in this way are likely to be lower than those obtained from full-depth tests. This is because the relationship between depth of water in the soakage pit, effective area of outflow and infiltration rate can vary with depth by a factor as large as two, even when the soil conditions are constant.

When it is not possible to carry out a full depth soakage test, the soil infiltration rate calculation should be based on the time for the water level to fall from 75% to 25% of the actual maximum water depth achieved in the test. The effective area of loss is then calculated as the internal surface area of the pit to 50% of the maximum depth achieved in the test, plus the base area of the pit.

Soakage trials should be undertaken where the drain will discharge to the soakaway. Full-depth tests with repeat determinations at locations along the line of trench soakaways are important where soil conditions vary or the soil is fissured since infiltration rates can vary enormously. A preliminary design length for the soakaway should be calculated from the first soakage trial pit result and, if the design length exceeds 10 m, a second trial pit should be excavated at the design length along the line of the soakaway. In all ground conditions, a second trial pit should be dug if the trench soakaway (designed on the basis of one trial pit) is longer then 25 m. If more than one trial pit is used, the final design should be based on the mean value of the infiltration test rates.

Construction
A small soakaway can be built by excavating a pit and filling it with very coarse material, such as broken brick, crushed rock or gravel[9]. A geomembrane or concrete blinding should be placed over the granular material to prevent topsoil being washed into the soakaway.

A large soakaway formed of perforated precast concrete rings should be installed within a square pit, with sides about twice the selected ring diameter to assist in the construction of the ring chamber, and the space between the rings and the soil backfilled with coarse granular material. Advantage may be taken of the oversizing of the soakaway pit by taking into account the additional volume of the voids in the granular backfill material below the discharge drain invert in calculating the design storage volume.

In all cases, the backfill should be separated from the surrounding soil by a suitable geotextile to prevent migration of fines into the soakaway. If migration from the surrounding soil occurs, it can cause ground settlement around the soakaway sufficient to affect the stability of adjacent buildings. The top surface of the coarse backfill should also be covered with a geotextile to prevent the ingress of material during and after surface reinstatement. Geotextile should not be wrapped around the outside of the ring units as it cannot be cleaned satisfactorily or removed when it has become blocked.

Maintenance

There have always been problems with the maintenance of soakaways. A rubble-filled soakaway may be difficult to find. All soakaways should be provided with an inspection access at the point of discharge of the drain to the soakaway. The access will identify the location and allow material to be cleared from the soakaway.

There is little monitoring of soakaway performance but this could be most informative about changes in soil infiltration rate and in warning of soakaway blockage. The inspection access should provide a clear view to the base of the soakaway, even when the soakaway is of the filled type (Figure 12.2). For small, filled soakaways, a 225 mm perforated pipe provides a suitable inspection well. Wet-well soakaways are easier to access for inspection and cleaning. Trench-type soakaways should have at least two inspection access points, one at each end of a straight trench, with a horizontal perforated or porous distributor pipe linking the ends along the top of the fill material (Figure 12.3). It may be convenient with a trench soakaway to have several drain discharge points along the length of the trench, each connected to the soakaway via an inspection access chamber.

12.3 Drains and sewers

A drain is a buried pipe that removes either foul water from sanitary pipework or rainwater and conveys it to a sewer. The drainage of a building and the disposal of sewage have a major impact on public health. The design of the drainage system should ensure that waterborne waste is carried away efficiently with minimal risk of blockage or leakage of effluent into the ground. A blocked drain can cause effluent leakage and flooding. The hydraulic features of the design are beyond the scope of this book and only the geotechnical aspects of the provision of drains and sewers, such as the procedures for burying the pipes to ensure that they are adequately protected and do not cause unnecessary subsidence of nearby foundations, are considered[10].

A drainage system normally consists of a network of pipes laid from a building to fall to a public sewer. Drains should be laid in

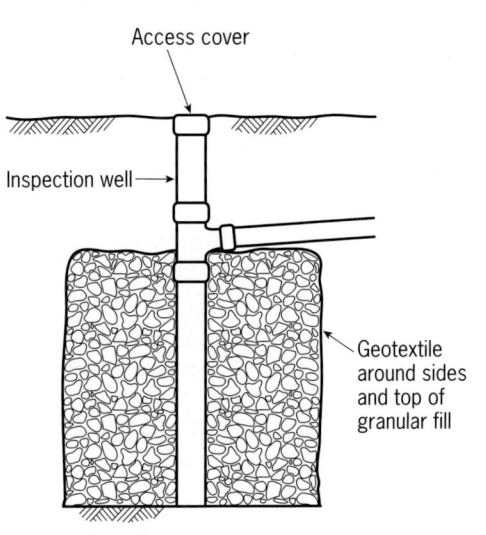

Figure 12.2 Small filled soakaway

ANCILLARY WORKS

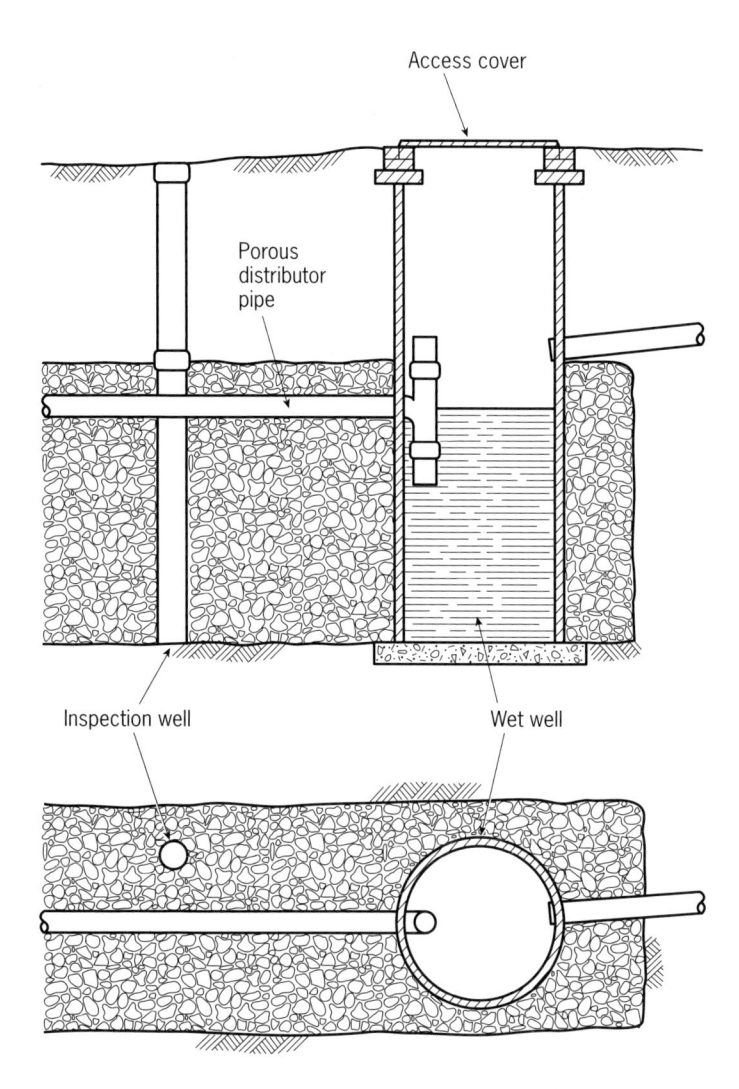

Figure 12.3 Trench-type soakaway with large wet well equipped with T-piece overflow to porous distributor pipe and separate inspection well

straight runs between access points at an approximately constant gradient. Access to drains is required for testing, inspection, maintenance and removal of debris. The traditional means of access is the inspection chamber, which becomes a manhole when it provides working space at drain level. Manholes are expensive but rodding eyes may be substituted, particularly at the head of shallow branch drains. A rodding eye consists of a vertical or inclined length of pipe with a suitable cover or cap at ground level and a curved junction with the drain at its base.

Bedding and backfilling
A drain is formed by placing a line of pipes in the bottom of a trench, which is then suitably backfilled. The load on a pipeline depends on the diameter of the pipe, the flexibility of the pipe (Box12.2), the depth at which it is laid, the trench width, the traffic, the foundation or other superimposed loading, and the prevailing site conditions[11]. The loading on the pipe is also influenced greatly by the type of backfill placed under and around the pipe.

Rigid pipes may be placed on the bottom of the trench or on a layer of granular material. Laying directly on the bottom of the trench is appropriate only where the trench bottom can be trimmed accurately using a shovel, such as in firm clay. It is not appropriate in soft silt and clay, soil containing large stones or flints, or in dense sand and gravel where a pick is required for excavation. The trench should be excavated to a depth slightly less than that required, so that hand trimming brings the bottom to the correct level. Over-dug places should be levelled using well-compacted fill. Where accurate hand trimming is not possible, the trench should be excavated below the pipe invert level to allow a minimum thickness of 100 mm of well compacted granular bedding material to be laid on the bottom of the trench[12].

To provide adequate protection for flexible pipes, the drain should be fully surrounded with suitably compacted, selected granular material. Total encasement in concrete is not recommended for flexible pipes. The pipe should be laid on a compacted layer of the selected material with a minimum thickness of 100 mm. The trench should be kept as narrow as practicable, but a clearance of at least 150 mm is needed on both sides of the pipe to allow the fill to

Box 12 2 Flexible and rigid pipes
Pipes and their joints may be either flexible or rigid. Flexibility is desirable in locations where ground movement is likely. Flexible joints enable rigid pipes to deform without fracturing. Flexible pipes depend particularly on the compaction of the sidefill material to prevent excessive distortion. For rigid pipes, there is often a choice between using a high strength pipe on a weak bedding or a weaker pipe on a stronger bedding. The choice will usually be determined by overall economy, although site conditions and material availability may also have to be considered. On small jobs, it may be preferable to choose a pipe with minimum bedding requirements rather than provide the supervision needed to ensure compliance with a more onerous bedding specification. Pipes should be firmly supported throughout their length and should not be laid on blocks or other supports that will create 'hard spots'.

ANCILLARY WORKS

be compacted. Selected material should be used at least until the pipe is completely covered. Where the general backfill contains stones or the trench is more than 2 m deep in poor ground, the selected material should be extended upwards to provide more cover to the pipe. The selected material should be placed and compacted equally on both sides of the pipe using hand rammers or light plate vibrators. Compaction may be unnecessary when using single sized aggregate[13].

In less stable soils, such as soft clays, silts or fine sands, and especially where there is a high water-table, it may be necessary to increase the thickness of the bedding and surround. In exceptionally soft ground with a high water-table or running sands, specialist advice may be required.

The trench will normally be backfilled using the excavated material. However, where the excavated material is difficult to compact and it is desirable to minimise settlement (for example where a road is to be constructed over the line of the trench), it may be preferable to import a more suitable material. With flexible pipes, the first 300 mm of general backfill should be selected to be free of stones larger than 40 mm. The general backfill material should be spread in layers no more than 300 mm thick and thoroughly compacted before placing the next layer. Heavy compactors should not be used until there is a minimum cover of 600 mm over the top of the pipe[14]. Trench sheeting should be removed as the work proceeds to avoid leaving unfilled voids at the sides of the trench. Pipes may be protected from traffic loading by covering with a reinforced concrete slab placed over a granular surround.

Pipes near or under buildings
Where pipes are placed close to an existing building, the foundations of the building could settle as a result of the excavation of the trench or of long-term settlement of the backfill material[15]. It is recommended that a pipe trench within 1 m of a loadbearing wall foundation should be filled with concrete to the level of the underside of the foundation (Figure 12.4); where the separation A between the trench and the foundation is more than 1 m, the trench should be filled with concrete to a level of $A - 150$ mm below the underside of the foundation. Pipes should be supported clear of the trench bottom by placing small wooden blocks or cradles under each pipe; concrete is then placed under each pipe to give continuous support. Where the pipes are flexibly jointed, the joints should be provided through the full thickness of the concrete at each pipe joint, using a thick compressible board.

A drain located under a building can cause major difficulties[16]. Consequently, where a drain has to pass under a building, particular care is needed and it should be surrounded by at least 100 mm of granular or other flexible filling and additional flexible joints may be needed on sites prone to excessive settlement. Concrete encasement should be provided

Ground level

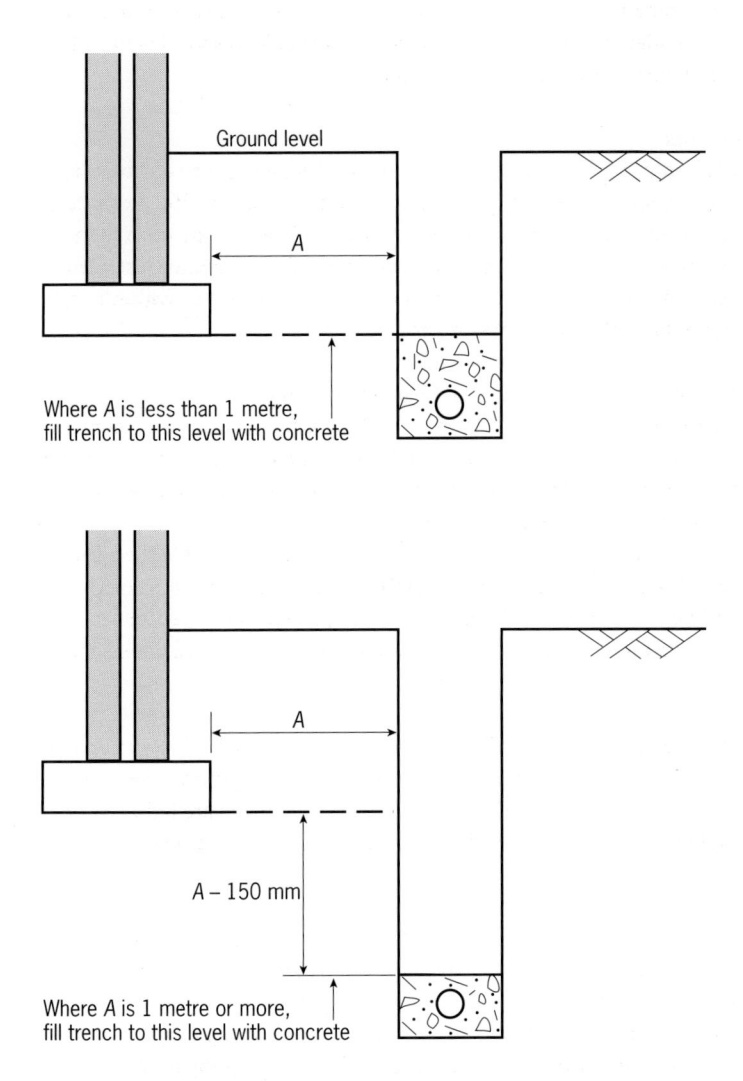

A

Where A is less than 1 metre,
fill trench to this level with concrete

A

A – 150 mm

Where A is 1 metre or more,
fill trench to this level with concrete

Figure 12.4 Placing drain close to building

ANCILLARY WORKS

where the crown of the pipe is within 300 mm of the underside of the floor slab and this encasement should be integral with the slab.

Where a drain passes through a wall or foundation, there is a major risk of differential settlement[17]. There are two approaches to the problem:
- provide an opening with an adequate clearance (at least 50 mm) all round the pipe;
- build a short length of pipe into the wall (with its joints at no more than 150 mm from the face of the wall or foundation), which is connected on each side to short lengths of pipe (less than 600 mm long) fitted with flexible joints.

12.4 Small embankments

Embankments are required in many civil engineering works, including the construction of highways and dams. These are specialised forms of construction with specific performance requirements and there is much technical literature[18]; such applications are outside the scope of this book. In this section, a brief description is given of some matters relevant to the use of embankments in landscaping works which are usually concerned with reducing the visual and acoustic impact of major roads on residential areas.

Geotechnical and economic parameters control embankment design:
- The geotechnical performance requirements of embankment barriers are quite simple; the embankment should be stable over a long period. Embankment instability could affect adjacent buildings. Small consolidation settlement of the fill or the foundation of the embankment will not usually be of concern.
- For economic reasons, it is desirable to use local materials; where possible the fill will be obtained from excavations at the site where the embankment is being constructed. Construction costs will be linked to the volume of fill to be placed so the steeper the slopes, the cheaper the embankment; but there may be environmental and maintenance considerations which make flatter slopes preferable.

The two major and inter-related elements of the design are the selection of fill material and the choice of the gradient of the embankment slopes. Coarse fill can generally be safely built to steeper slopes than can clay fill. However, where embankments are built on soft ground, the low strength of the foundation may be the factor controlling embankment stability.

Embankments should be designed to provide an adequate factor of safety against slope instability. A landscaping embankment is unlikely to be more than 5 m high; where a small embankment is built on a firm, level foundation, using well-graded and well-compacted coarse fill, slopes of 1 vertical to 2 horizontal will usually be satisfactory (Figure 12.5). If a small embankment is built on a firm, level foundation using a well-compacted sandy clay fill with $120 > c_u > 50$ kPa, slopes of 1 vertical to 3 horizontal might be adopted.

Figure 12.5 Small embankment

Where the embankment is higher than 5 m, or is being built on soft or sloping ground, or is being built of heavily overconsolidated clay, or there is some unusual feature, it is recommended that specialist geotechnical advice is sought. The minimum acceptable width of the top of the embankment is likely to be governed by the practicalities of construction with earthmoving machinery. It is unlikely to be less than 3m.

The method of deposition and spreading of the fill depends on the available earthmoving plant. The fill should be placed in layers sufficiently thin to permit adequate compaction through the full depth of the layer. Many different types of compactor can be used in embankment construction, including pneumatic-tyred rollers, vibrating smooth-wheeled rollers and vibrating sheepsfoot rollers. Each type of equipment is produced in a wide range of sizes: there are small, pedestrian-operated vibrating rollers and large, towed, vibrating rollers. The selection of compaction plant depends on the type of fill to be compacted, the scale of the operation and the availability of the equipment[19].

12.5 Small retaining walls

Retaining walls are used to support the ground at slopes steeper than those at which it would be naturally stable. The design of large retaining walls, which often form part of major civil engineering works, requires specialist knowledge and experience and is not covered here[20].

Small retaining walls, typically not more than 2 m high, can provide near flat terraces on a sloping site for gardens or roads. While some simple empirical guidance can be provided for the stable construction of a range of common types of bonded brickwork and blockwork earth retaining walls[21], in many situations it will be necessary to engage the services of a suitably experienced civil or structural engineer for the design and construction of a retaining wall[22]. Figure 12.6 shows a simple, small retaining wall, which may or may not be reinforced. The stability of the wall depends primarily on self-weight and there are a number of possible modes of failure (Box 12.3).

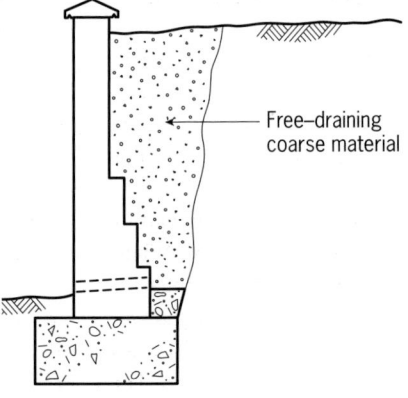

Figure 12.6 Small retaining wall

> **Box 12.3 Failure of small retaining walls**
>
> The retaining wall may break by a horizontal shear failure in the brickwork. There are three other primary modes of failure for a small retaining wall where the wall remains intact:
>
> **Overturning.** The wall should have adequate resistance against rotation. This can be calculated by taking moments about the most critical point, normally the toe of the wall.
>
> **Sliding.** The wall should have adequate resistance against translation, that is the sliding resistance of the base of the foundation of the wall should be greater than the lateral force from the soil that is pushing the wall.
>
> **Bearing failure.** The vertical thrust on the base of the retaining wall should not exceed the bearing resistance of the soil. The calculation of the bearing resistance should be based on the calculations described in Section 6.5, with due allowance for the inclination and eccentricity of the loading. An adequate factor of safety is required against each of these failure modes and the structural strength of the wall should be adequate to ensure that the wall is not over-stressed. There are well established methods to calculate the stability of retaining walls, but calculations are not always necessary for small walls and typical sections of masonry wall are accepted as suitable for the relatively small disturbing forces that act on the wall.

Sloping ground on which the wall is constructed should be stable in the short and the long-term; before excavating for the foundations for a retaining wall, the stability of the slope should be investigated and assessed by a geotechnical engineer (Section 7.6). Excavation of the ground prior to the construction of a retaining wall needs to be undertaken with care to avoid ground collapse[23]. Faces higher than 1.2 m should be sloped back to a safe angle. The risk of instability of an exposed face in a clay soil increases with time so the time for which it is left unsupported should be minimised. It is prudent to excavate the ground progressively and construct the wall in panels.

Particular care should be taken when setting out the wall footings. With the exception of walls with piers, the bases of retaining walls should be constructed centrally on their foundations. Appropriate reinforcement is required for the connection between the wall and its concrete foundation, at which movement can occur.

The construction of a wall may cause a change in the moisture equilibrium in a clay soil (Sections 4.8 and 7.5). Where there are large trees, the soil may be desiccated and the construction of a wall may cause some reversal of this desiccation by cutting through roots and by introducing a pathway for moisture to get into the soil. Construction is therefore likely to be followed by a period of swelling which, in a severely desiccated clay, may continue for many years. The changes that take place in the soil are largely reversible; during periods of dry weather, the clay may shrink away from the wall allowing a gap to open up and, during subsequent wet periods, the soil will swell back to more or less its original volume. However, backfill material or general debris may fall down the cracks and prevent the soil re-occupying its original space. Each cycle of swelling will increase the earth pressure on the wall and may cause it to move slowly forward. The provision of compressible material between the back face of the wall and the clay surface can reduce this effect.

The build-up of water pressure behind a retaining wall can be prevented by backfilling the space immediately behind the wall with lightly compacted, free-draining material, such as coarse aggregate, clean gravel or crushed stone. It may be necessary to incorporate a geotextile filter fabric to prevent this material from becoming clogged with fine material eroded from the soil. The drainage zone has to be able to discharge through weep holes located near the base of the wall; they should be at least 75 mm diameter, spaced horizontally at not more than 1 m intervals. After construction, the owner should be advised to keep the weep holes clear and free from obstruction.

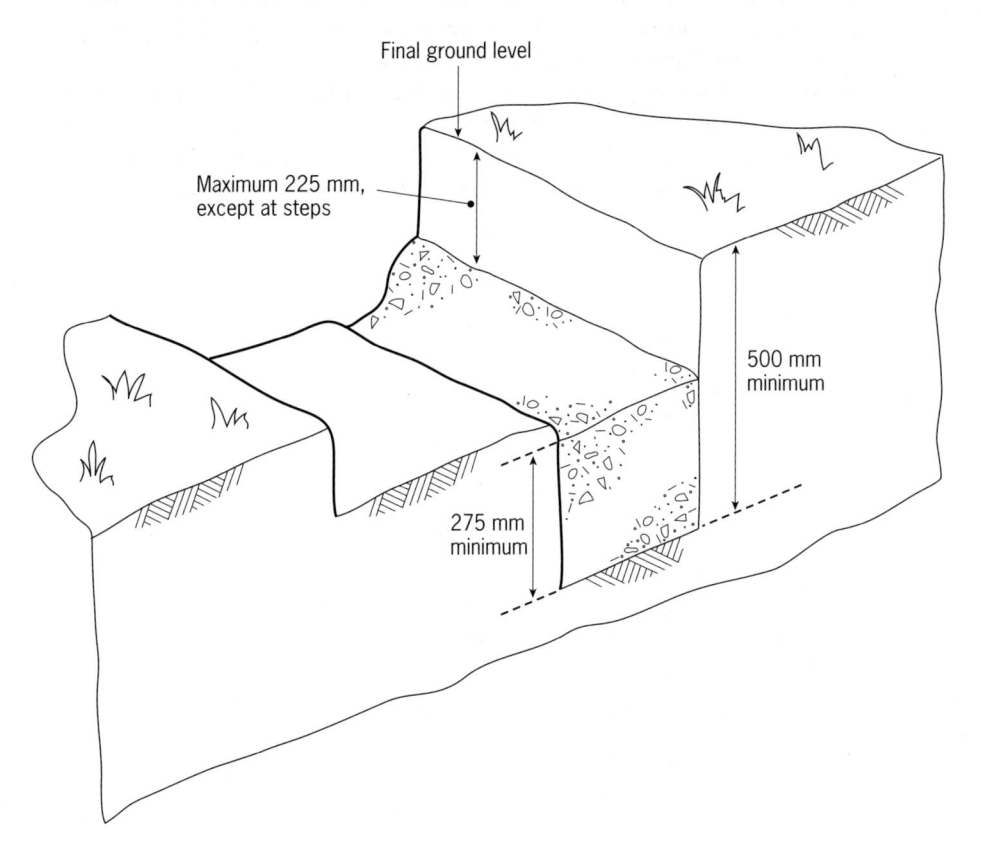

Figure 12.7 Foundation of freestanding wall

ANCILLARY WORKS

12.6 Freestanding walls

Simple freestanding walls, not forming part of a building, are widely used for boundary demarcation, landscaping, screening, security, and noise barriers. They are exposed to the weather and wind loading on both faces. A small part of a wall becoming unstable can lead to progressive collapse of the whole wall with a risk of serious injury to people. The stability of freestanding walls depends to a considerable degree on the provision of suitable foundations, as well as on their resistance both to wind load and to deterioration of their fabric[24].

The foundation requirements for freestanding walls are generally less onerous than those for buildings. For walls not forming part of a building and not exceeding 2 m in height, a foundation depth of 0.5 m is adequate for most ground conditions. When excavating the foundation trench, it can be helpful to remove an additional strip of topsoil to provide a working surface approximately level with the top of the footing (Figure 12.7). For higher walls in clay soils, the foundation depth should be increased to at least 0.75 m. Consideration should also be given to deeper foundations where walls are to be founded on highly shrinkable soils and in close proximity to large trees or where large trees have been recently removed (Section 7.5). For many applications, it should be possible to found a freestanding wall at, say, half the depth recommended for house construction.

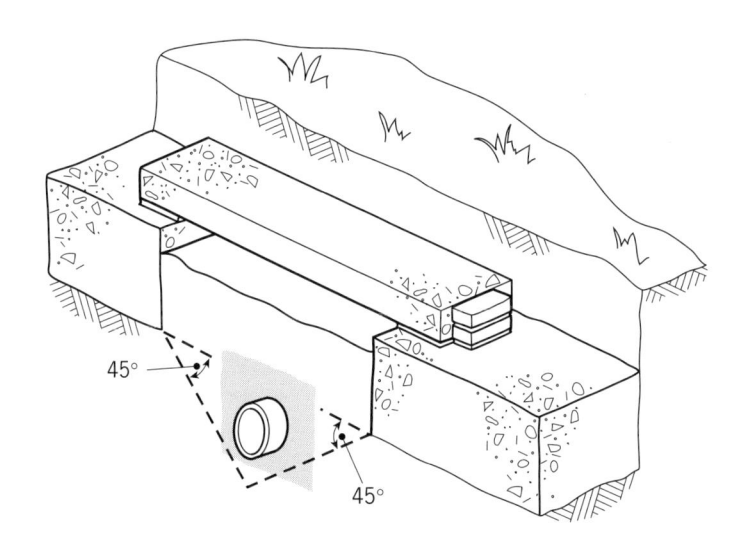

Figure 12.8 Bridging over drain

271

Foundation width is more likely to be governed by the need to resist the overturning moment generated by wind pressure rather than to reduce the bearing pressure. Typical foundation widths for brick and blockwork walls range from 0.6 m for walls 1 m high to 0.8 m for walls 1.8 m high[25]. Some consideration of bearing resistance may be necessary for construction on soft clays and loose sands and specialist advice may be required.

Where possible, construction on made ground should be avoided unless the fill has been properly compacted. Where construction on made ground is unavoidable, any local soft spots, or pockets of organic material should be removed and backfilled using lean mix concrete. It may be necessary to accept a certain amount of settlement and foundations should be reinforced longitudinally to limit the distortion to the wall.

When building a wall near trees, it is important not to sever tree roots; not only could this de-stabilise the tree, it could also cause clay swelling. A bridge, in the form of a reinforced concrete lintel, should be provided over any drains or large tree roots (Figure 12.8). Drains may be damaged if the pipework falls below any 45° line drawn from the lower edge of the foundation. Sufficient room (200 mm all round) should be allowed for tree roots to expand in girth without damaging the wall.

Notes
1. Problems involving hardcore are described in more detail in BRE Digest 276. The increased use of precast concrete foundation systems may mean that ground supported floor slabs become a less common feature of domestic buildings.

2. This is a form of sulfate disintegration from which even sulfate-resisting cement is not immune.

3. Pye and Miller (1990) described the chemical and biochemical weathering of pyritic mudrocks in a shale embankment. They concluded that engineering problems associated with volume loss due to carbonate dissolution, or volume increase due to gypsum formation, are only likely to occur in mudrocks which contain both finely divided pyrite and calcite or other readily soluble carbonate.

4. Settlement of ground floor slabs was a frequent structural defect in new houses where solid floor construction had been used and as a consequence NHBC required that a suspended floor be used where the depth of filling material needed is more than 600 mm at any point. '*Ground bearing slabs are not acceptable where fill exceeds 600 mm in depth*' (NHBC Standards Chapter 5.1).

5. Further details of the design and construction procedures for soakaways is in Digest 365 *Soakaway design*. See also BS EN 752-4 *Drain and sewer systems outside buildings* Part 4 *Hydraulic design and environmental considerations*.

6. Approved Document H3 of the Building Regulations *Surface water drainage* states that infiltration devices should not be built:
(a) within 5 m of a building or road or in areas of unstable land;
(b) in ground where the water-table reaches the bottom of the device at any time of the year;
(c) where the presence of any contamination in the run-off could result in pollution of groundwater source or resource.
It should be sufficiently far from any other soakaway so that the overall soakage capacity of the ground is not exceeded.

ANCILLARY WORKS

7. The design procedure outlined in this section of the book is an abbreviated form of that given in BRE Digest 365. See also BS EN 752-4 *Drain and sewer systems outside buildings* Part 4 *Hydraulic design and environmental considerations*.

8. See BRE Digest 365. A useful notation for identifying storm rainfall is *MX-D*, where D is the duration in minutes and X is the return period in years. For all parts of the UK, *M5-60* min (the rainfall in a storm of 60 minutes duration with a return period of five years) can be assumed to be 20 mm. To adjust this figure to storms of different duration and to different return periods, it is necessary to apply the correction factors which can be found in BRE Digest 365. Examples of the calculation of design values for R, the total rainfall in the design storm, are in Digest 365. A computer program is available for performing the design calculations.

9. NHBC Standards Chapter 5.3 *Drainage below ground* specifies a particle size range of 10 to 150 mm.

10. Approved Document H of the Building Regulations *Drainage and waste disposal* deals with many commonly encountered situations concerning drains and sewers in the vicinity of buildings. The 2002 edition, which came into effect on 1 April 2002, contains the following sections:
H1 Foul water drainage
H2 Wastewater treatment systems and cesspools
H3 Rainwater drainage
H4 Building over sewers
H5 Separate systems of drainage
H6 Solid waste storage
Further information is in NHBC Standards Chapter 5.3.

11. Approved Document H1 of the Building Regulations *Foul water drainage* gives detailed requirements. See also NHBC Standards Chapter 5.3.

12. Approved Document H1 states that a single size granular material complying with the requirements of BS EN 1610:1998 *Construction and testing of drains and sewers* Annex B Table B.15 should be used.

13. Bedding and cover of pipes are described in sections 2.41-2.45 and Tables 8, 9 and 10 of Approved Document H1.

14. See NHBC Standards Chapter 5.3 *Drainage below ground level*.

15. See Approved Document H1 section 2.25, and NHBC Standards Chapter 5.3.

16. See Approved Document H1 section 2.23.

17. See Approved Document H1 section 2.24, and NHBC Standards Chapter 5.3.

18. The Highways Agency *Specification for highway works* contains a detailed specification for the compaction of earthworks (Series 600); see also Parsons (1992). CIRIA Report 161 *Small embankment reservoirs* (Kennard et al, 1996) provides a comprehensive guide to the planning, design, construction and maintenance of small embankment reservoirs for amenity use.

19. Extensive research on the compaction of soils and granular materials carried out at the Transport Research Laboratory has been summarised by Parsons (1992). Suitable types of compaction plant for the main types of fill are described in BS 6031:1981 *Code of practice for earthworks*; guidance is also given on many practical aspects of earth moving, such as site clearance, wet weather, freezing conditions and haul roads. A detailed specification for layer thickness for different types of fill and different types of compaction plant is in the Highways Agency *Specification for highway works*. However, this specification has a more exacting situation in mind than that of a landscaping barrier and therefore some relaxation of the specification should be tolerable. Some general guidance is given by Trenter and Charles (1996).

20. BS 8002: 1994 *Code of practice for earth retaining structures* should be referred to for the design of large retaining walls.

21. BRE Good Building Guide 27 *Building brickwork or blockwork retaining walls* provides advice on the dimensioning and locating of wall footings and on the provision of structural reinforcement to small walls.

22. The services of a suitably qualified and experienced civil or structural engineer should be obtained for the design and construction of a retaining wall in the following circumstances:
❐ higher than about 1.7 m above the top of the foundation;
❐ supporting soil heaps, stored materials or buildings close to wall;
❐ supporting vehicles or traffic on backfill close to wall;
❐ retaining very wet earth, peat or water; for example, a garden pond;
❐ retaining soil with a slope adjacent to the wall steeper than 1:10 vertical to horizontal;
❐ supporting a fence of any type other than a simple guard-rail;
❐ forming part of, or adjoining, a building;
❐ not constructed of bricks or blocks;
❐ dry masonry construction;
❐ in area of mining subsidence or other unstable ground;
❐ water-table lies within 0.5 m of underside of foundation.

23. BS 6031: 1981 *Code of practice for earthworks* and BS 8004 *Code of practice for foundations* give guidance on safety issues.

24. BRE Good Building Guides GBG14 *Building simple plan brick or blockwork freestanding walls* and GBG19 *Building reinforced, diaphragm and wide plan freestanding walls* give practical advice on the construction of safe, freestanding walls.

25. Detailed guidance for foundation widths suitable to resist wind overturning forces in locations with different degrees of exposure are given in BRE Good Building Guides GBG14 and GBG19.

GLOSSARY

Some of the terms listed in the glossary have additional meanings which are outside the scope of the book; these are not included.

Active earth pressure
Earth pressure acting on a retaining structure when the structure moves away from the retained soil.
Activity
Ratio of *plasticity index* to *clay fraction*.
Aggressive conditions
Environmental conditions that cause damage, usually from chemicals in the air or the ground, to construction materials such as concrete.
Allowable bearing pressure
Maximum allowable loading intensity at the base of a foundation, taking into account ultimate bearing capacity and the required margin against failure, the amount and kind of settlement expected and the ability of the structure to accommodate settlement.
Alluvium
Soil transported by, and subsequently deposited from rivers; *fine soil* is deposited from still or slow moving water by sedimentation. Sometimes called alluvial soil.
Angle of friction
An alternative expression for *angle of shear resistance*.
Angle of repose
The steepest angle to the horizontal that a heap of loose coarse material can stand.
Angle of shear resistance ϕ'
Parameter that relates shear strength to normal stress, usually expressed in terms of effective stress. Also termed angle of shearing resistance or angle of friction. See *Mohr-Coulomb failure criterion*.
Angular distortion
Ratio of the *differential settlement* (δs) between two points on a structure and the distance (L) between them, less the tilt of the whole structure. Sometimes termed relative rotation.
Anisotropic
Exhibiting different physical properties in different directions. Soil is usually anisotropic.
Approved Document
Document associated with the UK Building Regulations.
Aquiclude
Relatively impermeable barrier to groundwater flow.
Aquifer
Permeable body of ground that is capable of storing a significant volume of water.
Artesian water
Groundwater that is under pressure and confined by relatively impermeable ground. Artesian water pressure often corresponds to a head of water which is above ground level.
Atterberg limits
Liquid limit and *plastic limit* are known as Atterberg limits.
Auger
Tool which is rotated to make a boring.
Axisymmetric condition
Condition of axial symmetry in which two of the *principal stresses* are equal. In a *triaxial compression test* a cylindrical specimen is loaded under axisymmetric conditions.
Backfill
Material used to fill an excavation or to cover buried services, or placed behind a retaining wall.
Backhoe
Hydraulic excavator with a bucket on the end of an arm; the bucket can also be used to tamp fill material. Sometimes called a backacter.

Band drain
Type of prefabricated vertical drain usually consisting of a plastic core surrounded by a geotextile sleeve. The core provides a flow path along the drain and supports the sleeve which in turn acts as a filter separating the core and the flow channels from the soil.
Basement
Storey of a building that is below ground level.
Base resistance
Resistance generated at the base of a pile.
Bearing capacity
Maximum applied pressure that the ground can support. See *ultimate bearing capacity*.
Bearing pressure
Contact pressure between the foundation and the ground.
Bearing resistance
Resistance generated at the interface between the foundation and the ground. It is the loading intensity at which the ground fails in shear. Sometimes termed *ultimate bearing capacity*.
Bedding
(a) In building and civil engineering, a layer of material laid on a surface in order to fill irregularities. Buried drains are usually laid on either a granular or a concrete bedding.
(b) Geological term describing layering within sedimentary strata.
Bedrock
Continuous, solid, undisturbed rock, which may be covered by soil deposits.
Berm
Horizontal ledge on the slope of an embankment or cutting.
Biodegradation
Decomposition of organic matter resulting from the activity of micro-organisms normally under anaerobic conditions (environment in which oxygen is absent).
Bioremediation
Remedial treatment for contamination utilising microbial degradation.
Borehole
A cylindrical hole formed in the ground by drilling. In site investigation, boreholes provide a means of obtaining information about the ground strata.
Boulder clay
Commonly used term for a glacial deposit; its use is not encouraged because such a deposit may not contain either boulders or clay. See *glacial till*.
Boulders
Very coarse soil particles which are larger than 200 mm are classified as boulders.
British Standard
A numbered publication of the British Standards Institution.
British Standards include Codes of practice (eg BS 8004: 1986 *Code of practice for foundations*) and test specifications (eg BS 1377: 1990 *Methods of test for soils for civil engineering purposes*). For European Standards, see *Euronorm*.
Brittle
See *strain softening*.
Brownfield land
Land that has been previously developed.
Bulk modulus K
Ratio of the applied stress to volumetric strain under conditions of hydrostatic stress

ie $K = \dfrac{\Delta p}{\Delta \varepsilon_v}$

where $p = \sigma_1 = \sigma_2 = \sigma_3$ and for an isotropic material $\Delta \varepsilon_v = 3\Delta \varepsilon_1 = 3\Delta \varepsilon_2 = 3\Delta \varepsilon_3$.
The modulus describes change in volume.

GLOSSARY

Capillarity
Rise of water in *ground* above the *water-table* due to capillary action which is associated with surface tension. *Pore water pressure* due to capillarity is negative.

Casing
Lengths of tubing inserted into a borehole to support the hole.

Chalk
Porous, soft, fine-grained *sedimentary rock*, predominantly composed of the calcareous skeletons of micro-organisms. There are extensive deposits in England.

Characteristic value of soil strength
Cautious estimate of the operational strength of the soil in the field at failure. In the load and resistance factor design method, a design value of soil strength is calculated by applying a *partial factor* to the characteristic value.

Clay
Very fine soil consisting mainly of *clay minerals*. Soil particles smaller than 0.002 mm are classified as the *clay fraction*, but it is preferable to restrict the term to clay particles formed from clay minerals.

Clay fraction
Fraction by weight of particles finer than 0.002 mm.

Clay minerals
Group of alumino-silicate minerals (eg kaolinite, illite and montmorillonite) with a characteristic sheet structure.

Coarse soil
Soil with more than about 65% of sand and gravel sizes. Gravel and sand are coarse soils. The terms *granular*, *cohesionless* and *coarse-grained* are often used to describe coarse soils.

Coarse-grained soil
See *coarse soil*.

Cobbles
Very coarse soil particles between 60 and 200 mm.

Coefficient of consolidation c_v
Coefficient which gives a measure of the rate of consolidation and which is defined as:

$$c_v = \frac{k}{m_v \gamma_w}$$

where:
k is the *coefficient of permeability*
m_v is the *coefficient of volume compressibility*
γ_w is the unit weight of water.

Coefficient of permeability k
Coefficient which describes the capacity of ground to allow the flow of liquid through it. For one-dimensional flow:
$k = q/(iA)$
where:
q is the flow rate
i is the hydraulic gradient
A is the area perpendicular to the flow.
See *permeability*.

Coefficient of secondary compression C_α
Coefficient which describes the rate of *secondary compression*; corresponds to the ratio of the change in height to the initial height of a laboratory consolidation test specimen over one log cycle of time during secondary compression. Sometimes based on change in *void ratio* rather than vertical strain.

Coefficient of uniformity U
Measure of shape of particle size distribution, expressed as ratio: D_{60}/D_{10} where:
D_{60} is the particle size such that 60% of the particles are finer
D_{10} is the particle size such that 10% of the particles are finer.
Coefficient of volume compressibility m_v
Measure of *compressibility* given by the ratio of the change in volume to the increase in vertical stress that has caused that change in confined compression.
Cohesion
Shear resistance mobilised between adjacent soil particles that adhere to each other without the need for any *normal stress*. The effective cohesion c' is the shear strength at zero normal effective stress. See *Mohr-Coulomb failure criterion*.
Cohesionless soil
Soil with zero *cohesion*, that is soil whose shear strength can be attributed to particle interlocking and friction. Coarse soil such as *sand*, *gravel* and non-plastic *silt* are cohesionless.
Cohesive soil
Soil with *cohesion*. *Fine soil*, particularly *clay*, is often described as cohesive, but this can be misleading since much of the apparent cohesive strength can be attributed to negative pore water pressure rather than true cohesion.
Collapse compression
Reduction in volume of a *partially saturated soil* which occurs due to an increase in *water content* without there necessarily being any increase in applied stress. Causes *collapse settlement* at the ground surface.
Collapse potential
Potential of soil for *collapse settlement* or *collapse compression*.
Collapse settlement
Settlement of the ground which occurs when a *partially saturated soil* undergoes a reduction in volume that is attributable to an increase in water content without there necessarily being any increase in applied stress. Sometimes referred to as *collapse compression*.
Colliery spoil
Waste from the deep mining of coal.
Compaction
Process of packing soil particles closer together by some mechanical means, such as rolling, ramming or vibration, thus reducing the volume of the voids and increasing the density of the soil.
Compaction curve
Relationship between the *dry density* and the *water content* of a soil for a particular compactive effort.
Compaction grouting
Grouting technique in which a bulb of grout is formed around the injection point and the surrounding ground is compressed.
Compensation grouting
Grouting technique which is used to prevent differential displacement of the ground occurring during tunnelling operations.
Compressibility
Ability of a material to decrease in volume and increase in density under pressure. See *coefficient of volume compressibility, compression index*.
Compression index
Measure of *compressibility* given by the slope of the curve in a plot of *void ratio* versus the logarithm of vertical *effective stress* in an *oedometer* test.

GLOSSARY

Conceptual ground model
Site investigation should provide the information about ground conditions required to form a conceptual ground model, including a three-dimensional stratigraphic model of the ground, a groundwater model and, where relevant, modifications caused by human activities and a soil contamination model.

Cone penetration test (CPT)
In-situ test in which a cone penetrometer is pushed into the ground and the penetration resistance is measured. The penetrometer includes a cone, a friction sleeve and connections to push rods. Some penetrometers have additional sensors and measuring systems.

Confined compression
Uniaxial compression in which there is only vertical compression, the horizontal strain being zero. This can be termed uniaxial strain (ie $\varepsilon_1 \neq 0$, $\varepsilon_2 = \varepsilon_3 = 0$) or one-dimensional consolidation.

Confining pressure
All-round, or cell pressure that is applied to a triaxial test specimen.

Consistency
Condition of clay in terms of firmness.

Consistency index I_C
Numerical difference between liquid limit and natural water content expressed as a ratio of plasticity index:

$$I_C = \frac{w_L - w}{I_P}$$

Consistency limits
See *Atterberg limits*.

Consolidation
Process of packing soil particles closer together by increasing the *effective stress* using some form of static loading, thus reducing the volume of voids and increasing the density of the soil. Consolidation of saturated clay is a time-dependent process which results from the slow expulsion of water from the soil pores.

Constant volume strength
Shear strength of a soil when it has been subjected to a shear displacement after peak strength sufficient to bring it to a state of constant volume at the critical void ratio. It can be described by the constant volume (or critical state) angle of shear resistance ϕ'_{cv}.

Constrained modulus D
Ratio of axial stress to axial strain in confined compression, usually expressed in terms of effective stress

$$D = \Delta\sigma'_1 / \Delta\varepsilon_1$$

where:

$\Delta\varepsilon_2 = \Delta\varepsilon_3 = 0$

Note that $D = 1/m_v$

Contaminant
Substance which is, or may become, harmful to people or buildings, including substances which are corrosive, explosive, flammable, radioactive or toxic.

Contaminated land
Land that contains substances which, when present in sufficient quantities or concentrations, are likely to cause harm, directly or indirectly, to man, to the natural environment or to the built environment. Part IIA of the Environmental Protection Act 1990: *'... land is only contaminated where it appears to the authority, by reason of substances in, on or under the land, that: (a) significant harm is being caused or there is a significant possibility of such harm being caused; or (b) pollution of controlled waters is being, or is likely to be, caused.'*
Harm is defined by reference to harm to health of living organisms or other interference with the ecological systems of which they form a part and, in the case of man, includes harm to property.

Course
Horizontal layer of bricks or blocks, including mortar laid with them. The term is also used for a layer of material such as damp-proofing.
Creep compression
Compression which occurs at constant water content and under constant *effective stress*.
Critical state strength
See *constant volume strength*.
Crown hole
Surface depression resulting from the collapse of strata into old mine workings.
Damp-proof course (DPC)
Strip of impervious material near bottom of a wall to exclude rising damp.
Damp-proof membrane (DPM)
Wide layer of impervious material beneath a *ground supported floor slab* or *raft foundation*.
Darcy's law
Law concerning the flow of *groundwater* through the ground which states that the flow q is proportional to the *hydraulic gradient i*. The coefficient of proportionality is known as the *coefficient of permeability* or hydraulic conductivity k.
$q = kiA$
where:
A is the area perpendicular to the direction of flow.
The law is valid for a wide range of soil types and hydraulic gradients.
Dead load
Load due to permanent construction, including weight of floors, walls, permanent partitions, roof and services.
Deflection ratio
Maximum vertical displacement (Δ) relative to the straight line connecting two points on a structure divided by the length between the two points (L).
Degree of saturation
Ratio of volume of water contained in the voids to the total volume of voids. Usually expressed as a percentage.
Density
Mass per unit volume.
Density index I_D
Degree of packing in *coarse soil*, such as *sand* or *gravel*; relates the in-situ density to the limiting conditions of maximum density and minimum density. Sometimes known as relative density.
Derelict land
Land that has been so damaged by industry, mining and urban development that it can no longer be put to beneficial use without treatment.
Derived value of soil strength
Value of soil strength obtained from test results by theory, correlation or empiricism. Together with information from other sources on the site, forms the basis for the selection of the *characteristic value* of soil strength.
Desiccation
Reduction in water content of a soil caused by *evapotranspiration*.
Desk study
Examination of existing information concerning a site, including historical records, geological maps, borehole records, and air photographs to determine ground conditions and previous land use.
Deviator stress
Stress difference on a body. In a *triaxial compression test*, it is the stress in excess of confining pressure due to applied axial load: the difference between the major and minor principal stress, $\sigma_1 - \sigma_3$.

GLOSSARY

Dewatering
Removal of *groundwater*, typically used to keep water out of a construction site.

Differential settlement
Settlement of one part of a building or other type of structure relative to another part of the same structure. While the total settlement of a structure may interfere with some aspect of its functions, such as connections to services, it is differential settlement that causes structural damage.

Dilatancy
Expansion of a dense soil when distorted by shearing.

Direct shear test
Shear strength test carried out in a shear box. The relative movement of two halves of a soil *specimen*, which is square in cross-section, is constrained to take place along a horizontal surface.

Disturbed sample
A *sample* which is obtained without attempting to retain the structure of the soil, sometimes described as a bulk sample. As the structure has been destroyed and there may have been significant changes in the water content, a disturbed sample should only be used for classification and compaction tests.

Downdrag
Downward loading on the shaft of a pile or other type of deep foundation through frictional forces induced by the soil settling relative to the foundation element. Also termed negative skin friction.

Drain
Artificial conduit which conveys water:
(a) in general building works, a buried pipe that removes foul water or rainwater from a building and conveys it to a sewer;
(b) in agriculture, a porous pipe that drains the surrounding ground, and is termed a field drain or agricultural drain;
(c) in civil engineering earthworks, permeable material, including granular fill and geosynthetics, that facilitate the removal of water from the surrounding ground, including horizontal drainage blankets, *sand drains* and *band drains*.

Drained shear strength
Shear resistance of soil in a drained condition in which any excess pore water pressure dissipates during shearing.

Drift geology map
Map showing superficial deposits such as *alluvium*, *head* and glacial materials; the underlying bedrock is not normally shown.

Drilling
Process by which a borehole is formed in the ground.

Durability
Ability of a construction material to remain serviceable over a long period.

Dynamic compaction
Ground treatment method in which deep compaction is effected by repeatedly dropping a heavy weight onto the ground surface from a great height.

Dynamic probing (DP)
In-situ test in which a solid cone is driven into the ground; the penetration resistance of the soil is measured in terms of the number of blows required to drive the cone a specified distance.

Earthfill
Fill composed of natural soil materials such as sand, silt and clay.

Earth pressure
Usually refers to the lateral pressure exerted by a soil mass against an earth supporting structure such as a retaining wall.

Eccentric load
Load applied at a point distant from the centre of the foundation and therefore applying a bending moment to the foundation.

Effective overburden pressure
Intensity of vertical *effective stress* at a specified depth below ground level, prior to construction.
Effective stress σ'
Stress carried by the soil particles.
Elasticity
Ability of a material to recover its original form and dimensions when the forces acting upon it are removed.
Elastic modulus
Ratio of stress to strain in a particular material. The strain may be a change in length (*Young's modulus*), in shear (*shear modulus*) or in volume (*bulk modulus*). Strictly only applies to a material with elasticity, but in practice it is much more widely used.
Electro-level
Gravity-sensing electrolytic transducer that produces an output voltage proportional to the tilt angle. The device can be installed at depth within a borehole or fixed to a structure to measure tilt.
Engineered fill
Fill which is selected, placed and compacted to an appropriate specification, so that it will exhibit the required engineering behaviour.
Engineering geology
Application of geological knowledge and principles to engineering works; closely related to *geotechnical engineering*.
Equipotential line
Line representing constant *head*.
Eurocode
The structural Eurocodes are European harmonised design codes. Eurocode 7 deals with geotechnical design.
Euronorm
European standards are designated EN or Euronorm.
Evaporation
Conversion of a liquid into a vapour. The combined process of evaporation and *transpiration* is termed *evapotranspiration*.
Evapotranspiration
Combined process of *evaporation* and *transpiration*.
Excavation
Hole, pit or trench formed in the ground by digging-out the soil or rock. The excavation may be made to reach a suitable *foundation* level for a building or for services, or for soil exploration (see *trial pit*).
Expansive soil
Clay soil that undergoes large volume changes during wetting and drying cycles. Also termed shrinkable clay.
Factor of safety
In foundation design, the ratio of the *bearing capacity* to the applied vertical stress. Also termed safety factor.
Field inspection vane
Lightweight, hand-held vane for assessing the *undrained shear strength* of a clay soil to about 3m depth.
Fill
Material deposited through human activity rather than by geological processes. Ground formed of fill is referred to as filled ground or, more commonly, made ground.
Fines
Particles finer than 0.06 mm (particles of silt and clay size).

GLOSSARY

Fine soil
Soil with more than about 35% of silt and clay sizes. Clay and silt are fine soils. The terms *cohesive* and *fine-grained* are sometimes used to describe fine soils.

Fine-grained soil
See fine soil.

Fissured clay
Clay with a system of fissures similar to the jointing system of a rock mass, but on a small scale.

Floor slab
Concrete slab, which may be either *ground supported* or *suspended*.

Flow line
Path of travel of a particle of *groundwater* as it flows through the ground.

Flow net
Net formed by *equipotential* and *flow lines*, which is used in the graphical solution of a seepage problem.

Footing
Widening at the base of a wall or a column to spread its weight over the ground. In earlier times the widened depth of a wall might have been in brickwork, today it is usually in concrete. Sometimes termed a *spread footing*.

Formation
Surface of the ground in a completed excavation before concreting.

Foundation
(a) Part of a building or other type of structure designed and constructed to be in direct contact with and transmitting loads to the ground.
(b) Ground underneath and supporting a building or other type of structure.

Foundation inspection
Examination of the ground on which footings are to be built by a building inspector to ensure compliance with the Building Regulations.

Friction angle
See *angle of shear resistance*.

Frost heave
Swelling of soil due to expansion of water on freezing causing upward movement of the ground.

Gabion
Large cuboid shaped basket formed of steel wire and filled with rock. Gabions are permeable and flexible, and can be used to form retaining structures.

Gap graded soil
Soil with a particle size distribution in which certain grain sizes are missing.

Geophysics
Science concerned with the physical properties and processes of the Earth.

Geosynthetic
Generic term for civil engineering materials such as geotextiles, geogrids, geomembranes and geocomposites that are used to modify or improve ground behaviour.

Geotechnical engineering
Branch of civil engineering concerned with ground and groundwater and their relation to the design and construction of engineering works.

Geotechnics
Science of ground behaviour and its application to building and other construction activities. Includes *soil mechanics* and *rock mechanics*.

Glacial till
Heterogeneous soil with a wide variety of particle sizes transported and deposited by glaciers. Sometimes termed *boulder clay*.

Grading
Measure of the *particle size distribution* of a soil.

Granular soil
See *coarse soil* and *cohesionless soil*.
Gravel
Coarse soil which is coarser than *sand*. Soil particles between 2 mm and 60 mm are classified as gravel.
Ground
Soil, rock and fill in place prior to construction works.
Ground beam
Reinforced concrete beam at or near ground level which acts as a foundation for a wall or floor of the building. It may span between *piers* or *piles* or be itself a *strip foundation*.
Ground engineering
See *geotechnical engineering*.
Ground floor
Floor which is the nearest to ground level. A concrete floor can be a *ground supported floor slab*. A suspended floor may be of timber or concrete.
Ground supported floor slab
Concrete floor slab supported by the ground. It is normally laid on a hardcore bed and a damp-proof membrane is provided.
Ground treatment
Controlled alteration of the state, nature or mass behaviour of ground materials in order to achieve an intended satisfactory response to existing or projected environmental and engineering actions.
Groundwater
Subsurface water; the term is sometimes restricted to subsurface water below the *water-table*.
Grouting
Controlled injection of fluid material into the ground to improve its physical characteristics.
Gypsum
White mineral used in gypsum plasters; causes sulfate attack on mortar and concrete.
Hardcore
Limited amount of selected fill put down as infill within the foundations of a building unit or beneath an oversite concrete slab.
Hazard
Situation or event which has the potential for harm, including human injury, damage to property, environmental and economic loss.
Head
(a) In fluid mechanics, the height of the water level in a standpipe piezometer in relation to some horizontal datum line.
(b) Geological term for an unstratified deposit on a sloping site thought to have been produced by creep of water saturated ground when subjected to alternate periods of freezing and thawing.
Heave
Volumetric expansion of ground beneath part or all of a building resulting in an upward displacement of the ground.
Hogging
Mode of deformation of a foundation or beam undergoing upward bending; the opposite of *sagging*.
Hydraulic conductivity
See *permeability*.
Hydraulic gradient
Loss of *head* per unit distance along a *flow line*. The critical hydraulic gradient corresponds to the situation where the upward flow of water reduces the effective stress to zero. See *quicksand*.
Hydrofracture grouting
Grouting technique in which the ground is fractured and the grout is forced into the fractures so that lenses and sheets of grout are formed.

GLOSSARY

Hydrogeology
Branch of geology concerned with physics, chemistry and environmental relationships of water within the Earth's crust.

Igneous rock
Rock which has come directly from the earth's mantle as magma (molten rock), which is forced up through the crust before cooling and solidifying. Basalt and granite are igneous rocks.

Imposed loads
Loads assumed to be produced by the intended use of the building or other type of structure, including the weight of movable partitions. Does not include *wind load*.

Inclinometer
Electronic instrument for measuring horizontal ground displacement. The inclinometer torpedo is lowered down a plastic access tube which is installed in a vertical borehole. Measurements of tilt on successive occasions are made at the same depth positions.

Isotropic
Exhibiting the same physical properties in all directions. Soils are not normally isotropic.

Jet grouting
Grouting technique in which high pressure erosive jets of grout break down the soil structure and partially replace the soil with the remainder mixed with the ground in-situ.

Land condition record (LCR)
Voluntary log-book for brownfield land, which is kept and maintained by the landowner and transferred with the land. It should provide a clear record of the physical and chemical nature of the land and give details of the use that the land has supported.

Landfill
Term often used specifically to describe domestic refuse.

Landslide
Large-scale downslope ground movement caused by slope instability. Also termed landslip.

Leachate
Solution formed when water percolates through a permeable medium such as domestic refuse or colliery spoil.

Leaching
Removal of substances by dissolving them in water.

Leakage
Concentrated, uncontrolled flow of water through a crack or defect (contrast with *seepage*).

Limestone
Sedimentary rock, usually composed mainly of calcium carbonate, often in the form of calcite. It may also contain considerable amounts of magnesium carbonate (dolomite).

Liquefaction
Phenomenon in which soil loses shear strength due to an increase in pore pressure, often triggered by dynamic loading such as an earthquake. Saturated sandy soils are particularly vulnerable.

Liquid limit w_L
Water content at which fine soil passes from the liquid to the plastic condition, as determined by the liquid limit test.

Liquidity index I_L
Numerical difference between natural water content and the plastic limit expressed as a ratio of plasticity index. $I_L = (w - w_P)/I_P$

Load
Weight carried by a structure or its foundation. Includes *dead load*, *imposed loads* and *wind load*.

Load and resistance factor design
Design method in which different *partial factors* are applied to various soil properties and loads.

Loadbearing wall
Wall carrying some load in addition to its own weight and wind force; supports structure above it including floors, roof and their loads.

London clay
Deposit of stiff fissured clay which is widely encountered in London and elsewhere in south-east England.
Longwall mining
A modern method of coal mining which involves the extraction of complete panels of coal by a long, continuously moving face.
Low-rise building
Building not more than three storeys in height.
Made ground
Ground formed by human activity. The material of which made ground is composed is termed *fill*. Made ground is sometimes described as filled ground.
Marine soil
Soil deposited underwater in ocean; usually uniform.
Marl
Calcareous *clay*; the term is applied to a wide variety of sediments and rocks with a considerable range of composition. Calcareous marls grade into clays, by reduction in the amount of lime, and into clayey limestones.
Masonry
A loadbearing structure built of individual blocks of stone, concrete or brick, usually laid in mortar. The term is often restricted to stonework.
Maximum dry density
For a specified compaction procedure, fine fill has an optimum water content at which maximum density is achieved. Maximum dry density and *optimum water content* of a soil refer to particular compaction procedure and can be misleading if taken out of the context of that procedure.
Metamorphic rock
Rock derived from pre-existing *sedimentary* or *igneous rock* through the action of extremely high pressure and temperature. Metamorphic rocks include gneiss, marble, schist and *slate*.
Mitigation
Limitation of undesirable consequences of situation or event.
Mohr circle
Graphical representation of the stresses acting on the various planes at a particular point in a soil.
Mohr-Coulomb failure criterion
Linear relationship between shear strength τ_f and normal stress σ' derived from the envelope of a sequence of *Mohr circles* representing different stress conditions at failure for a given soil. In terms of effective stress:
$\tau_f = c' + \sigma' \tan \phi'$
where:
c' is the effective *cohesion*
ϕ' is the *angle of shear resistance*.
Moisture content
See *water content*.
Mudstone
Sedimentary rock formed mainly of *clay* or *silt* size particles. Sometimes the term is considered to refer only to rock that it is not laminated or easily split into thin layers, but the term can also be used to include all *shale* type sedimentary rocks.
Needle
Beam placed through a hole in a wall to transfer the weight of the wall above to *piers* or *piles* under its ends.
Negative skin friction
See *downdrag*.

GLOSSARY

Non-engineered fill
Fill which has arisen as a by-product of human activity, usually involving the disposal of waste material; it has not been placed with a subsequent engineering application in view.

Normal consolidation line
Nearly linear relationship between void ratio (or vertical strain) and the logarithm of vertical effective stress for a normally consolidated soil undergoing one-dimensional consolidation.

Normally consolidated soil
Soil which has never experienced vertical effective stress greater than the current vertical effective stress.

Normal stress σ_n
Force or load per unit area on a plane normal to the direction of the force. In geotechnical engineering normal stress is usually compressive and therefore compression is reckoned as positive.

Oedometer
Laboratory consolidation press in which a test specimen is subjected to one-dimensional consolidation.

One-dimensional consolidation
See *confined compression*.

Opencast mining
Mining carried out by excavation from the ground surface. Large of strips of land are successively excavated and the overburden is cast into the excavation.

Optimum water content w_{opt}
Water content at which the *maximum dry density* is achieved using a specified compactive procedure.

Organic content
Content of *organic matter* content in a soil expressed as a percentage of the original dry weight.

Organic matter
Technically any material that contains carbon is organic. When applied to soils, geotechnical engineers and engineering geologists use the term in a more restricted sense to describe materials recently derived from plants or animals.

Organic soil
Soil with a high content of *organic matter*.

Outcrop
Exposed area of bedrock at ground surface.

Overburden
Overlying deposits.
(a) In geotechnical engineering it usually refers to all material overlying the point of interest in the ground; hence the overburden pressure at that point is the vertical stress attributable to the weight of the overlying ground.
(b) In mining engineering it refers to the material overlying a mineral deposit; hence the overburden ratio is the proportion of overburden thickness to mineral deposit.

Overburden pressure
See *effective overburden pressure, total overburden pressure*.

Overconsolidated soil
Soil which has experienced a vertical effective stress greater than the current vertical effective stress.

Overconsolidation ratio (OCR)
Ratio of maximum past vertical effective stress (preconsolidation pressure) to the current vertical effective stress in the ground.

Oversailing
Brick course overhanging or projecting beyond the course below.

Pad foundation
Isolated foundation or footing consisting of a simple circular, square or rectangular slab, which is usually provided to support a structural column.

Partial factor
Factor that is applied to a soil property or to a load in the load and resistance factor design method.
Partially saturated soil
In a partially saturated soil, the voids of the soil are not completely filled with water, but also contain air. Soil above the water-table is generally considered to be partially saturated, although in some situations it could have a high degree of saturation, or even be fully saturated. Also termed partly saturated or unsaturated soil.
Particle density ρ_s
Mass per unit volume of the solid particles of a soil. In the past, the term *specific gravity* was used to describe the ratio of the density of the soil solids to the density of water; when particle density is expressed in Mg/m^3, it is numerically equal to the specific gravity.
Particle size
Equivalent diameter of soil particles.
Particle size distribution
Distribution of particle sizes within a soil can be presented as a particle size distribution curve in which the percentage of particles finer than a given size is plotted against the logarithm of particle size. Sometimes called a grading curve.
Partition wall
Non-loadbearing internal wall between the rooms of a building.
Partly saturated soil
See *partially saturated soil*.
Passive earth pressure
Earth pressure acting on a retaining structure when the structure is pushed into the retained soil.
Pathway
Route by which a *contaminant* moves from its source to a *receptor*.
Peagravel
Rounded, uniform, medium size gravel, used as bedding and surround for buried drains.
Peak strength
Maximum shear resistance of a soil.
Peat
Type of organic soil.
Penetration test
In-situ test in which a device is pushed or driven into the ground while the resistance of the soil to penetration is recorded (eg standard penetration test, cone penetration test).
Penetrometer
Device which is pushed or driven into the ground to measure the resistance of the soil.
Percentage air voids
Volume of air voids in soil expressed as a percentage of the total voids.
Perched water-table
Water-table which is isolated from *groundwater* at deeper levels by an impermeable stratum.
Permeability
Capacity of ground to allow flow of liquid through it; its magnitude is described by the *coefficient of permeability*. Also known as hydraulic conductivity.
Permeameter
Laboratory device for measuring the *coefficient of permeability* of a soil.
Permeation grouting
Grouting technique in which grout is injected into coarse soil or fractured rock to fill voids and fissures and produce a solidified mass.
Phreatic surface
See *water-table*.
Phreatic zone
Subsurface water below the *water-table*.

GLOSSARY

Pier
(a) Wide column or short wall that forms a foundation which is more massive than a pile. It can be used as an element of an underpinning system.
(b) Thickened section of a wall provided at intervals to give lateral support.

Piezometer
Device for measuring the pore water pressure at a specific point in the ground. A piezometer tip consists of a porous element which allows water to pass into the instrument. The tip is sealed into the ground so that it is isolated from *groundwater* pressures elsewhere. There are many types of piezometer and recording system.

Pile
Timber, steel or concrete post or column forming a slender structural foundation element, which is usually installed vertically in the ground, to provide vertical or lateral support for a structure.

Pinning up
Ramming suitable mortar or concrete into the gap between *underpinning* and an existing foundation.

Piping
Type of internal erosion in a water retaining embankment in which movement of soil particles carried by the water erodes a channel or 'pipe' through the soil, leading to sudden collapse.

Plane strain
Strain condition in which strain in one direction is zero.

Plasticity
Ability of a material to deform continuously without rupture and without the strain and deformation being recovered when the forces acting on the material are removed. Clay soil changes its mechanical behaviour with change of water content and exhibits plastic behaviour within a range of water content from the *plastic limit* to the *liquid limi*t.

Plasticity index I_p
Range of water content over which a clay soil deforms plastically ($I_p = w_L - w_P$)

Plastic limit w_P
Water content at which fine soil becomes too dry to be in a plastic condition, as determined by the plastic limit test. Below this lower bound of plastic behaviour, the soil loses its intact behaviour and begins to break up into discrete pieces.

Plump hole
See *crown hole*.

Pocket penetrometer
Very small, lightweight, hand-held penetrometer used for assessing the unconfined compressive strength of clay, and hence the undrained shear strength, in sides of pits. Since it only tests very small volume of soil, it can give misleading results and is not a substitute for laboratory tests.

Poisson's ratio υ
Ratio of lateral strain to axial strain in uniaxial compression with no lateral restraint.

ie $\upsilon = \dfrac{\Delta \varepsilon_3}{\Delta \varepsilon_1}$

where:
$\Delta \sigma_3 = 0$

Pollarding
Cutting off the top of a tree to produce a close rounded head of young branches.

Poorly graded soil
See *uniformly graded soil*.

Pores
Spaces between soil particles, filled with liquid (usually water, termed *pore* water) and/or gas (usually air). Also described as interstices or *voids*.

Pore water
Water partially or completely occupying the pores in a soil.

Pore water pressure
Pressure in water contained in pores within the soil.
Porosity *n*
Ratio of volume of voids to total volume occupied by the soil. Usually multiplied by 100 to give a percentage.
Preconsolidation pressure σ'_p
The maximum vertical effective stress to which soil ever has been subjected.
Preloading
(a) Ground treatment method in which consolidation is achieved by temporarily loading the ground, usually with a surcharge of fill.
(b) Ground may have been preloaded during its geological history, in which case it will be *overconsolidated*.
Pressure
See *stress*.
Pressure bulb
Zone of ground under a loaded area where the load applies a significant vertical stress.
Primary consolidation
Reduction in volume of low permeability, fine soil caused by expulsion of water from soil pores and transfer of load from excess *pore water pressure* to soil particles.
Principal stress
At a point in the ground or in a soil test specimen, the stress normal to one of the three mutually perpendicular planes on which shear stresses are zero. The three mutually perpendicular directions are called the three principal axes of stress. The major principal stress is σ_1, the intermediate principal stress is σ_2, and the minor principal stress is σ_3.
Proctor curve
Compaction curve obtained using the standard laboratory Proctor compaction test.
pulverised-fuel ash (pfa)
Waste product from power stations which burn pulverised coal composed of nearly-spherical particles predominantly in the coarse sand and fine silt range.
Quick clay
Clays which are highly sensitive to disturbance are commonly found in Scandinavia and Eastern Canada.
Quicksand
Sand through which water flows upwards reducing the vertical *effective stress* to zero. Sand in this condition has no *shear strength* and hence no *bearing resistance*.
Raft foundation
Type of *foundation* which is continuous in two directions, usually covering an area equal to or greater than the base area of the structure. It is usually a *slab* of reinforced concrete supported by the ground.
Rapid impact compaction
Ground treatment method in which *compaction* of the ground is effected by repeated impacts of a weight onto a plate which remains in contact with the ground surface.
Receptor
People, other forms of life, ecological system, groundwater, or buildings and services, which could be adversely affected by a hazard, particularly contamination.
Relative compaction C_R
Ratio of in-situ dry density to maximum dry density achieved with specified degree of compaction in standard laboratory compaction test.
Relative density
See *density index*.

GLOSSARY

Residual strength
Shear strength of a soil when subjected to very large shear displacement after peak strength has been mobilised. It can be described by the residual angle of shear resistance ϕ'_r; in sands this is usually equal to the constant volume value ϕ'_{cv}, but in some clays it may be much smaller than ϕ'_{cv}.

Resistance
Capacity of material to withstand loads without failure.

Retaining structure
Structure designed to retain the ground and resist lateral displacement.

Retaining wall
Wall designed to retain the ground and resist lateral displacement.

Risk
Combination of probability or frequency of the occurrence of a defined hazard and some measure of the magnitude of the consequences of the occurrence.

Risk management
Overall application of policies, processes and practices dealing with risk.

Risk register
File in which risk information is stored which includes a description of the risk, an assessment of its likelihood and consequence, and any remedial actions. Can be used in the management of geotechnical and geoenvironmental risk on a building development.

Rock
Naturally occurring assemblage of minerals, consolidated, cemented or otherwise bonded together by strong and permanent cohesive forces to form a hard and rigid deposit of generally greater strength and stiffness than *soil*.

Rockfill
Fill produced by blasting or ripping rock strata; rockfill may contain large rock fragments.

Rockhead
Surface between overlying soil and solid *bedrock* below.

Rock mechanics
Application of principles of mechanics and hydraulics to engineering problems dealing with rock as an engineering material.

Safety factor
See *factor of safety*.

Sagging
Mode of deformation of a foundation or beam undergoing downward bending; the opposite of *hogging*.

Sample
A portion of the ground obtained from an exploratory borehole or trial pit. A sample may be either a *disturbed sample* or an *undisturbed sample*.

Sand
Coarse soil finer than gravel. Soil particles between 0.06 mm and 2 mm are classified as sand.

Sand drain
Vertical drain formed by filling a borehole with sand to enable the surrounding silt or clay soil to drain and consolidate more quickly.

Sandstone
Sedimentary rock formed mainly of quartz grains of sand size cemented with calcite or iron compounds.

Saturated soil
When all the voids are filled with water, the soil is saturated. Soil below the *water-table* is generally considered to be saturated, although the *degree of saturation* could be slightly less than 100%.

Secondary compression
Time dependent reduction in volume of fine soil caused by adjustment of soil structure after *primary consolidation* has been completed.

Sedimentary rock
Rock formed by settlement from water of mineral and rock particles, followed by compression under high pressure and temperature, and possibly chemical action. *Sandstone, mudstone, shale, limestone* and *chalk* are sedimentary rocks.
Seepage
Slow uniform flow of water through a porous medium such as a type of soil or a building material (contrast with *leakage*).
Sensitivity
Ratio of *undrained shear strength* of an undisturbed clay test specimen to the *undrained shear strength* of the same *specimen* after remoulding at constant water content.
Serviceability limit state
State corresponding to conditions beyond which specified service criteria for a structure or structural element are no longer met.
Set
Penetration of a driven *pile* for each blow of the drop hammer driving the pile into the ground.
Settlement
Downward movement of the ground or of a structure and its foundation resulting from movement of the ground below it. The terms settlement and subsidence are both widely used to describe downward movement of the ground and of buildings.
(a) In non-technical usage, the terms settlement and subsidence are virtually interchangeable; insofar as there is any distinction, settlement tends to imply a more gradual movement whereas subsidence is often associated with the ground caving-in.
(b) In geotechnical engineering, the usual term for downward movement of buildings or the ground on which they are founded is settlement. Subsidence is sometimes used to describe widespread movements, particularly mining subsidence.
(c) British house insurers provide cover for subsidence rather than for settlement and, for insurance purposes, settlement is regarded as movement due to the distribution or redistribution of loading and stresses within the various elements of a building which normally occurs in the early stages of the life of a building, whereas subsidence is defined as the downward movement of a building and its foundation caused by loss of support of the ground beneath the foundations.
Shaft friction
See *skin friction*.
Shaft resistance
Resistance generated on vertical face of shaft of pile.
Shale
Laminated *sedimentary rock* formed mainly of *clay* or *silt*-size particles which have been compressed by the weight of overlying rocks.
Shallow foundation
Generally taken to be a foundation where the depth below finished ground level is less than 3 m. Where the ratio of embedment depth to foundation width is large, say $(D/B) > 2.5$, the foundation may need to be designed as a deep foundation.
Shear box
See *direct shear test*.
Shear modulus G
Ratio of shear stress to shear strain under conditions of simple shear. The modulus describes the change in shape.
Shear resistance
See *shear strength*.
Shear strength τ_f
Stress at which a soil fails in shear.
Shear stress
Force per unit area on a plane parallel to the direction of the shear force.

GLOSSARY

Shrinkable clay
Clay soil that undergoes large volume changes during wetting and drying cycles.
Shrinkage limit
Water content below which soil ceases to shrink with further loss of water.
Shrinkage potential
Volume change potential of a clay soil associated with reduction of water content.
Silt
Fine soil, coarser than clay and finer than sand. Soil particles between 0.002 mm and 0.06 mm are classified as silt.
Sinkhole
Steep sided depression formed when a void migrates to the ground surface. The feature can be caused by the solution of carbonate rocks such as limestone and chalk. Sinkholes can also occur in earth embankment dams due to internal erosion.
Site reconnaissance
See *walkover survey*.
Skin friction
Frictional force generated on shaft of *pile*. See *shaft resistance*.
Slab
Large flat rectangular thin area of rigid material, usually concrete. A floor slab may be supported by the ground (*ground-supported floor slab*) or may be suspended clear of the ground between supporting walls or beams (*suspended floor slab*). A slab of reinforced concrete supported by the ground may form a *raft foundation*.
Slab-on-grade
Concrete slab supported by the ground that may form a *raft foundation* (USA).
Slate
(a) Geological term for fine-grained, clayey *metamorphic rock* that splits readily into thin slabs having great tensile strength and durability.
(b) In building, the term is applied to any rectangular sheet of roofing material.
Sleeper wall
Wall built to carry the joists of a timber ground floor.
Slip
Downslope movement caused by slope instability is referred to as a slip or a *slide*.
Slip indicator
Simple device for detecting horizontal ground displacement at a *slip surface*.
Slip surface
Surface along which slope failure occurs.
Soakaway
Pit, which may be empty or filled with coarse material, into which surface water can drain and then soak away into the surrounding ground.
Soil
Assemblage of mineral and/or organic particles that can be separated by gentle mechanical means such as agitation in water. Civil engineers regard most superficial deposits as soil, whereas soil scientists, geographers and geologists use the term in an agricultural sense.
Soil mechanics
Application of principles of mechanics and hydraulics to engineering problems dealing with soil as an engineering material.
Solid geology map
Map showing the *bedrock* geology as it would appear if the superficial deposits were removed.
Specific gravity
See *particle density*.

Specimen
Part of a soil or rock sample used for a laboratory test. A test specimen is usually prepared from an *undisturbed sample* for laboratory strength or compressibility test. In practice laboratory test specimens are often referred to as samples.

Spread footing
Foundation wider than wall or column it supports.

Stability
Resistance to collapse, failure or sliding, which depends on the shear strength of the material.

Standard penetration test (SPT)
Commonly used in-situ test in which the penetration resistance is measured by driving a split spoon sampler into the base of a borehole.

Stiffness
Resistance of a material such as soil to deformation.

Stone columns
Ground treatment technique in which poor ground is stiffened by the installation of columns of stone. See *vibrated stone columns*.

Strain ε
Intensity of deformation measured as ratio of change in dimension to original dimension (eg ratio of change in length to original length).

Strain softening
Reduction in strength with increasing strain. Such behaviour is sometimes termed *brittle*.

Strength
(1) Strength of a material, such as soil, is the maximum stress which can be sustained before failure.
(2) Strength of a structural element of a building is its ability to resist the loads that are applied to it.

Stress σ
Intensity of loading, measured as load (or force) per unit area. Similar to *pressure*, but pressure is more usually applied to fluids. However, loading applied by a *foundation* to the ground is described as bearing pressure. The stress on a plane within a body of soil has two components; compressive stress acting normal to the plane which reduces the volume of the soil and *shear stress* acting parallel to the plane which distorts the soil (changes the shape of the body of soil).

Stress bulb
See *pressure bulb*.

Strip footing
Traditional shallow strip *foundation* for low-rise building.

Strip foundation
Shallow *foundation* normally provided for a load-bearing wall.

Structural fill
See *engineered fill*.

Subsidence
See *settlement*.

Substructure
Part of structure that is below ground level, including *foundation* and *basement*.

Subsurface water
All underground water.

Suction
Negative pore pressures in ground above the water-table result in attractive forces between soil particles; this attraction is termed soil suction.

Superstructure
Part of structure that is above ground level.

Suspended floor slab
Floor slab suspended clear of ground between supporting walls or beams.

GLOSSARY

Swallow hole
See *sinkhole*.
Swelling
Expansion of a clay soil due to reduction of *effective stress*, resulting from either reduction of *total stress* or absorption of water, can cause *heave* of the ground and the building.
Till
See *glacial till*.
Tilt
Rigid body rotation of a whole structure.
Total overburden pressure
Intensity of vertical *total stress* at a specified depth below ground level, prior to construction.
Total stress
Actual stress in soil mass due to applied pressure. In a saturated soil, total stress is the sum of *effective stress* and *pore water pressure*.
Transpiration
Extraction of moisture by vegetation.
Tremie pipe
Pipe used to place concrete under water to prevent segregation of the concrete.
Trench
Long narrow excavation in the ground.
Trench-fill foundation
Strip foundation formed by filling a trench with concrete.
Trial pit
Open excavation dug to examine ground conditions, recover samples or carry out field tests.
Triaxial compression test
Compression test under *axisymmetric conditions* in which a cylindrical soil specimen enclosed in an impermeable membrane is subjected to a confining pressure and then brought to failure by increasing the axial load (ie σ_1 is increased while $\sigma_2 = \sigma_3$ is kept constant.)
Ultimate bearing capacity
Value of loading intensity for a foundation at which the resistance of the ground to displacement of the foundation is fully mobilised.
Ultimate limit state
State associated with collapse or other similar forms of structural failure.
Unconfined compression
Uniaxial compression in which there is no lateral restraint. This can be termed uniaxial stress ie $\sigma_1 \neq 0$, $\sigma_2 = \sigma_3 = 0$.
Unconfined compressive strength q_u
Compressive strength of a test specimen subjected to *unconfined compression*. $q_u = 2c_u$
Underpinning
New permanent support for a wall, which transfers loads to a deeper level and which is placed beneath old foundations without removing the building.
Undisturbed sample
A *sample* obtained completely intact and with the structure of the soil not modified in any way would be truly undisturbed. In practice, some disturbance is unavoidable and the term is used to describe samples where disturbance is minimised and no change in soil characteristics of practical significance has occurred. High-quality undisturbed samples are required for laboratory strength and compressibility tests.
Undrained shear strength c_u
Shear resistance of soil in an undrained condition.
Uniformity coefficient
See *coefficient of uniformity*.

Uniformly graded soil
Soil with a majority of soil particles which are very nearly the same size.
Unit weight
Weight per unit volume.
Unsaturated soil
See *partially saturated soil*.
Uplift force
(a) Structures with foundations beneath the *water-table* are subject to an uplift force from the *pore water pressure* acting on the *foundation*.
(b) *Foundations* on clay can be subjected to an uplift force due to *swelling of the clay*.
Vadose zone
Subsurface water above the *water-table*.
Vane test
Shear strength test in which a relative rotational movement takes place between a cylindrical volume of soil and the surrounding material. The test is carried out by rotating a vane, comprising four blades set at right angles to each other in the shape of a cruciform. The measurement of the torque required to rotate the vane in a *clay* soil gives a measure of the *undrained shear strength*. Sometimes termed a *shear vane test*.
Vibrated concrete columns
Formed in a similar manner to *vibrated stone columns*, but using concrete instead of stone.
Vibrated stone columns
Deep vibratory ground treatment achieved by penetration into the ground of a large depth vibrator, sometimes called a vibrating poker; usually the cylindrical hole formed by the vibrator is backfilled in stages with stone, forming stone columns. Often termed *vibro*.
Vibro
See *vibrated stone columns*.
Virgin compression line
Nearly linear relationship between void ratio and the logarithm of vertical effective stress for soil undergoing one-dimensional consolidation in an oedometer after an initial recompression stage. See *normal consolidation line*.
Void ratio e
Ratio of volume of *voids* to volume occupied by solid particles of soil.
Voids
Spaces between soil particles. Also described as *pores*.
Walkover survey
Visual survey of a site carried out to obtain information on ground conditions and land use; a thorough walkover survey should be carried out at an early stage.
Water content
Weight of water expressed as a fraction or percentage of the weight of solid particles is the gravimetric water content (w) of a soil, commonly termed moisture content. Volumetric water content (w_v) is used in determining shrinking and swelling clay behaviour.
Water shrinkage factor
Ratio of the water deficiency in a soil layer to the vertical ground movement that occurs in that layer.
Water-table
Surface of the standing water level, that is the surface approximated by the level to which water fills open borings which penetrate a short distance into saturated ground. Sometimes termed *groundwater* level or *phreatic* surface.
Weathering
Process by which soil, rock or building materials are degraded by physical and chemical processes through the action of surface agencies such as rain, frost, temperature changes, pollution and plants.
Weephole
Small drain hole for water.

GLOSSARY

Well-graded soil
Soil with a wide and even distribution of particle sizes.
Wind load
Load on a building or other type of structure due to the effect of wind pressure or suction.
Young's modulus E
Ratio of the axial stress to the axial strain in uniaxial compression with no lateral restraint

$$E = \frac{\Delta\sigma_1}{\Delta\varepsilon_1}$$

where:
$\Delta\sigma_2 = \Delta\sigma_3 = 0$

REFERENCES

The references are listed in alphabetical order by author.
BRE publications are listed on pages 307 and 308; British, European and International standards are listed on pages 309 and 310; websites are listed on page 311.

Alonso E E and Delage P [eds] (1995). *Unsaturated soils.* Proceedings of 1st International Conference, Paris, 3 volumes. Balkema, Rotterdam.

Association of Geotechnical and Geoenvironmental Specialists (1998a). *Code of conduct for site investigation.* AGS, Beckenham, Kent.

Association of Geotechnical and Geoenvironmental Specialists (1998b). *Guidelines for good practice in site investigation.* AGS, Beckenham, Kent.

Association of Geotechnical and Geoenvironmental Specialists (1998c). *AGS Guide: the selection of geotechnical soil laboratory testing.* AGS, Beckenham, Kent.

Association of Geotechnical and Geoenvironmental Specialists (2000). *Guidelines for combined geoenvironmental and geotechnical investigations.* AGS, Beckenham, Kent.

Atkinson J (1993). *An introduction to the mechanics of soils and foundations.* McGraw-Hill, London. 337pp.

Atkinson M F (2000). *Structural defects reference manual for low-rise buildings.* Spon, London. 240pp.

Atkinson M F (2003). *Structural foundations manual for low-rise buildings.* Second edition. Spon, London. 264pp.

Biddle P G (1998). *Tree root damage to buildings.* 2 volumes. Willowmead, Wantage.

Blyth F G H and de Freitas M H (1984). *Geology for engineers.* Seventh edition. Butterworth-Heinemann, Oxford. 336pp.

Bolton M D (1986). The strength and dilatancy of sands. *Geotechnique*, vol 36, no 1, March, pp 65-78.

Boscardin M D and Cording E J (1989). Building response to excavation-induced settlement. *ASCE Journal of Geotechnical Engineering*, vol 115, no 1, pp 1-21.

Broch E and Franklin J A (1972). The point load strength test. *International Journal of Rock Mechanics and Mining Science*, vol 9, pp 669-697.

Bromhead E N (1992). *The stability of slopes.* Second edition. Blackie, Glasgow. 411p.

Brown R W (1992). *Foundation behaviour and repair: residential and light commercial.* Second edition. McGraw Hill, New York. 271pp.

Bruce D A, Bruce M E C and Dimillio A F (1999). Dry mix methods: a brief overview of international practice. *Dry mix methods for deep soil stabilisation* (eds H Bredenberg, G Holm and B B Broms). Proceedings of International Conference, Stockholm, pp 15-25. Balkema, Rotterdam.

Budhu M (2000). *Soil mechanics and foundations.* Wiley, New York. 586pp.

Burford D (1988). Heave of tunnels beneath the Shell Centre, London, 1959-1986. *Géotechnique*, vol 38, no 1, pp 135-137.

Burford D, Crilly, M S and Handley V (1999). Monitoring and repair of a building damaged by ground movements using the Hoopsafe system. *Structural faults and repair 99.*

Burford D and Charles J A (1991). Long term performance of houses built on opencast ironstone mining backfill at Corby, 1975-1990. *Ground movements and structures*, volume 4 (ed J D Geddes). Proceedings of 4th International Conference, Cardiff, July 1991, pp 54-67. Pentech Press, London, 1992.

REFERENCES

Burland J B (1973). Shaft friction of piles in clay – a simple fundamental approach. *Ground Engineering*, vol 6, no 3, pp 30-42.

Burland J B and Burbidge M C (1985). *Settlement of foundations on sand and gravel.* Proceedings of Institution of Civil Engineers, Part 1, vol 78, December, pp 1325-1381.

Burland J B and Hancock R J R (1977). Underground car park at the House of Commons, London: geotechnical aspects. *Structural Engineer*, vol 55, no 2, pp 87-100.

Burland J B, Standing J R and Jardine F M (2001). *Building response to tunnelling – case studies from construction of the Jubilee Line Extension, London.* Vol 1: the project, vol 2: case studies. Report SP200, CIRIA, London.

Burland J B and Wroth C P (1975). Review Paper: Settlement of buildings and associated damage. *Settlement of structures.* Proceedings of British Geotechnical Society Conference, Cambridge, April 1974, pp 611-654. Pentech Press, London, 1975.

Card G B (1995). *Protecting development from methane.* CIRIA Report 149, London. 190pp.

Casagrande A (1936). *The determination of the pre-consolidation load and its practical significance.* Proceedings of 1st International Conference on Soil Mechanics and Foundation Engineering, Harvard, vol 3, pp 60-64.

Chandler R J [ed] (1991). *Slope stability engineering: developments and applications.* Proceedings of International Conference, Isle of Wight. Thomas Telford, London. 443pp.

Chandler R J and Gutierrez C I (1986). The filter-paper method of suction measurement. *Geotechnique*, vol 36, no 2, pp 265-268.

Chandler R J, Crilly M S and Montgomery-Smith G (1992). A low-cost method of assessing clay desiccation for low-rise buildings. *Civil Engineering*, Proceedings of Institution of Civil Engineers, vol 92, May, pp 82-89.

Charles J A (1996). The depth of influence of loaded areas. *Geotechnique*, vol 46, no 1, pp 51-61.

Charles J A and Skinner H D (2001). The delineation of building exclusion zones over hughwalls. *Ground Engineering*, vol 34, no 2, February, pp 28-33.

Charles J A and Watts K S (1996). *The assessment of the collapse potential of fills and its significance for building on fill.* Proceedings of Institution of Civil Engineers, Geotechnical Engineering, vol 119, January, pp 15-28.

Charles J A and Watts K S (2002). *Treated ground: engineering properties and performance.* Report C572. CIRIA, London. 168pp.

Charles J A, Burford D and Watts K S (1986). Improving the load carrying characteristics of uncompacted fills by preloading. *Municipal Engineer*, vol 3, no 1, pp 1-19.

Charles J A, Burford D and Hughes D B (1993). Settlement of opencast mining backfill at Horsley 1973-1992. *Engineered fills.* Proceedings of conference held in Newcastle-upon-Tyne, September 1993, pp 429-440. Thomas Telford, London.

Charles J A, Driscoll R M C, Powell J J M and Tedd P (1996). Seventy-five years of building research: geotechnical aspects. Proceedings of Institution of Civil Engineers, *Geotechnical Engineering*, vol 119, July, pp 129-145.

Charles J A, Skinner H D and Watts K S (1998). The specification of fills to support buildings on shallow foundations: the '95% fixation'. *Ground Engineering*, vol 31, no 1, January, pp 29-33.

Cheney J E (1988). 25 years' heave of a building constructed on clay after tree removal. *Ground Engineering*, vol 21, no 5, pp 13-27.

Clark R G (1998). Theme lecture – Realism in remediation. *Environmental geotechnics* (ed P S Seco e Pinto). Proceedings of 3rd International Congress, Lisbon, vol 4, pp 1257-1272. Balkema, Rotterdam.

Clayton C R I (2001). Managing geotechnical risk: time for change? Proceedings of Institution of Civil Engineers, *Geotechnical Engineering*, vol 149, no 1, January, pp 3-11.

Clayton C R I, Matthews M C and Simons N E (1995). *Site investigation*. Blackwell Science, Oxford.

Clevenger W A (1956). Experiences with loess as a foundation material. *ASCE Journal of Soil Mechanics and Foundations Division*, vol 82, no SM3, paper 1025.

Coduto D P (1998). *Geotechnical engineering: principles and practice*. Prentice Hall, New Jersey. 759pp.

Cooling L F and Ward W H (1948). *Some examples of foundation movements due to causes other than structural loads.* Proceedings of 2nd International Conference on Soil Mechanics and Foundation Engineering, Rotterdam, vol 2, pp 162-167.

Crilly M S, Driscoll R M C and Chandler R J (1992). *Seasonal ground and water movement observations from an expansive clay site in the UK.* Proceedings of 7th International Conference on Expansive Soils, Dallas, vol 1, pp 313-318.

Curtin W G, Shaw G, Parkinson G I and Golding J M (1994). *Structural foundation designers' manual*. Blackwell, Oxford. 377pp.

Cutler D F and Richardson I B K (1989). *Tree roots and buildings.* Second edition. Longman, London.

Davies M C R and Schlosser F [eds] (1997). *Ground improvement geosystems - densification and reinforcement.* Proceedings of 3rd International Conference, London. Thomas Telford, London. 499pp.

Davis R O and Selvadurai A P S (1996). *Elasticity and geomechanic*s. Cambridge University Press, Cambridge. 201pp.

DEFRA/Environment Agency (2002). Contaminated Land Research Reports CLR 7, 8, 9 and 10.

Degen W S (1997). 56m deep vibrocompaction at German lignite mining area. *Ground improvement geosystems - densification and reinforcement*. Proceedings of 3rd International Conference, London, pp 127-133. Thomas Telford, London.

Department of the Environment (1994). *Landsliding in Great Britain.* HMSO, London. 361pp.

Department of the Environment, Transport and the Regions (2000a). Planning policy guidance notes PPG 3: *Housing* and PPG 14: *Development on unstable land.* The Stationery Office, London.

Department of the Environment, Transport and the Regions (2000b). *Our towns and cities: the future – delivering an urban renaissance.* Urban White Paper.

Driscoll R (1983). The influence of vegetation on the swelling and shrinking of clay soils in Britain. *Geotechnique,* vol 33, pp 93-105.

Driscoll R (2004). Counting to seven: Eurocode 7 – geotechnical design is coming. *Ground Engineering,* vol 37, no 3, March, pp 42-43.

Driscoll R and Chown R (2001). Shrinkage and swelling of clays. *Problematic soils.* Proceedings of Symposium at Nottingham Trent University, pp 53-66. Thomas Telford, London.

Driscoll R M C and Crilly M S (2000). *Subsidence damage to domestic buildings: lessons learned and questions remaining.* FBE report 1. Foundation for the Built Environment, Garston.

Dunnicliff J (1988). *Geotechnical instrumentation for monitoring field performance.* Wiley, New York. 577pp.

Eaglestone F and Apted J (1988). *Building subsidence – liability and insurance.* BSP Professional Books, Oxford. 304pp.

REFERENCES

Eakin W R G and Crowther J (1985). Geotechnical problems on land reclamation sites. *Municipal Engineer*, vol 2, October, pp 233-245.

Elson W K (1984). *Design of laterally loaded piles.* CIRIA Report 103, London.

Environment Agency and NHBC (2000). *Guidance for the safe development of housing on land affected by contamination.* R&D publication 66. Stationery Office, London. 86pp.

Evans D, Jefferis S A, Thomas A O and Cui S (2001). *Remedial processes for contaminated land – principles and practice.* Report C548, CIRIA, London.

Findlay J D, Brooks N J, Mure J N and Heron W (1997). Design of axially loaded piles – United Kingdom practice. *Design of axially loaded pies – European practice* (eds F De Cock and C Legrand), pp 353-376. Balkema, Rotterdam.

Fleming W G K, Weltman A J, Randolph M F and Elson W K (1992). *Piling engineering.* Second edition. Blackie, Glasgow.

Frank R, Bauduin C, Driscoll R, Kavvadas M, Ovesen N K, Orr T and Schuppener B (2004). *Designers' Guide to EN1997-1 Eurocode 7: Geotechnical design – General rules.* Thomas Telford, London.

Fredlund D G and Rahardjo H (1993). *Soil mechanics for unsaturated soils.* Wiley, New York. 560pp.

Freeman TJ, Burford D and Crilly MS (1992). Seasonal foundation movements in London clay. *Ground movements and structures* (ed JD Geddes). Proceedings of 4th International Conference, Cardiff, July 1991, pp 485-501. Pentech Press, London, 1992.

Gadd K M (1951). *From Ur to Rome.* Revised edition. Ginn and Company, London. 280pp.

Garvin S, Hartless R, Smith M A, Manchester S and Tedd P (1999). *Risks of contaminated land to buildings, building materials and services: a literature review.* Environment Agency R&D Technical Report P331. WRc, Swindon.

Gazetas G, Tassoulas J L, Dobry R and O'Rourke M (1985). Elastic settlement of arbitrarily shaped foundations embedded in half-space. *Geotechnique,* vol 35, no 3, pp 339-346.

Giroud J P (1970). Stresses under linearly loaded rectangular area. *ASCE Journal of Soil Mechanics and Foundations Division*, vol 96, no SM1, pp 263-268.

Greenwood D A (1987). Underpinning by grouting. *Ground Engineering*, vol 20, no 3, April, pp 21-32.

Handy R L (1995). The day the house fell, ASCE. Quoted in *Geotechnical engineering – principles and practices* by D P Coduto, Prentice Hall, New Jersey, 1998.

Harris M R, Herbert S M and Smith M A (1995). *Remedial treatment for contaminated land, volume 3: site investigation and assessment.* Special Publication 103, CIRIA, London.

Harris M R, Herbert S M and Smith M A (1995). *Remedial treatment for contaminated land, volume 4: classification and selection of remedial measures.* Special Publication 104, CIRIA, London.

Harris M R, Herbert S M and Smith M A (1995). *Remedial treatment for contaminated land, volume 5: excavation and disposal.* Special Publication 105, CIRIA, London.

Harris M R, Herbert S M and Smith M A (1995). *Remedial treatment for contaminated land, volume 6: containment and hydraulic measures.* Special Publication 106, CIRIA, London.

Harris M R, Herbert S M and Smith M A (1995). *Remedial treatment for contaminated land, volume 7: ex-situ remedial measures for soils, sludges and sediments.* Special Publication 107, CIRIA, London.

Harris M R, Herbert S M and Smith M A (1995). *Remedial treatment for contaminated land, volume 8: ex-situ remedial measures for contaminated groundwater and other liquids.* Special Publication 108, CIRIA, London.

Harris M R, Herbert S M and Smith M A (1995). *Remedial treatment for contaminated land, volume 9: in-situ methods of remediation.* Special Publication 109, CIRIA, London.

Harris M R, Herbert S M, Smith M A and Mylrea K (1998). *Remedial treatment for contaminated land, volume 12: policy and legislation.* Special Publication 112, CIRIA, London.

Hatem D J [ed] (1998). *Subsurface conditions – risk management for design and construction management professionals.* Wiley, New York. 465pp.

Hawkins A B [ed] (1997). *Ground chemistry implications for construction.* Proceedings of International Conference, Bristol, 1992. Balkema, Rotterdam, 1997. 658pp.

Hawkins A B (1998). Engineering significance of ground sulphates. *Geotechnical site characterisation* (eds P K Robertson and P W Mayne). Proceedings of 1st International Conference, Atlanta, vol 1, pp 685-692. Balkema, Rotterdam.

Hawkins R G P and Shaw H S (2004). *The practical guide to waste management law.* Thomas Telford, London.

Head K H (1992). *Manual of soil laboratory testing*, Volume 1, Soil classification and compaction tests, Pentech Press, London. Second edition. 388pp.

Head K H (1994). *Manual of soil laboratory testing*, Volume 2, Permeability, shear strength and compressibility tests. Second edition. Wiley, New York. 440pp.

Head K H (1998). *Manual of soil laboratory testing*, Volume 3, Effective stress tests. Wiley, Chichester. Second edition. 428pp.

Health and Safety Commission (1992). *Management of Health and Safety at Work Regulations,* Approved Code of Practice. HMSO, London.

Health and Safety Commission (1995). *A guide to managing health and safety in construction.* HSE Books.

Health and Safety Commission (1995). *Managing construction for health and safety*, Construction (Design and Management) Regulations 1994 Approved Code of Practice L54. HSE Books.

Health and Safety Commission (2001). *Managing health and safety in construction.* Construction (Design and Management) Regulations 1994 Approved Code of Practice and Guidance HSG224. HSE Books.

Health and Safety Executive (1985). *The Abbeystead explosion.* A report of the investigation by the Health and Safety Executive into the explosion on 23 May 1984 at the valve house of the Lune/Wyre water transfer scheme at Abbeystead. HMSO, London.

Highways Agency (1998). *Manual of contract documents for highway works*: volume 1 Specification for highway works. The Stationery Office, Norwich.

Hudson J P and Harrison J A (1997). *Engineering rock mechanics – an introduction to the principles.* Pergamon, Oxford.

Humpheson C, Simpson B and Charles J A (1992). Investigation of hydraulically placed PFA as a foundation for buildings. *Ground movements and structures* (ed J D Geddes). Proceedings of 4th international conference, Cardiff, July 1991, pp 68-88. Pentech Press, London, 1992.

Institution of Civil Engineers (1987). *Specification for ground treatment.* Thomas Telford, London.

Institution of Civil Engineers (1987). *Specification for ground treatment - notes for guidance.* Thomas Telford, London.

Institution of Civil Engineers (1999). *Specification for cement-bentonite cut-off walls.* Thomas Telford, London.

REFERENCES

Institution of Civil Engineers Site Investigation Steering Group (1993). *Site investigation in construction.* Thomas Telford, London.
Part 1: Without site investigation ground is a hazard.
Part 2: Planning, procurement and quality management of site investigation.
Part 3: Specification for ground investigation.
Part 4: Guidelines for the safe investigation by drilling of landfills and contaminated land.

Institution of Structural Engineers (2000). *Subsidence of low-rise buildings.* Second edition. ISE, London. 176pp.

Jennings J E and Knight K (1957). *The additional settlement of foundations due to a collapse of structure of sandy subsoils on wetting.* Proceedings of 4th International Conference on Soil Mechanics and Foundation Engineering, London, vol 1, pp 316-319.

Kalinski R J and Kelly W E (1993). Estimating water content of soils from electrical resistivity. *Geotechnical Testing Journal*, vol 16, no 3, pp 323-329.

Kennard M F, Hoskins C G and Fletcher M (1996). *Small embankment reservoirs.* Report 161, CIRIA, London. 447 pp.

Knipe C V, Lloyd J W, Lerner D N and Greswell R (1993). *Rising groundwater levels in Birmingham and the engineering implications.* Special Publication 92, CIRIA, London.

Kwan J, Rudland D and Nesbit N (2001). Risk assessment and remediation of contaminated land – training material. *Geoenvironmental engineering – geoenvironmental impact management* (eds R N Yong and H R Thomas). Proceedings of 3rd British Geotechnical Association Conference, Edinburgh, pp 72-78. Thomas Telford, London.

Lambe T W and Whitman R V (1979). *Soil Mechanics, SI Version.* John Wiley, New York. 553pp.

Lancellotta R (1995). *Geotechnical Engineering.* Balkema, Rotterdam. 436pp.

Law Society (1998). *Coal mining searches - England and Wales: guidance notes and directory.*

Law Society of Scotland (1999). *Coal mining searches - guidance notes and directory.*

Lea F M (1971). *Science and building.* HMSO, London. 203pp.

Leach B A and Goodger H K (1991). *Building on derelict land.* Special Publication 78, CIRIA, London. 232pp.

Longworth T I (2004). Assessment of sulfate-bearing ground for soil stabilisation for built development. *Ground Engineering*, vol 37, no 5, May, pp 30-34.

Loxham M, Orr T and Jefferis S A (1998). Soil contamination and remediation. *Environmental geotechnics* (ed P S Seco e Pinto). Proceedings of 3rd International Congress, Lisbon, vol 3, pp 1039-1055. Balkema, Rotterdam.

Lunne T, Robertson P K and Powell J J M (1997). *Cone penetration testing in geotechnical practice.* Blackie, London. 312pp.

Lupini J F, Skinner A E and Vaughan P R (1981). The drained residual strength of cohesive soils. *Geotechnique*, vol 31, no 2, June,pp 181-213.

Mair R J (1994). Report on session 4: displacement. *Grouting in the ground.* Proceedings of International Conference (ed A L Bell), Institution of Civil Engineers, pp 375-383. Thomas Telford, London.

Marshall D, Patch D and Dobson, M (1997). *Root barriers and building subsidence.* Arboricultural Practice Note 4. Farnham: Arboricultural Advisory and Information Service.

Martin W S (1996). *Site guide to foundation construction. - a handbook for young professionals.* Special Publication 136, CIRIA, London.

Matthews M C, Hope V S and Clayton C R I (1996). The use of surface waves in the determination of ground stiffness profiles. Proceedings of Institution of Civil Engineers, *Geotechnical Engineering,* vol 119, no 2, April, pp 84-95.

Mayne P W, Jones J S and Dumas J C (1984). Ground response to dynamic compaction. *ASCE Journal of Geotechnical Engineering,* vol 110, no 6, June, pp 757-774.

McCann D M and Green C A (1996). *Geophysical surveying methods in a site investigation programme.* Proceedings of International Conference on *Advances in Site Investigation Practice,* London, March 1995, pp 687-700. Thomas Telford, London, 1996.

McDowell P W, Barker R D, Butcher A P, Culshaw M G, Jackson P D, McCann D M, Skipp B O, Matthews S L and Arthur J C R (2002). *Geophysics in engineering investigations.* CIRIA Report, C562, London.

Meyerhof G G (1965). Shallow foundations. *ASCE Journal of Soil Mechanics and Foundations Division,* vol 91 no SM2, pp 21-31.

Meyerhof G G (1976). Bearing capacity and settlement of pile foundations. *ASCE Journal of Geotechnical Engineering Division,* vol 102, no GT3, pp 197-228.

Mitchell J M and Jardine F M (2002). *A guide to ground treatment.* Report C573, CIRIA, London. 246pp.

Moseley M P and Kirsch K [eds] (2004). *Ground improvement.* Second edition. Spon, London, 344pp.

Musson R M W and Winter P W (1996). *Seismic hazard of the UK.* Report for Department of Trade and Industry (AEA Technology, Warrington).

Musson R M W and Winter P W (1997). Seismic hazard maps for the UK. *Natural Hazards,* vol 14, pp 141-154.

NHBC (various dates). *NHBC Standards.* National House-Building Council, Amersham.
Chapter 4.1: Land quality – managing ground conditions (September 1999, amendments April 2001).
Chapter 4.2: Building near trees (April 2003).
Chapter 4.4: Strip and trench fill foundations (September 1999).
Chapter 4.5: Raft, pile, pier and beam foundations (September 1999).
Chapter 4.6: Vibratory ground improvement techniques (September 1999).
Chapter 5.1: Substructure and ground bearing floors (September 1999, amendments April 2002).
Chapter 5.2: Suspended ground floors (September 1999, amendments April 2002).
Chapter 5.3: Drainage below ground (September 1999, amendments April 2003).

Nixon P J (1978). Floor heave in buildings due to the use of pyritic shales as fill material. *Chemistry and Industry,* 4 March, pp 160-164.

Ove Arup and Partners (1993). *Preliminary study of UK seismic hazard and risk.* Report for Department of the Environment.

Parliamentary Office of Science and Technology (1998). *A brown and pleasant land – household growth and brownfield sites.* Report 117, July. House of Commons, London

Parsons A W (1992). *Compaction of soils and granular materials: a review of research performed at the Transport Research Laboratory.* HMSO, London. 323pp.

Picard L (2000). *Dr Johnson's London: life in London 1740-1770.* Weidenfeld and Nicolson, London. 362pp.

Pilyugin A I (1967). *Settlement of loess foundations under canal embankments.* Proceedings of 3rd Asian Conference on Soil Mechanics and Foundation Engineering, pp 29-32.

Poulos H G and Davis E H (1974). *Elastic solutions for soil and rock mechanics.* Wiley, New York.

Powell J H (1998). *A guide to British stratigraphical nomenclature.* Report SP149, CIRIA, London.

REFERENCES

Price G, Longworth T I and Sullivan P J E (1994). Installation and performance of monitoring systems at the Mansion House. Proceedings of Institution of Civil Engineers, *Geotechnical Engineering,* vol 107, pp 77-87.

Privett K D, Matthews S C and Hodges R A (1996). *Barriers, liners and cover systems for containment and control of land contamination.* Special Publication 124, CIRIA, London.

Proctor R R (1933). Fundamental principles of soil compaction. *Engineering News-Record*, vol 111, no 9, 31 August, pp 245-248.

Pye K and Miller J A (1990). Chemical and biochemical weathering of pyritic mudrocks in a shale embankment. *Quarterly Journal of Engineering Geology*, vol 23, pp 365-381.

Rawlings C G, Hellawell E E and Kilkenny W M (2000). *Grouting for ground engineering*. Report C514, CIRIA, London.

Robson P (1990). *Structural appraisal of traditional buildings*. Gower Technical.

Schmertmann J H (1970). Static cone to compute static settlement over sand. *ASCE Journal of Soil Mechanics and Foundations Division*, vol 96, no SM3, pp 1011-1043.

Schofield A and Wroth C P (1968). *Critical state soil mechanics.* McGraw Hill, London.

Simpson B, Blower T, Craig R N and Wilkinson W B (1989). *The engineering implications of rising groundwater levels in the deep aquifer beneath London*. Special Publication 69, CIRIA, London.

Skempton A W (1959). Cast in-situ bored piles in London clay. *Geotechnique*, vol 9, pp 153-173.

Skempton A W (1986). Standard penetration test procedures and the effects in sands of overburden pressures, relative density, particle size, ageing and overconsolidation. *Geotechnique*, vol 36, no 3, pp 425-447.

Skempton A W and Bjerrum L (1957). A contribution to the settlement analysis of foundations on clay. *Geotechnique*, vol 7, no 4, pp 168-178.

Skempton A W and Macdonald D H (1956). *The allowable settlement of buildings.* Proceedings of Institution of Civil Engineers, part 3, vol 5, pp 727-768.

Skinner H D, Watts K S and Charles J A (1997). Building on colliery spoil: some geotechnical considerations. *Ground Engineering*, vol. 30, no 5, June, pp 35-40.

Slocombe B C, Bell A L and Baez J I (2000). The densification of granular soils using vibro methods. *Geotechnique*, vol 50, no 6, December, pp 715-725.

Taunton P and Adams R (2001). Impact management during remediation of a former landfill in an urban environment. *Geoenvironmental engineering – geoenvironmental impact management* (eds R N Yong and H R Thomas). Proceedings of 3rd British Geotechnical Association conference, Edinburgh, pp 251-256. Thomas Telford, London.

Terzaghi K (1936). *Relation between soil mechanics and foundation engineering.* Proceedings of 1st International Conference on Soil Mechanics and Foundation Engineering, Harvard, vol 3, pp 13-18.

Terzaghi K (1939). Soil mechanics - a new chapter in engineering science (45th James Forrest Lecture). *Journal of Institution of Civil Engineers*, vol 12, pp 106-141.

Terzaghi K (1951). *The influence of modern soil studies on the design and construction of foundations.* Proceedings of Building Research Congress, London, 1951, division 1, part 3, pp 139-145.

Thomas P R (1991). *Geological maps and sections for civil engineers*. Blackie, Glasgow. 106pp.

Thorburn S and Littlejohn G S (1993). *Underpinning and retention*. Second edition. Blackie, London.

Tomlinson M J (2001). *Foundation design and construction.* Seventh edition. Prentice Hall, Harlow. 569pp.

Trenter N A and Charles J A (1996). A model specification for engineered fills for building purposes. *Geotechnical Engineering,* Proceedings of Institution of Civil Engineers, vol 119, no 4, October, pp 219-230.

Urban Task Force (1999). *Towards an urban renaissance.* Spon, Andover, Hampshire.

Van Impe W F (1989). *Soil improvement techniques and their evolution.* Balkema, Rotterdam. 125pp.

Watts K S, Charles J A and Butcher A P (1989). Ground improvement for low-rise housing using vibro at a site in Manchester. *Municipal Engineer*, vol 6, no 3, pp 145-157.

Watts K S and Charles J A (1993). Initial assessment of a new rapid impact ground compactor. *Engineered fills.* Proceedings of International Conference, Newcastle-upon-Tyne, September, pp 399-412. Thomas Telford, London.

Watts K S and Charles J A (1999). Settlement characteristics of landfill wastes. Proceedings of Institution of Civil Engineers, *Geotechnical Engineering*, vol. 137, October, pp 225-233.

Westcott F J, Lean C M B and Cunningham M L (2001a). *Piling and penetrative ground improvement methods on land affected by contamination: interim guidance on pollution prevention.* Environment Agency, National Groundwater and Contaminated Land Centre Project NC/99/73.

Westcott F J, Smith J W N and Lean C M B (2001b). Piling on land affected by contamination: environmental impacts, regulatory concerns and effective solutions. *Geoenvironmental engineering – geoenvironmental impact management.* Proceedings of 3rd British Geotechnical Association conference, Edinburgh, pp 103-108. Thomas Telford, London.

Whittaker B N and Reddish D J (1989). *Subsidence – occurrence, prediction and control.* Elsevier, Amsterdam. 528pp.

Williams G M and Aitkenhead N (1991). Lessons from Loscoe: the uncontrolled migration of landfill gas. *Quarterly Journal of Engineering Geology*, vol 24, pp 191-207.

Wood A A and Griffiths C M (1994). Debate: Contaminated sites are being over-engineered. Proceedings of Institution of Civil Engineers, *Civil Engineering*, vol 102, no 3, pp 97-105.

Xanthakos P P, Abramson L W and Bruce D A (1994). *Ground control and improvement.* Wiley, New York. 910pp.

Younger P L (1993). Possible environmental impact of the closure of two collieries in County Durham. *Journal of Institution of Water and Environmental Management*, vol 7, pp 521-531.

REFERENCES

BRE Publications

Digests
240 Low-rise buildings on shrinkable clay soils: Part 1
241 Low-rise buildings on shrinkable clay soils: Part 2
242 Low-rise buildings on shrinkable clay soils: Part 3
251 Assessment of damage in low-rise buildings
276 Hardcore
298 The influence of trees on house foundations in clay soils
313 Mini-piling for low-rise buildings
315 Choosing piles for new construction
318 Site investigation for low-rise building: desk studies
322 Site investigation for low-rise building: procurement
343 Simple measuring and monitoring of movement in low-rise buildings Part 1: cracks
344 Simple measuring and monitoring of movement in low-rise buildings Part 2: settlement, heave and out-of-plumb
348 Site investigation for low-rise building: the walk-over survey
352 Underpinning
361 Why do buildings crack?
365 Soakaway design
381 Site investigation for low-rise building: trial pits
383 Site investigation for low-rise building: soil description
386 Monitoring building and ground movement by precise levelling
395 Slurry trench cut-off walls to contain contamination
403 Damage to structures from ground-borne vibrations
411 Site investigation for low-rise building: direct investigations
412 Desiccation in clay soils
427 Low-rise buildings on fill
   Part 1: classification and load-carrying characteristics
   Part 2: site investigation, ground movement and foundation design
   Part 3: engineered fill
471 Low-rise building foundations on soft ground
472 Optimising ground investigation
475 Tilt of low-rise buildings: with particular reference to progressive foundation movemen
479 Timber piles and foundations
Special Digest 1 Concrete in aggressive ground (in four parts)

Reports
BR 104 A review of routine foundation design practice
BR 184 Foundation movement and remedial underpinning in low-rise buildings
BR 211 Radon: guidance on protective measures for new dwellings
BR 212 Construction of new buildings on gas-contaminated land
BR 255 Performance of building materials in contaminated land
BR 292 Cracking in buildings
BR 332 Floors and flooring
BR 391 Specifying vibro stone columns
BR 414 Protective measures for housing on gas-contaminated land
BR 424 Building on fill: geotechnical aspects
BR 440 Foundations, basements and external works
BR 447 Brownfield sites: ground-related risks for buildings
BR 458 Specifying dynamic compaction
BR 465 Cover systems for land regeneration: thickness of cover systems for contaminated land
BR 470 Working platforms for tracked plant

GEOTECHNICS FOR BUILDING PROFESSIONALS

Information Papers
15/85 The effect of a rise of water table on the settlement of opencast mining backfill
16/86 Preloading uncompacted fills
2/87 Fire and explosion hazards associated with the redevelopment of contaminated land
3/89 Subterranean fires in the UK – the problem
5/89 The use of 'vibro' ground improvement techniques in the United Kingdom
4/93 A method of determining the state of desiccation in clay soils
5/97 Building on fill: collapse compression on inundation
18/01 Blastfurnace slag and steel slag: their use as aggregates

Information Sheets
A simple guide to in-situ ground testing:
  IS 1 What is it and why do it?
  IS 2 Cone penetration testing
  IS 3 Flat dilatometer testing
  IS 4 Dynamic probing
  IS 5 Pressuremeter testing
  IS 6 Large-diameter plate loading tests
  IS 7 Geophysical testing

Good Building Guides
  3 Damp-proofing basements
 14 Building simple plan brick or block-work freestanding walls
 19 Building reinforced, diaphragm and wide plan freestanding walls
 25 Buildings and radon
 27 Building brickwork or blockwork retaining walls
 39 Simple foundations for low-rise housing:
     Part 1: Site investigation
     Part 2: 'Rule of thumb' design
     Part 3: Groundworks: getting it right
 53 Foundations for low-rise building extensions
 59 Building on brownfield sites:
     Part 1: Identifying the hazards
     Part 2: Reducing the risks

Good Repair Guides
  1 Cracks caused by foundation movement
  2 Damage to buildings caused by trees
  3 Repairing damage to brick and block walls

REFERENCES

British, European and International Standards

BS 1377:1990 Methods of test for soils for civil engineering purposes
    Part 1 General requirements and sample preparation
    Part 2 Classification tests
    Part 3 Chemical and electro-chemical tests
    Part 4 Compaction-related tests
    Part 5 Compressibility, permeability and durability tests
    Part 6 Consolidation and permeability tests in hydraulic cells and with pore pressure measurement
    Part 7 Shear strength tests (total stress)
    Part 8 Shear strength tests (effective stress)
    Part 9 In-situ tests
BS 5837:1991 Guide for trees in relation to construction
BS 5930:1999 Code of practice for site investigations
BS 6031:1981 Code of practice for earthworks
BS 8002:1994 Code of practice for earth retaining structures
BS 8004:1986 Code of practice for foundations
BS 8006:1995 Code of practice for strengthened/reinforced soils and other fills
BS 8081:1989 Code of practice for ground anchorages
BS 8103-1:1995 Structural design of low-rise buildings: Code of practice for stability, site investigation, foundations and ground-floor slabs for housing
BS 8301:1985 Code of practice for building drainage
BS 10175:2001 Investigation of potentially contaminated sites - Code of practice

BS EN 752-4:1998 Drain and sewer systems outside buildings: Hydraulic design and environmental considerations
BS EN 1610:1998 Construction and testing of drains and sewers

prEN 1997-1:2003 Eurocode 7: Geotechnical design – Part 1: General rules
prEN 1997-2:2003 Eurocode 7: Geotechnical design – Part 2. Ground investigation and testing

BS EN ISO 14688-1:2002 Geotechnical investigation and testing – Identification and classification of soil – Part 1: Identification and description
BS EN ISO 14688-2:2004 Geotechnical investigation and testing – Identification and classification of soil – Part 2: Principles for a classification
BS EN ISO 14689-1:2003 Geotechnical investigation and testing – Identification and classification of rock – Part 1: Identification and description

Geotechnical investigation and testing - field testing
prEN ISO 22476-1:2004 Part 1: Electrical cone and piezocone penetration tests (CPT and CPTU)
prEN ISO 22476-2:2003 Part 2: Dynamic probing
prEN ISO 22476-3:2003 Part 3: Standard penetration test

CEN ISO/TS 17892 Geotechnical investigation and testing – Laboratory testing of soil*
17892-1:2003   Part 1: Determination of water content
17892-2:2003   Part 2: Determination of density of fine grained soil
17892-3:2003   Part 3: Determination of particle density – pycnometer method
17892-4:2003   Part 4: Determination of particle size distribution
17892-5:2003   Part 5: Incremental loading oedometer test
17892-6:2003   Part 6: Fall cone test
17892-7:2004   Part 7: Unconfined compression test on fine-grained soils
17892-8:2004   Part 8: Unconsolidated undrained triaxial test
17892-9:2004   Part 9: Consolidated triaxial compression tests on water saturated soils
17892-10:2004 Part 10:Direct shear tests
17892-11:2003 Part 11:Determination of permeability by constant and falling head
17892-12:2004 Part 12:Determination of Atterberg limits

BS EN 1536:2000   Execution of special geotechnical works – Bored piles
BS EN 1537:2000   Execution of special geotechnical works – Ground anchors
BS EN 1538:2000   Execution of special geotechnical works – Diaphragm walls
BS EN 12063:1999   Execution of special geotechnical works – Sheet pile walls
BS EN 12699:2001   Execution of special geotechnical works – Displacement piles
BS EN 12715:2000   Execution of special geotechnical works – Grouting
BS EN 12716:2001   Execution of special geotechnical works – Jet grouting
prEN 14199:2001   Execution of special geotechnical works – Micropiles
prEN 14475:2002   Execution of special geotechnical works – Reinforced fill
prEN 14490:2002   Execution of special geotechnical works – Soil nailing
prEN 14679:2003   Execution of special geotechnical works – Deep mixing
prEN 14731:2003   Execution of special geotechnical works – Ground treatment by deep vibration

* *The European Standards for the laboratory testing of soil are being prepared as Technical specifications. With a technical specification (TS), unlike an EN, conflicting national standards may continue to exist.*

Websites

British Geological Survey	www.bgs.ac.uk
British Geotechnical Association	www.britishgeotech.org.uk
British Standards Institution	www.bsi-global.com
Building Research Establishment	www.bre.co.uk
Coal Authority	www.coal.gov.uk
Construction Industry Research & Information Association	www.ciria.org.uk
Environment Agency	www.environment-agency.gov.uk
Health and Safety Executive	www.hse.gov.uk
Institution of Civil Engineers	www.ice.org.uk
Institution of Structural Engineers	www.istructe.org.uk
National House-Building Council	www.nhbc.co.uk
Office of the Deputy Prime Minister	www.odpm.gov.uk
Royal Institution of Chartered Surveyors	www.rics.org
The Stationery Office	www.thestationeryoffice.com

INDEX

Information on some of the terms listed in the index can also be found in the glossary.

aggressive ground 129-30, 138, 171-4
  see also chemican hazards
air voids, volume of 26
  see also compaction; density and densification
alluvial soil 20, 21
Atterberg limits 29, 30, 65, 90, 91

backfill
  boreholes 152
  brownfield land 139, 168
  drains and sewers 264–5
  foundation damage 221, 225, 227
  freestanding walls 272
  mines 35, 67–8, 155, 168, 206, 221
  retaining walls 270
  soakaways 258, 26
  see also fills; vibrated stone and concrete columns
band drains *see* vertical drains
base resistance 118–9, 127
basements
  contaminated land 174, 175
  excavation for 150, 218
  foundation depths 114
  ground assessment 83
  groundwater level changes 221, 228
  infilling 139
  raft foundations 109
bearing pressures and resistance 92, 114–20, 127, 135, 140, 190, 269
  see also soil strength
bedding, drainage 264–5
bioremediation 208
boreholes
  backfilling 152
  foundation movement investigation 225–6, 227, 228
  ground assessment 64–5, 82, 86, 87–9, 92, 247, 249
  mine and cavity location 157–8, 159
  vane shear tests 97, 98
boulder clay *see* glacial till
brass screws 231
British Standards *see* codes and standards
brownfield land 163–88
  backfill 139, 168
  fills 166–9, 173, 179, 180, 181–2, 207, 208
  ground treatment 139, 182, 183, 193, 194
  groundwater level changes 171, 207
  hazards 76, 78, 83, 164, 170–8, 182–3

precast concrete foundations 111
  see also contaminated land; fills
building extensions 128–9
Building Regulations 7, 8–13
buildings
  damage 214–5
  repair/strengthening 238–9
  weight 218
  see also cracks; foundations

cavities *see* mine workings; natural cavities
chalk 34
  frost heave 222, 254
  natural cavities 85, 156, 157, 222
  soakaways 156, 257
  trees and 148, 228
chemical hazards 129–130, 164, 178, 214
  brownfield land 174–5, 182–3
  colliery spoil 223, 255
  cracks 129, 174, 214
  fills 138, 223, 255
  floor slabs 169, 174, 182, 223, 255–6
  foundations 174–5, 182, 223
  ground beams 227
  groundwater 84, 171–2, 183, 221, 223, 228
  hardcore 254, 255–6
  soil stabilisation 204
  see also contaminated land
clay soils 29–32
  compaction 69–70
  compressibility 54, 55–6
  consistency classification 62
  consolidation 20–1, 22, 31, 52, 190, 197–8
  deformation 218
  differential settlement 147, 250
  earth pressure 47
  erosion 20–1, 29, 31, 226
  foundation depths 108, 135, 140, 145, 148, 218–9, 227
  freestanding walls 271
  frost 29, 135
  ground investigation 79, 80, 85, 89, 94, 226
  ground treatment 190, 197, 198, 204
  groundwater level changes and 221, 228
  landslides 61, 227, 229–30
  moisture content 29–30, 63, 64–5, 66, 69, 70
  organic matter 32–3
  particle sizes 23
  permeability 29, 49, 52, 142, 190
  piled foundations 110, 120

311

raft foundations 147
retaining walls 269–70
settlement 32, 50, 125, 143, 198, 226, 228
shallow foundations 63, 128, 135, 229
shrinkage and swelling 63–6, 108, 110, 120, 218–1, 221
slope creep 150
standard penetration tests 92
strength 50, 61–2, 91, 116, 117, 118
strip footings and foundations 106, 108
underpinning in 243, 248, 250
void ratio 25
see also desiccation; fine soils; fissured clay; glacial till; heave; laminated clay soils; soft ground; trees; undrained shear strength
coarse soils 23, 26–8, 31
  deformation 51–2, 218
  foundation design 57, 116, 117, 118, 119, 126–7, 137
  ground investigation 89, 91, 92, 94
  groundwater 49–50, 221
  heave 218
  permeability 27, 50
  settlement 116, 219
  see also gravel; sand
codes and standards 13–14, 15, 20, 209
cohesive soils 23, 94
collapse, mines and cavities 67, 155–6, 158, 173, 221, 222, 248
collapse compression 37, 66–8, 221, 228
  diagnosis 226–7
  minimising risk 189–90, 200
  vibrated stone columns 192–3
colliery spoil 168
  chemical hazards 223, 255
  collapse compression 68
  hardcore 254, 255, 256
  mine shaft infill 208
  underground fires 172, 175
compaction 68–70, 91, 199–200, 201, 265
  see also compression; consolidation; density and densification; dynamic compaction; impact compaction
compressibility 32, 33, 51–7, 135, 190
  see also volume compressibility coefficient
compression
  fills 37, 66–8, 221, 226–7
  ground 8, 51, 56, 58–60, 126, 135
  see also collapse compression; consolidation; secondary compression; settlement; soil strength; stress; virgin compression line
compression index 55
cone penetration tests 28, 94–7, 119, 126
consistency index 30
  see also Atterberg limits

consolidation
  clay soils 20–1, 22, 31, 52, 190, 197–8
  coefficient of 56, 125
  fine soils 91, 203
  see also compaction; compression; normally consolidated soils; overconsolidated soil; pore water pressure; settlement
constrained modulus 51, 54, 56, 126
contaminated land 163, 170–8
  excavations 201, 207, 208
  ground investigation 83, 164, 176–8
  ground treatment 206–9
  monitoring 209
  regulations 10, 11–13
  see also brownfield land; chemical hazards
contracts 80–1
corseting 239
cracks
  buildings 107, 111, 112–3, 129, 174, 214
  investigation 214, 215–9, 222, 224, 225, 230–2
  soils 64, 142, 150, 151, 223
creep compression *see* secondary compression
crown holes *see* natural cavities
cut-off walls 166, 181, 207

deep foundations 57, 105–6, 118–20, 127–8, 145, 150
  see also piled foundations
deflection ratio 112–13
deformation 44–5, 51–7, 218
  Building Regulations 10, 11
  foundation design and problems 117, 121, 124, 215
  mine collapses 155
  retaining walls 153
  rock 34
  soil strength and 61
  see also bearing pressures and resistance; distortion; effective stress; hogging; sagging; settlement
density and densification 25–6, 28, 68, 69, 189–190, 192
  see also compaction; particle density
density index 28, 69, 91, 92, 93, 94, 137
derelict land *see* brownfield land
desiccation
  drainage 148, 221
  ground movement 148, 150, 227
  overconsolidation 21, 55, 62
  testing for 64–5
  trees and 135, 140, 142–8, 218–21, 228, 237–8, 248–9
  see also moisture content; shrinkage and swelling; soil suction

INDEX

desk studies 79, 80–3, 85, 177, 224
dewatering 47, 201, 208, 228
differential settlement
  acceptable ground movements 111–13
  brownfield land 173, 183
  causes 105, 136, 173, 183, 218, 221, 267
  clay soils 147, 250
  foundation damage 83, 128–9, 215, 218, 221, 222
  shrinkable clays 147, 250
  soft ground 135, 136
  treatment 192, 238, 239
  see also distortion; foundation movement
direct investigation 86–90, 177, 225–6
  see also boreholes; samples and sampling; testing; trial pits
direct shear tests 58, 91
displacement transducers 231
distortion
  effects 111, 215, 216, 224, 232
  foundation design 105, 111, 112–13, 117
  freestanding walls 272
  see also deformation; effective stress
distortion surveys 224
domestic refuse 35, 56, 105, 138, 167, 172, 175, 184
  see also landfill
downdrag 119–20, 139, 147, 159, 248
drainage 148, 262, 264–7
  brownfield land 175, 183
  desiccation and 148, 221
  fine soils 202
  foundations 106–7, 220–1, 228, 241, 265–7, 271
  ground treatment 153, 197, 202–3, 236, 237
  natural cavities and 156
  regulations 11
  retaining walls 151, 270
  unstable slopes 150–1, 202
  see also groundwater flow; permeability; soakaways; vertical drains
drainage blankets 50, 203
dynamic compaction 194–5
dynamic loading 137, 190
dynamic probing (DP) 93–4

earth pressure 46–7, 51, 270
earthfill 200, 201
effective stress 45–7
  collapse compression 66
  foundation design 116, 118, 119, 125, 126, 152
  groundwater flow 49
  partially saturated soils 63
  testing 59, 60–1
  see also deformation; distortion; liquefaction;
overconsolidation ratio; preconsolidation pressures
elastic theory 120–4
embankments 50, 51, 67, 91, 153, 267–8
environmental factors 194, 196, 227
  see also contaminated land; trees
erosion 27, 28, 49, 51, 219–221
  clay soils 20–1, 29, 31, 226
  fills 50
  trenches 203
  underpinning and 248
  see also piping
Eurocodes see codes and standards
excavations
  basements 150, 218
  contaminated land 201, 207, 208
  drainage trenches 264–5
  foundations 83, 145, 148, 150, 173
  freestanding walls 271
  ground movement and foundation damage 222, 223, 227, 232, 236, 239
  ground treatment 190, 199–201
  groundwater and 37, 47, 83, 201, 228
  retaining walls 236, 269
  soakaways 261
  soil softening 219–20
  underground fire control 176
  underpinning work 241, 243, 247, 248
  see also trial pits

factors of safety 13, 117, 127
fills 35–7, 137–9
  brownfield land 166–9, 173, 179, 180, 181–2, 207, 208
  chemical hazards 138, 223, 255
  compaction 199–200, 201, 265
  compression and compressibility 37, 55, 56, 66–8, 218, 221, 226–7
  definition 20
  embankments 267, 268
  erosion 50
  foundations 37, 83, 109, 136–7, 139, 182
  freestanding walls 272
  grading 27
  ground investigation 82, 83, 85
  ground treatment 139, 189–190, 192–3, 195, 196, 197, 198–201
  groundwater 50, 201, 221
  heave 109, 169, 180, 200, 254
  mine workings 35, 67–8, 155, 168, 199, 205–6, 221
  overburden pressure 122
  settlement 35, 37, 68, 189, 200, 226, 265
  shrinkage and swelling 138, 169
  soakaways 257

313

underpinning in 247, 248
unstable slopes 152–3
see also backfill; collapse compression; colliery spoil; compaction; hardcore; landfill; rubble
filter paper tests 63
fine soils 23, 28–32
  behaviour 50, 51–2, 221
  compaction 68–70
  consolidation 91, 203
  drainage 202
  foundation design 115, 116, 124–6, 127
  grouting 238
  permeability 23, 49, 50, 91
  soil suction 63
  see also chalk; clay soils; cohesive soils; silt soils
fires 172, 175–6, 181
fissured clay 32, 87, 92
floor slab piling 245, 247
floor slabs 111
  chemical attack 169, 174, 182, 223, 255–6
  damage 105, 147, 227
  drainage pipes 267
  hardcore 253, 254, 255–6
  underpinning 243, 250
  ventilation 139
  vibrated stone columns 194
  see also raft foundations
foundation depths 114, 123–4
  building extensions 128–9
  clay soils 108, 135, 140, 145, 148, 218–9, 227
  coarse soils 57
  freestanding walls 271
  frost 106, 109, 114, 135, 222, 227
  overburden pressure and 57
  strip footings and foundations 106, 108, 123, 135, 136
  trees 108, 140, 145, 148
  trench-fill foundations 108
  underpinning 247, 248, 250–1
  see also deep foundations; shallow foundations
foundation design 37, 57, 104–62, 215, 218, 241
foundation movement 104–5, 145–7, 213–35, 265
  see also ground movement; heave; settlement
foundations 105–11
  brownfield land 182–4
  chemical hazards 174–5, 182, 223
  drainage 241, 271
  excavations 83, 145, 148, 150, 173
  fills 37, 83, 109, 136–7, 139, 182
  freestanding walls 270–2
  retaining walls 269
  see also deep foundations; floor slabs; pad foundations; piled foundations; raft foundations; shallow foundations; strip footings and foundations; trench-fill foundations; underpinning

freestanding walls 271–2
friction angle 92, 119
frost
  building damage 214
  Building Regulations 10
  chalk 222, 254
  clay soils 29, 135
  foundation depths 106, 109, 114, 135, 222, 227
  rock 35
  silt soils 28, 222
  see also heave

gas migration
  hazards 172, 175
  prevention 180, 181, 183–4
  sources and routes 166, 175, 176, 184, 221
gasworks 165, 166
glacial till 20, 21, 30, 55, 140
glossary 275 97
graticule 230
gravel 27
  deformation 218
  foundation design 57, 106, 119, 127, 137
  hardcore 254
  particle sizes 23, 27
  permeability 49
  trees and 228
  see also coarse soils
gravimetric water content *see* moisture content
ground beams 109, 111, 159, 227, 247, 250
ground chemistry *see* chemical hazards
ground investigation 75–103, 151–2, 157–8, 224–9, 247, 249
  contaminated land 83, 164, 176–8
  see also testing
ground movement
  acceptable 111–113, 114
  brownfield land 173
  Building Regulations 10
  desiccation and 148, 150, 227
  difficult ground 134–62
  excavations 222, 223, 227, 232, 236, 239
  groundwater level changes and 67, 68, 207, 221, 227, 228
  monitoring 233
  see also collapse; compression; deformation; foundation movement; heave; landslides; retaining walls; settlement; shrinkage and swelling; unstable slopes
ground treatment 189–212, 236–52
  brownfield land 139, 182, 183, 193, 194
  drainage 153, 197, 202–3, 236, 237
  shallow foundations and 136–7
  testing 94, 210
  see also bioremediation; fills; grouting; soil

INDEX

stabilisation
ground types 19–42
groundwater 37–9
  chemical attack potential 84, 171–2, 183, 221, 223, 228
  coarse soils 49–50, 221
  excavations and 37, 47, 83, 201, 228
  fills and 50, 201, 221
  ground assessment 82, 84, 85, 228
  mines and natural cavities 34, 155, 156, 221
  slope stability 223
groundwater flow 47–51, 52, 125, 181
  see also drainage; permeability
groundwater level changes 221, 228
  bearing resistance 116
  brownfield land 171, 207
  ground movement 67, 68, 207, 221, 227, 228
  measurement 154
  overconsolidation 21, 31, 55
  see also dewatering; downdrag; drainage
groundwater levels 38
  excavation and refilling 201
  foundation design and repair 115, 116, 123, 241
  soakaways 257, 260
  soft natural ground 135
groundwater pollution
  brownfield land 164, 165, 166, 172–3, 183
  regulation 12
  soakaways 258
  treatment 207, 208
grouting 153, 159, 176, 204–6, 207, 208, 238
gypsum 254, 256

hardcore 147, 169, 253–7
  see also rubble
head *see* hydraulic gradient
health and safety 7, 9, 12–13, 87, 191, 195, 201
  see also brownfield land; contaminated land; risk assessment and management
heave
  clay soils 29, 32, 85, 201, 218, 219, 221, 227, 237, 238
  coarse soils 218
  compaction grouting 205
  fills 109, 169, 180, 200, 254
  foundation depths 114
  foundation movement and damage 105, 174, 215, 217, 221, 222–3, 237
  groundwater level changes 207
  insurance risk 14, 217, 223
  investigation 223, 224, 226, 227, 230, 233
  load removal 218
  silt soils 28, 222
  tree removal 85, 219, 227, 237, 238
  underpinning 239, 241, 243, 245, 249

see also chemical hazards; frost; ground movement
hogging 112, 113, 215
hydraulic gradient 47–9, 52, 125

impact compaction 190, 194–6
in-situ testing 91–9, 191
inclined piling 244, 246
inclinometers 154, 233
infiltration 38, 259–61
insurance aspects 14–15, 217, 218, 223, 236

laminated clay soils 21–2, 32
land condition records 178
landfill 12, 35, 166, 172, 180, 181
  see also gas migration
landslides 149–50, 223
  Building Regulations 10
  clay soils 61, 227, 229–30
  foundation damage diagnosis 226
  ground assessment 82, 84, 85
  insurance risk 14, 217
  planning guidance 8
  underpinning 248
  see also unstable slopes
leaching and leachates 166, 171–2, 179, 208, 254
liquefaction 99, 137
liquidity index 30
load and resistance factor design 117
loadbearing walls 106, 193, 250, 265
loads 10, 114, 116, 118, 120, 247
  see also bearing pressures and resistance; dynamic loading; overburden pressure; preloading; stress
loose soils 21, 28, 137

made ground *see* fills
marine soil 20, 21
marl *see* clay soils
mass concrete underpinning 240–1, 247, 250–1
maximum dry density 69, 70, 201
mine workings 154–9
  collapse 67, 155–6, 158, 173, 221, 222, 248
  fills 35, 67–8, 155, 168, 199, 205–6, 221
  ground investigation 82, 83, 85, 227
  groundwater 34, 155, 156, 221
  planning guidance 8
  raft foundations 109, 158, 159
  see also colliery spoil; natural cavities
mini-piling underpinning 243–7, 251
mixing *see* soil stabilisation
Mohr-Coulomb failure criterion 60, 61
moisture content 24–5
  clay soils 29–30, 63, 64–5, 66, 69, 70
  earthfills 200, 201

ground assessment 29–30, 90–1, 259
groundwater treatment 207
see also density and densification; desiccation; dewatering; shrinkage and swelling; undrained shear strength; water content

monitoring
  contaminated land 209
  gas migration 184
  ground and foundation movement 154, 204, 210, 222, 229–33, 237
  mine workings 158
  soakaways 260, 262
mudstone 34, 68

natural cavities 8, 34, 85, 154–9, 176, 222, 227
  see also mine workings
natural ground 135–7, 190, 193, 197
  see also soft ground; soils
needles 240, 243, 244, 245–6
negative skin friction *see* downdrag
NHBC standards 15
normally consolidated soils 20–1, 55, 92, 125, 127

oedometer tests 52–6, 67, 91
opencast mining backfill 35, 67, 168, 221
optimum water content 69, 70, 201
organic fills 139, 193
organic soils 32–3, 116, 135
  see also soft ground
overburden pressure
  eroded soils 20
  foundation design 57, 115, 119, 121, 122
  overconsolidated soils 55
  undrained shear strength tests 62
overconsolidated soils 20–1, 31, 55
  bearing resistance 140
  desiccation and 21, 55, 62
  earth pressure 47
  embankments 268
  lamination 22
  settlement calculations 125
  suction profiles 65
  water content 66
overconsolidation ratio (OCR) 21
overstressing 214, 269

pad foundations 109, 123, 182, 227, 240
partial factors 13–14, 117
partial underpinning 239, 250–1
partially saturated soils 23, 26, 37, 38, 43, 47, 62–70
particle density 25, 90, 91
particle shapes, coarse soils 27–8
particle sizes 22–3, 26–7, 30–1, 90, 91, 97, 255

peat 33, 56, 79, 116, 135–6, 221, 228
permeability 47–50
  chemical solutions 130
  clay soils 29, 49, 52, 142, 190
  coarse soils 27, 50
  fine soils 23, 49, 50, 91
  laminated clay soils 21–2
  organic soils 33
  soft ground 135
  see also drainage; groundwater flow
phreatic surfaces *see* groundwater levels
pier-and-beam underpinning 241–3, 247, 250, 251
pier foundations 109, 139
piezometers 154
pile-and-beam underpinning 243, 247, 248, 249, 250, 251
piled foundations 110
  bearing resistance 118–20, 127
  brownfield land 179, 180, 183
  clay soils 110, 120
  coarse soils 57, 127
  difficult ground 139, 145, 147, 152, 159, 191, 227
  ground investigation 86
  groundwater level changes 221
  settlement 110, 118, 119, 120, 127–8
  see also deep foundations; mini-piling underpinning
piled-raft underpinning 243, 247
pipes *see* drainage
piping 50, 51
planning guidance and constraints 8
plasticity index 29, 30, 31, 66, 98
pollution *see* contaminated land; groundwater pollution
pore water pressure 45–7, 48, 50, 95, 125, 255
  see also consolidation; drainage
precast concrete foundations 111
preconsolidation pressures 21, 54, 55
preloading 190, 197–9
pressure bulbs 120
Proctor compaction test 70, 201
procurement 80–1
pulverised-fuel ash 35, 204, 206, 255

quality control 209–210
quick soils 49

raft foundations 109–10
  chemical attack 182
  clay soils 147
  damage investigation 227
  difficult ground 135–6, 137, 139, 150, 152, 215
  distortion and 113, 215

INDEX

fills 109, 182
mine workings 109, 158, 159
settlement estimation 125
shrinkable clays 147
significant depth 123
vibrated stone columns 193
see also floor slabs; piled-raft underpinning
rapid impact compaction 190, 194, 196
refraction surveys 99
regulations 6–18
relative compaction 70
relative density *see* density index
remedial treatment *see* ground treatment
residual soils 20, 22
residual strength 58, 61
retaining walls 268–70
  drainage 151, 270
  earth pressures 47
  excavations 236, 269
  ground movement prevention 83, 153, 236
  underpinning as 241
  water-table and 228
  see also water-retaining structures
risk assessment and management 12, 176–9, 191
  see also health and safety
rock 29, 33–5, 117, 149, 153, 154
  see also mudstone; residual soils; sandstone
rubble 164, 223, 254, 255–6
  see also fills; hardcore

safety *see* factors of safety; health and safety; risk assessment and management
sagging 112, 113, 215
samples and sampling 85, 88, 89–90, 92, 225–6
  see also testing
sand 27
  deformation 218
  foundations 57, 106, 119, 127, 137
  groundwater level changes 221
  grouting 238
  laminated clay soils 21–2
  liquefaction 137
  particle sizes 23
  permeability 49, 50
  rapid impact compaction 196
  settlement 127, 228
  shrinkage and swelling 221
  standard penetration tests 92, 93
  strength 61
  trees and 228
  underpinning 243
  void ratio 25
  see also coarse soils
sandstone 34, 55, 68
saturated soils
  behaviour 44–62, 137, 140–2
  definition 23, 26, 38
  moisture content 24, 29
  preloading 197
  see also partially saturated soils
scrapyards 165, 166
secondary compression 57, 126, 218
seismic measurement 99
sensitivity 29
services
  brownfield land 174–5, 176, 180, 182–4
  natural cavity collapses 156
  settlement and 111, 219
  shallow foundations 106
  soft ground 135
  unstable slopes 151
  see also drainage
settlement
  brownfield land 166, 167, 168
  Building Regulations 10
  clay soils 32, 50, 125, 143, 198, 226, 228
  coarse soils 116, 219
  compression and 51, 56
  embankments 267
  fills 35, 37, 68, 189, 200, 226, 265
  fine soils 50
  foundation damage 215, 218, 219, 226, 228, 230, 233
  foundation treatment 247
  freestanding walls 272
  ground treatment 189–90, 193, 194, 196, 198, 204, 205
  groundwater level changes 59, 67, 68, 207, 221, 228
  hardcore 255, 256–7
  insurance risk 14–15, 217, 223
  loose natural ground 137
  monitoring 210
  organic soils 33
  peat 79, 228
  performance specification 209
  piled foundations 110, 118, 119, 120, 127–8
  planning guidance 8
  sand 127, 228
  services and 111, 219
  shallow foundations 83, 85, 116, 120, 124–7
  silt soils 228
  soakaways 261
  soil strength and 50
  strip footings and foundations 227
  trees and 85, 142, 143
  see also collapse; compression; consolidation; differential settlement; downdrag; strain
settlement estimation
  allowable bearing pressure 117

317

building damage 113
consolidation coefficient 56
elastic theory 120–8
engineered fill 200
penetration tests 92, 96
stress-strain relationship 45
sewage works 166–7
sewers *see* drainage
shaft resistance 118, 119, 127
shallow foundations 105–6
  bearing resistance 115–17
  building extensions 128
  clay soils 63, 128, 135, 229
  coarse soils 126–7, 137
  damage 140, 147, 150, 219, 221, 222
  differential settlement 83
  drains and 107
  fills 83, 136–7, 139
  fine soils 124–6
  ground treatment 136–7
  partially saturated soils 38, 43
  settlement 83, 85, 116, 120, 124–7
  soil behaviour 44
  trees and 85, 219, 229
  trial pits 87
  see also strip footings and foundations
shear resistance, angle of *see* friction angle
shear strength 33, 57–62, 97–8, 137
  see also direct shear tests; effective stress; undrained shear strength
shrinkage and swelling
  building materials 214
  Building Regulations 10
  fills 138, 169
  hardcore 256
  soils 63–6, 108, 110, 120, 218–9, 221
  see also desiccation; heave; moisture content; trees; water shrinkage factor
shrinkage potential 66, 238
silt soils 21, 28
  bearing resistance 116
  frost heave 28, 222
  groundwater level changes 221, 228
  organic matter 32–3
  particle sizes 23
  permeability 49
  settlement 228
  shrinkage and swelling 221
  strip foundation widths 106
  underpinning 243
  see also fine soils; laminated clay soils; soft ground
sinkholes *see* natural cavities
site investigation 75–6, 176–8, 257
  see also ground investigation

slabs *see* floor slabs; raft foundations
slags 35, 169, 180, 254, 256
slip indicators 154
slopes *see* unstable slopes
soakaways 148, 151, 156, 183, 257–62, 263
soft ground 135–7, 190, 193, 265, 267, 268
  see also clay soils; organic soils; silt soils
soil softening 61, 219–21
soil stabilisation 136, 152–4, 203–4, 208, 238
  see also ground treatment
soil strength 50, 58–60, 61, 89, 90, 97–8
  clay soils 50, 61–2, 91, 116, 117, 118
  see also bearing pressures and resistance; shear strength
soil suction 63, 64–65
  see also desiccation
soils 19–33, 43–74, 226–7
  cracking 64, 142, 150, 151, 223
  erosion 27, 28, 49, 51, 219–221
  see also coarse soils; cohesive soils; fine soils; natural ground; organic soils; overconsolidated soils; saturated soils
solution features *see* natural cavities
specific gravity *see* particle density
specification and quality control 209–10
spread footings 86, 106
  see also strip footings and foundations; trench-fill foundations
standard penetration tests 28, 92–3, 97, 126
steel rulers 230
strain 44–5
  see also compression; deformation
strain softening 61
stress 44–5, 120–3
  see also effective stress; overstressing
stress bulbs 120
strip footings and foundations 106–8, 114
  bearing resistance 115–17
  Building Regulations 10
  chemical attack 182
  clay soils 106, 108
  difficult ground 135, 139, 159
  foundation depths 106, 108, 123, 135, 136
  settlement 227
  vertical stress distribution 122
  vibrated stone columns 193, 194
  see also shallow foundations
subsidence *see* settlement
sulfate attack *see* chemical hazards

tell-tales 230–1
testing
  codes and standards 13, 20
  compaction 69–70, 91
  compression 58–60, 126

318

INDEX

desiccation 64–5
effective stress 59, 60–1
ground assessment 90–9, 130, 191
ground treatment 94, 210
liquefaction 137
moisture content 29–30
permeability 50
soil properties 20, 89, 90–1
soil suction 63
see also cone penetration tests; direct shear tests; oedometer tests; Proctor compaction test; samples and sampling; standard penetration tests; triaxial compression tests; vane tests
tilt 105, 111–3, 135, 147, 215, 217
tilt-meters 154, 233
trees
  desiccation and 135, 140, 142–8, 218–21, 228, 237–8, 248–9
  foundation damage 139, 218–9, 220, 224, 227, 228, 229
  foundation depths 108, 140, 145, 148
  freestanding walls 271, 272
  ground assessment 82, 84, 85
  piled foundations 120, 147
  removal 220, 225, 227, 230, 236, 237, 248, 249
  retaining walls 269
  settlement and 85, 142, 143
  shallow foundations 85, 219, 229
  trench-fill foundations 145–7
  underpinning and 248–9
  unstable slopes 151, 154
trench-fill foundations 106, 108, 114, 145–7, 182, 227
trial pits
  foundation movement investigation 225
  ground assessment 79, 87, 89, 98, 178
  soakaways 259, 260–1
triaxial compression tests 58–60, 91, 125

ultimate bearing capacity *see* bearing pressures and resistance
underground features *see* mine workings; natural cavities
underground fires 172, 175–6, 181
underpinning 236, 239–52
undrained shear strength 61–2
  measurement 91, 92, 94, 97–8
  settlement prediction from 125
  shaft resistance and 119
  soft ground 135, 190, 193
  spread foundations 86
uniformity coefficient 27
unit weight 26, 46
unstable slopes 148–54, 223
  drainage 150–1, 202

ground assessment 85
groundwater flow 49
  monitoring 229–30
  planning guidance 8
  retaining walls 269
  see also embankments; landslides

vane tests 89, 91, 97–8
varved clay soils *see* laminated clay soils
ventilation 139, 184
vertical drains 50, 136, 190, 197–8, 203
vertical stress *see* effective stress; overburden pressure; preconsolidation pressures
vibrated stone and concrete columns 136, 179, 183, 190, 191–4, 210
vibration, building damage 222
virgin compression line 54, 55
void ratio 25, 53–4
volume change *see* shrinkage and swelling
volume compressibility coefficient 54–5, 56, 125, 140
volumetric water content 25
see also moisture content

walk-over surveys 84–6, 177
walls *see* cut-off walls; freestanding walls; loadbearing walls; retaining walls
water content *21-30, 50-2, 62-70, 89-91, 138-143, 200-1*
see also moisture content
water features 82, 85
water-retaining structures 50, 51
see also embankments; retaining walls
water shrinkage factor 142
water-table *see* groundwater levels
weight *see* buildings, weight; moisture content; unit weight
wind-deposited soils 20, 21
window sampling 89–90, 226

319